HCI Beyond the GUI

The Morgan Kaufmann Series in Interactive Technologies

Series Editors: Stuart Card, PARC; Jonathan Grudin, Microsoft; Jakob Nielsen, Nielsen Norman Group

HCI Beyond the GUI

Design for Haptic, Speech, Olfactory, and Other Nontraditional Interfaces

Edited by

Philip Kortum

AMSTERDAM • BOSTON • HEIDELBERG • LONDON
NEW YORK • OXFORD • PARIS • SAN DIEGO
SAN FRANCISCO • SINGAPORE • SYDNEY • TOKYO
Morgan Kaufmann Publishers is an imprint of Elsevier

Publisher: Denise E. M. Penrose
Publishing Services Manager: George Morrison
Project Manager: Marilyn E. Rash
Assistant Editor: Mary E. James
Copyeditor: Barbara Kohl
Proofreader: Dianne Wood
Indexer: Ted Laux
Cover Design: Jayne Jones
Cover Direction: Alisa Andreola
Typesetting/Illustration Formatting: SPi
Interior Printer: Sheridan Books
Cover Printer: Phoenix Color Corp.

Morgan Kaufmann Publishers is an imprint of Elsevier.
30 Corporate Drive, Suite 400, Burlington, MA 01803

This book is printed on acid-free paper.

Library of Congress Cataloging-in-Publication Data
HCI beyond the GUI: design for haptic, speech, olfactory and other nontraditional
   interfaces/edited by Philip Kortum.
     p. cm. — (The Morgan Kaufmann series in interactive technologies)
Includes bibliographical references and index.
ISBN-13: 978-0-12-374017-5 (alk. paper) 1. Human-computer interaction. 2. Graphical
   user interfaces (Computer systems) I. Kortum, Philip.
QA76.9.H85H397 2008
005.4'37—dc22   2007051584

For information on all Morgan Kaufmann publications, visit our Web site at *www.mkp.com* or *www.books.elsevier.com*.

Printed in the United States
08 09 10 11 12   5 4 3 2 1

Working together to grow
libraries in developing countries

www.elsevier.com | www.bookaid.org | www.sabre.org

ELSEVIER   BOOK AID
             International   Sabre Foundation

Contents

Preface

The computer revolution and the graphical user interfaces (GUIs) it ushered in has helped define the work of a generation of human factors professionals. The advent of the Internet established the standard GUI as one of the primary interfaces that both users and designers must deal with. Yet, despite the ubiquity of the GUI, nontraditional interfaces abound, and are in fact significantly more common than we might first think. From the oft-reviled interactive voice response system to the small-screen interfaces on our cell phones, these nontraditional interfaces play a huge role in our everyday lives.

This book was born out of a desire to collect the fundamental wisdom that might be needed to do the human factors work on a variety of non-GUI interfaces into a single reference source for practicing human factors professionals and to give students of psychology and engineering an opportunity to be exposed to the human factors for the multitude of non-GUI interfaces that they will most likely be working on in the real world.

It is my hope that this book serves both of these groups. First, the chapters are structured so as to provide the seasoned human factors professional with a ready reference source for those occasions when the project demands an interface that is outside the common GUI. The inclusion of the design guidelines and the online case studies was specifically intended to give the practicing human factors professional useful, practical advice on implementation. Second, the book has also been designed to be used as a teaching text for upper-division undergraduates and graduate students, serving as an introduction to the many fascinating interfaces that exist beyond the realm of the well-covered GUI. The discussion of the underlying technologies, the current implementations and the fundamental human factors of the interface have been written to help the student understand the "nuts and bolts" of each interface and gain an appreciation of the role of the human factors engineer in its design.

Acknowledgments

As with any such endeavor, there are many people who played an important role in helping the project come to fruition. First, thanks to my friends and colleagues who contributed to the book—without their dedicated efforts and expertise, this book would not exist.

I would also like to thank my editors at Morgan Kaufmann, Denise Penrose, Mary James, and Asma Palmeiro, for their unending patience in helping to make this book a reality. Arnie Lund, Caroline Jarrett, Gavin Lew, Christine Alverado, and Randolph Bias provided enormously helpful reviews, and the book is better for their substantial and copious comments on the early versions. Finally, I would like to thank Michael Riley, my first human factors professor at the University of Nebraska, for sparking my love of human factors as a discipline.

Dedication

To Rebecca.

Contributors

Aaron Bangor, AT&T Laboratories, Austin, TX (bangor@labs.att.com)
Bangor is a principal member of technical staff at AT&T Labs, Inc., in Austin. He has worked on a wide variety of user interface designs, including applications that have multiple interfaces of different modalities. He earned a Ph.D. in human factors engineering from Virginia Tech and is a certified human factors professional. Bangor serves on the Texas Governor's Committee on People with Disabilities. He is also active in the Human Factors and Ergonomics Society, including part editor of the forthcoming American national standard: *Human Factors Engineering of Software User Interfaces.* (Chapter 11)

Paulo Barthelmess, Adapx, Seattle, WA (Paulo.Barthelmess@naturalinteraction. com)
Barthelmess is a research scientist working with collaboration technology at Adapx. His research interests are in human-centered multimodal systems, exploring intelligent interfaces to facilitate the work of co-located or distributed groups of people. His current focus is on supporting collaborative document-centered annotation work using digital paper. Barthelmess has an extensive software engineering background, having worked in industry in many capacities for over 20 years. He received a Ph.D. in computer science from the University of Colorado at Boulder. (Chapter 12)

Virginia Best, Boston University, Boston, MA (ginbest@cns.bu.edu)
Best studied medical science at the University of Sydney, and received her Ph.D. in 2004 after specializing in human auditory spatial perception. She then worked as a research associate at Boston University, where she examined the role of spatial hearing in realistic multiple-source environments, and developed an interest in how spatial hearing is affected by hearing impairment. In 2008, she will continue her work on hearing impairment as a research fellow at the University of Sydney. (Chapter 2)

Jeff Brandt, AT&T Laboratories, Austin, TX (brandt@labs.att.com)
Brandt began his career with the AT&T Labs Human Factors Group in 1996, ensuring that new products and services are useful to and usable by AT&T's customers. He manages the Austin Human Factors Laboratory facilities and performs interface design and usability testing for Internet Protocol Television applications. Past projects include disaster recovery, privacy management, outgoing call control, voice dial, unified communications, and bill formatting. Brandt holds 5 patents and has 43 patents pending. He earned the M.S. in industrial engineering from the University of Washington and B.S. in cognitive/experimental psychology from Oregon State University. (Chapter 7)

Derek Brock, Intelligent Systems Section, Navy Center for Applied Research
   in Artificial Intelligence, U.S. Naval Research Laboratory, Washington, DC
   (derek.brock@nrl.navy.mil)
Brock is a computer scientist at the U.S. Naval Research Laboratory's Center for Applied Research in Artificial Intelligence. His work involves the application of auditory display, cognitive architectures, and models of human language use to the design of collaborative interfaces for desktop, immersive, mobile, and robotic systems. He holds B.S. and M.S. degrees in computer science and computer graphics and multimedia systems from George Washington University. Brock is a member of the Acoustical Society of America (ASA), Cognitive Science Society, Association for the Advancement of Artificial Intelligence (AAAI), and International Community for Auditory Display (ICAD). (Chapter 5)

Christopher Frauenberger, Department of Computer Science, Queen Mary,
   University of London, London, UK (frauenberger@dcs.qmul.ac.uk)
Frauenberger is a Ph.D. student in the Interaction Media Communication Group at the Department of Computer Science, Queen Mary, University of London. His research focuses on alternative modes of interacting with technology with a special interest in the design of auditory displays. Since 2006, he is a member of the board of the International Community for Auditory Display and contributes toward establishing audio and sound as a highly efficient alternative for human–computer interaction designers. (Chapter 5)

Erik Granum, Department of Media Technology and Engineering Science at
   Aalborg University, Aalborg, Denmark (eg@vision.auc.dk)
Granum is a professor of information systems and head of the Department of Media Technology and Engineering Science at Aalborg University, Denmark. His interests cover pattern recognition, continually operating vision systems, motion analysis, color vision, multimedia interfaces, visualization, virtual reality,

and creative use of media technology. He has been coordinator and partner of a range of national and international research projects and networks in computer vision, media technologies, and virtual reality. He was a major contributor in the establishment of a multimedia and virtual reality center at Aalborg University, and pursues interdisciplinary educations and research. (Chapter 3)

Abhishek Gupta, Rice University, Houston, TX (abhi@rice.edu)
Gupta received the bachelor of technology (honors) degree in mechanical engineering from the Indian Institute of Technology, Kharagpur, and the M.S. degree in mechanical engineering from Rice University in 2004, where he is currently a doctoral student. His current research interests include design and control of haptic interfaces, nanorobotic manipulation with haptic feedback, and robot-assisted training and rehabilitation in virtual environments. (Chapter 2)

Thomas Hermann, Neuroinformatics Group, Faculty of Technology, Bielefeld University, Bielefeld, Germany (thermann@techfak.uni-bielefeld.de)
Hermann studied physics and received a Ph.D. in computer science at Bielefeld University in 2002. He is a research professor at Bielefeld University where he launched the research on sonification. Hermann serves as member of the International Community for Auditory Display (ICAD) board of directors and is German delegate and vice chair of the EU COST Action IC0601 (SID, sonic interaction design). He is initiator and organizer of the International Workshop on Interactive Sonification and guest editor of an *IEEE Multimedia* special issue on interaction sonification. His research fields are sonification, data mining, and human–computer interaction. (Chapter 5)

Susan L. Hura, SpeechUsability, Cumming, GA (susan.hura@speechusability.com)
Hura is the founder of SpeechUsability, a consultancy focused on improving customer experience by incorporating user-centered design practices in speech technology projects. She founded the usability program at Intervoice, and prior to that worked a member of the human factors team at Lucent Technologies. As faculty member at Purdue University, she cofounded a multidisciplinary team researching novel approaches to speech recognition. Hura holds a Ph.D. in linguistics from the University of Texas at Austin. She served as co-chair of SpeechTEK 2007 and 2008, and is a member of the board of directors of AVIOS. (Chapter 6)

Hiroo Iwata, Graduate School of Systems and Information Engineering, University of Tsukuba, Tsukuba, Japan (iwata@kz.tsukuba.ac.jp)
Iwata is a professor in the Graduate School of Systems and Information Engineering, University of Tsukuba. His research interests include haptic interfaces,

locomotion interfaces, and spatially immersive displays. Iwata received the B.S., M.S., and Ph.D. degrees in engineering from the University of Tokyo. He is a founding member of the Virtual Reality Society of Japan. (Chapter 9)

Philip Kortum, Rice University, Houston, TX (pkortum@rice.edu)
Kortum is currently a faculty member in the Department of Psychology at Rice University in Houston. Prior to joining Rice, he worked for almost a decade at SBC Laboratories (now AT&T Laboratories) doing human factors research and development in all areas of telecommunications. Kortum continues to do work in the research and development of user-centric systems in both the visual (web design, equipment design, and image compression) and auditory domains (telephony operations and interactive voice response systems). He received his Ph.D. from the University of Texas at Austin. (Chapter 1)

Marcia O'Malley, Rice University, Houston, TX (omalleym@rice.edu)
O'Malley received the B.S. degree in mechanical engineering from Purdue University, and the M.S. and Ph.D. degrees in mechanical engineering from Vanderbilt University. Her current research interests include nanorobotic manipulation with haptic feedback, haptic feedback and shared control between robotic devices and their human users for training and rehabilitation in virtual environments, and educational haptics. She is co-chair of the ASME Dynamic Systems and Controls Division Robotics Technical Committee, and a member of the IEEE Technical Committee on Haptics. (Chapter 2)

Chris Masterton, Optimal Interfaces, Cary, NC (chris@optimalinterfaces.com)
Masterton has been a practicing interaction designer and usability specialist for more than 8 years. His broad user interface design experience includes large e-commerce websites for clients like IBM and Lloyds of London; interactive sites for Tribal DDB, Tourism British Columbia, Ontario Tourism; mobile phone interface design for Nokia and Motorola; and usability testing for DirectTV, Clorox, Intrawest, and the University of Minnesota, among others. In 1997, Chris achieved his bachelor's degree in cognitive science with a certificate in computing science from Simon Fraser University. For the past 7 years, Chris has also been the instructor for user interface design at the University of British Columbia's software engineering continuing studies program. (Chapter 10)

Dan Mauney, HumanCentric, Vancouver, BC, Canada (dmauney@ humancentrictech.com)
Mauney is a 14-year veteran in the wireless telecommunications human factors profession. He has developed a broad view and understanding of the wireless

telecommunications market by working directly for a major North American wireless operator (SBC Wireless, now AT&T), a major handset manufacturer (Nokia), a content provider (Mobileum), a wireless accessory manufacturer (Jabra Corporation), and currently for a service provider specializing in the wireless telecommunications field (HumanCentric Technologies). At HumanCentric Technologies, Mauney leads a team of human factors professionals specializing in helping clients with small screen design and evaluation. He holds a Ph.D. and M.S. in industrial engineering and human factors from Virginia Tech. (Chapter 10)

James T. Miller, AT&T Laboratories, Austin, TX (miller@labs.att.com)
Miller is a principal member of the Technical Staff at AT&T Labs, Inc. He is primarily responsible for the development, testing, and evaluation of web pages that present consumer and business products for sale and that provide online support for those products. In addition, he is also responsible for the development of interactive voice response systems, including some that use speech recognition. Miller earned his Ph.D. from the University of Colorado at Boulder. (Chapter 11)

Thomas B. Moeslund, Laboratory of Computer Vision and Media Technology, Aalborg University, Aalborg, Denmark (tbm@cvmt.dk)
Moeslund is an associate professor at the Computer Vision and Media Technology lab at Aalborg University, Denmark. He obtained his M.S. and Ph.D. degrees in 1996 and 2003, respectively, both from Aalborg University. He is actively involved in both national and international research projects, and is currently coordinating a national project and work package leader in an international project. His primary research interests include visual motion analysis, pattern recognition, interactive systems, computer graphics, multimodal systems, and machine vision. Moeslund has more than 70 publications in these areas. (Chapter 3)

John Neuhoff, Department of Psychology, The College of Wooster, Wooster, OH (jneuhoff@wooster.edu)
Neuhoff is a member of the board of directors for the International Community for Auditory Display (ICAD). He plays the saxophone and teaches auditory display and cognitive science at The College of Wooster. His work has been published in *Nature, Science,* the *Proceedings of the National Academies of Science,* and he has edited a book on ecological psychoacoustics. He has received grants from the National Science Foundation, and the National Institute for Occupational Safety and Health. His saxophone career has yet to blossom. (Chapter 5)

Michael Nielsen, Laboratory of Computer Vision and Media Technology, Aalborg University, Aalborg, Denmark (mnielsen@cvmt.dk)

Nielsen is an assistant professor in the study of media at Aalborg University. His Ph.D. thesis was focused on three-dimensional reconstruction-based sensors in precision agriculture, and he has also worked with gesture interfaces and shadow segmentation. Research interests include aspects of media technology such as interface design, games, camera-based interfaces, color and light theory, and shadow segmentation. (Chapter 3)

Sharon Oviatt, Adapx, Seattle, WA (sharon.oviatt@adapx.com)
Oviatt is a distinguished scientist at Adapx and president of Incaa Designs. Her research focuses on human-centered interface design and cognitive modeling, communication technologies, spoken language, pen-based and multimodal interfaces, and mobile and educational interfaces. She has published over 120 scientific articles in a wide range of venues, including work featured in recent special issues of *Communications of the ACM, Human–Computer Interaction, Transactions on Human–Computer Interaction, IEEE Multimedia, Proceedings of IEEE,* and *IEEE Transactions on Neural Networks.* She was founding chair of the advisory board for the International Conference on Multimodal Interfaces and General Chair of the ICMI Conference in 2003. In 2000, she was the recipient of a National Science Foundation Creativity Award for pioneering work on mobile multimodal interfaces. (Chapter 12)

S. Camille Peres, Psychology Department, University of Houston-Clear Lake, Houston, TX (peresSC@uhcl.edu)
Peres is currently an assistant professor in psychology at the University of Houston-Clear Lake. Her research is generally focused on the cognitive mechanisms associated with the acquisition of new skills, and specifically mechanisms associated with acquisition and use of efficient methods, optimal designs for interactive auditory displays, and incorporation of simulations in the teaching of statistics. Peres received her Ph.D. in psychology from Rice University with a focus on human–computer interaction. (Chapter 5)

Sharif Razzaque, Computer Science, University of North Carolina, Chapel Hill, NC (sharif@cs.unc.edu)
Razzaque is a research scientist for InnerOptic Technology, where he develops augmented-reality surgical tools for solving spatial coordination problems faced during surgery. He received his Ph.D. in computer science at the University of North Carolina at Chapel Hill for work in virtual environment locomotion interfaces. He has previously worked on haptic interfaces, physiological monitoring,

medical imaging, collaborative satellite–engineering tool development at Lockheed Martin, and cochlear implants at the University of Michigan. (Chapter 4)

Barbara Shinn-Cunningham, Departments of Cognitive and Neural Systems and
   Biomedical Engineering, Director of CNS Graduate Studies, Boston University,
   Boston, MA (shinn@cns.bu.edu)
Shinn-Cunningham is an associate professor in cognitive and neural Systems and biomedical engineering at Boston University. Her research explores spatial hearing, auditory attention, auditory object formation, effects of reverberant energy on sound localization and intelligibility, perceptual plasticity, and other aspects of auditory perception in complex listening situations. Shinn-Cunningham is also engaged in collaborative studies exploring physiological correlates of auditory perception. She received the M.S. and Ph.D. in electrical engineering and computer science from the Massachusetts Institute of Technology. (Chapter 5)

Tony Stockman, Department of Computer Science, University of London,
   London, UK (tonys@dcs.qmul.ac.uk)
Stockman is a senior lecturer at Queen Mary, University of London, and a board member of the International Community for Auditory Display (ICAD). He first employed data sonification to assist in the analysis of physiological signals during his doctoral research in the mid-1980s. He has over 30 years of experience as a consultant and user of assistive technology and has published over 30 papers on auditory displays and data sonification. (Chapter 5)

Moritz Störring, ICOS Vision Systems, Belgium (moritz.störring@icos.be)
Störring studied electrical engineering at the Technical University of Berlin, and at the Institut National Polytechnique de Grenoble, France, and received the PhD from Aalborg University, Denmark. As an associate professor at Aalborg University, his research interests included physics-based color vision, outdoor computer vision, vision-based human–computer interaction, and augmented reality. In 2006, Störring moved to industry where he is focused on automatic visual inspection of electronic components and intellectual property rights IPR. (Chapter 3)

Louise Valgerður Nickerson, Department of Computer Science, Queen Mary,
   University of London, London, UK (lou@dcs.qmul.ac.uk)
Valgerður Nickerson is a Ph.D. student at Queen Mary, University of London, in the Department of Computer Science. Her work focuses on developing auditory overviews using nonspeech sound for the visually impaired and for mobile and wearable computing. She holds a B.A. in French and Italian Language and

Literature from the University of Virginia, and a M.S. in advanced methods in computer science from Queen Mary. (Chapter 5)

Mary C. Whitton, Computer Science, University of North Carolina, Chapel Hill, NC (whitton@cs.unc.edu)
Whitton is a research associate professor in the Department of Computer Science, University of North Carolina at Chapel Hill. She has been working in high-performance graphics, visualization, and virtual environments since she cofounded the first of her two entrepreneurial ventures in 1978. At UNC since 1994, Whitton's research focuses on what makes virtual environment systems effective and on developing techniques to make them more effective when used in applications such as simulation, training, and rehabilitation. She earned M.S. degrees in guidance and personnel services (1974) and electrical engineering (1984) from North Carolina State University. (Chapter 4)

Yasuyuki Yanagida, Department of Information Engineering, Faculty of Science and Technology, Meijo University, Nagoya, Japan (yanagida@ccmfs.meijo-u.ac.jp)
Yanagida is a professor in the Department of Information Engineering, Faculty of Science and Technology, Meijo University. He received his Ph.D. in mathematical engineering and information physics from the University of Tokyo. Yanagida was a research associate at the University of Tokyo and a researcher at Advanced Telecommunication Research Institute International before he moved to Meijo University. His research interests include virtual reality, telexistence, and display technologies for various sensory modalities. (Chapter 8)

1

CHAPTER

Introduction to the Human Factors of Nontraditional Interfaces

Philip Kortum

1.1 STRUCTURE OF THE BOOK

As human factors professionals, we are trained in the art of interface design. However, more and more of that training has centered on computer interfaces. More specifically, it has focused on the graphical user interfaces (GUIs) that have become ubiquitous since the advent of the computer.

While the GUI remains the most common interface today, a host of other interfaces are becoming increasingly prevalent. *HCI Beyond the GUI* describes the human factors of these nontraditional interfaces. Of course, the definition of a "nontraditional" interface is rather arbitrary. For this book, I attempted to select interfaces that covered all of the human senses, and included nontraditional interfaces that are widely used, as well as those that are somewhat (if not totally) neglected in most mainstream education programs. Many of these interfaces will evoke a strong "wow" factor (e.g., taste interfaces) since they are very rare, and commercial applications are not generally available. Others, such as interactive voice response interfaces, may not seem as exciting, but they are incredibly important because they are widely deployed, and generally very poorly designed, and it is likely that every human factors professional will be asked to work on one of these during the course of her career. This book brings together the state of the art in human factors design and testing of 11 major nontraditional interfaces, and presents the information in a way that will allow readers who have limited familiarity with these interfaces to learn the fundamentals and see how they are put into action in the real world.

Each chapter in the book is structured similarly, covering the most important information required to design, build, and test these interfaces. Specifically, each chapter will address the following aspects.

Nature of the interface: Each chapter begins with a description of the fundamental nature of the interface, including the associated human perceptual capabilities (psychophysics). While the details of these discussions may seem unimportant to the practitioner who simply wants to build an interface, an understanding of pertinent human strengths and limitations, both cognitive and perceptual, is critical in creating superior interfaces that are operationally robust.

Interface technology: As with any interface, technology is often the limiting factor. Some of the interfaces described in this book use very mature technology, while others are on the cutting edge of the research domain. In either case, a detailed description of the technologies used and their appropriate implementations are provided so that the practitioner can specify and construct basic interfaces.

Current implementations of the interface: This section describes how and where the interface is used today. Examples of successful implementations for each interface are given, as well as examples of failures (where appropriate), which can be very instructive. Another topic that is included in this section is a discussion of the interface's application to accessibility. Many of these interfaces are of special interest because certain implementations provide crucial interfaces for people with physical or cognitive disabilities. For example, Braille is a low-tech haptic interface that allows blind users to read. This section briefly discusses the benefits of using the technology to assist individuals who have physical or cognitive impairments, and provides examples of special implementations of the technology for such users. If use of the interface has any special adverse consequences for the disabled population, these are noted as well.

Human factors design of the interface: This section will tell you, as the human factors designer, what you should be considering as you embark on the design or evaluation of a given nontraditional interface. It discusses when to select a particular interface, the data required to build the interface, and details on what a human factors professional would need to know in order to specify such an interface for use.

Techniques involved in testing the interface: Special interfaces usually require special testing methodologies. This section describes special testing considerations for the interface, special technology or procedures that might be required, and methods of data analysis if they are sufficiently different from standard analysis methods. Even if standard testing measures are used, a description of these and when they should be applied is included to guide the practitioner. Special attention is paid to the concept of iterative testing if it is applicable to the specific interface.

Design guidelines: For experienced designers, guidelines can appear to be too simplistic and inflexible to be of any practical value. However, for the beginning designer, they serve as an invaluable way to generate a first-generation design while leveraging the knowledge of expert designers. It is in this spirit that the Design Guidelines section of each chapter provides some of the most important lessons that should be applied. The guidelines presented for each interface are not meant to be exhaustive and inclusive. Rather, the goal of this section is to list the top 5 to 10 items that an expert would pass along to someone who was looking for important advice about the human factors implementation of the interface.

Case study of a design: This section presents a case study of the human factors specification/implementation/evaluation of the interface over its life cycle. Where practical, the case study is a real-world implementation. For certain interfaces, however, proprietary considerations dictated changes in names, dates, and identifying details to mask the identity of the interface. In some cases, the example has been made stronger though the use of multiple implementations rather than a single, life cycle case study.

Future trends: Since the focus is on nontraditional interfaces, most are still evolving as technology changes and as users (and designers!) become more familiar and comfortable with their use. This section describes the future of the interface in the next 10 to 20 years. Where is the interface headed? How will current implementations change? Will current implementations survive or be supplanted by new innovations? What is the end state of the interface when it is fully mature? In this section, the authors are given a chance to speculate how a particular interface will mature over time, and what users can look forward to.

The authors of certain chapters, particularly those focused on interfaces that use sound, have provided access to examples that you can listen to by visiting the book's website at www.beyondthegui.com. This website also contains case studies for each interface. These case studies provide examples of how the interfaces have been implemented, and how human factors contributed to those implementations.

1.2 NONTRADITIONAL INTERFACES

Scarcity of implementation was not the primary factor in determining the interfaces to be included in this book, as many of them are nearly ubiquitous. Others, such as taste interfaces, are quite rare. Further, even though the name of the book is *HCI Beyond the GUI*, several chapters do, in fact, deal with GUIs, but in a form that most designers have little experience with (see, for instance, Chapter 10 on small-screen design). The 11 interfaces selected for inclusion represent the most important nontraditional interfaces that a human factors professional should know and understand.

1.2.1   Haptic User Interfaces

Haptic interfaces use the sensation of touch to provide information to the user. Rather than visually inspecting a virtual three-dimensional object on a computer monitor, a haptic display allows a user to physically "touch" that object. The interface can also provide information to the user in other ways, such as vibrations. Of course, the gaming industry has led the way in introducing many of these nontraditional interfaces to the general public. Various interface technologies have heightened the realism of game play and make the game easier and more compelling. One of the early interfaces to take advantage of haptics can be found in Atari's Steel Talons sit-down arcade game (Figure 1.1).

The game was fun to play because the controls were reasonably realistic and the action was nonstop; however, unlike other contemporary first-person shooter

FIGURE   Atari's Steel Talons helicopter simulation, circa 1991.

1.1   While the graphics were unremarkable (shaded polygons), the game employed a haptic interface in the player's seat (as indicated by the arrow) that thumped the player (hard!) when the helicopter was being hit by ground fire. The added interface dimension caused the player to react in more realistic ways to the "threat" and made the information more salient. *Source:* Retrieved from *www. mame.net.*

games, Atari integrated a haptic feedback mechanism that was activated when the user's helicopter was "hit" by enemy fire. Other contemporary games used sounds and changes in the graphical interface (flashing, bullet holes) to indicate that the user was taking enemy fire. Atari integrated what can best be described as a "knocker" in the seat of the game. Similar in sound and feel to the device in pinball machines that is activated when the user wins a free game, this haptic interface was both effective and compelling. Although the information provided to players was identical to that presented via sound and sight, they reacted to it differently—their response was evocative of the fight-or-flight response seen in the real world, and they were more reluctant to just "play through" the warnings, as is so often seen in strictly visual games.

By selecting the right interface type for the information that must be presented, the designers created a better interface. Once the sole purview of high-end simulators and arcade games, haptics can now be found in home game consoles as well (e.g., Nintendo's Rumble Pac). Although generally not as sophisticated or realistic as their more expensive counterparts, the use of vibration in the hand controller provides the player with extra information about the environment and game play that was not previously available.

Other examples of compelling implementations of the haptic interface can be found in interfaces as diverse as automobile antilock braking feedback systems and threat identification systems for soldiers. Chapter 2 will address the entire spectrum of haptic interfaces, from simple implementations, such as vibrating mobile phone ringers, to some of the most sophisticated virtual-touch surgical simulators.

1.2.2   Gesture Interfaces

Gesture interfaces use hand and face movements as input controls for a computer. Although related to haptic interfaces, gesture interfaces differ in the noted absence of machine-mediated proprioceptive or tactile feedback. The simplest form of gesture interfaces can be found in motion-activated lights—the light interprets the user's motion as the signal that it should turn itself on. Other commercial implementations of gesture interfaces have recently begun to make their way into the game world as well.

In 2001, Konami released a game called MoCap Boxing. Unlike earlier versions of boxing games that were controlled with joysticks and buttons, Konami's game required the player to actually box. The player donned gloves and stood in a specified area that was monitored with infrared motion detectors. By moving and boxing, the player could "hit" the opponent, duck the opponent's hits, and protect his body by simply replicating the moves a real boxer would make. Figure 1.2 shows the game in action.

This technology, too, has found its way into the home with the recent introduction of Nintendo's Wii system. Unlike other current home gaming

FIGURE

1.2

Konami's gesture interface game, MoCap Boxing.

Unlike previous generations of sports games, the user does not use buttons to code his intentions. Instead, he dons boxing gloves and moves in a motion capture area (the mat the user is standing on) to control the interface. The end effect is a fairly realistic game that is intuitive (and tiring!) to use. (Courtesy of Konami.)

systems, Wii makes extensive use of the gesture interface in a variety of games, from bowling to tennis, allowing the user to interact in a more natural style than previously when interaction was controlled via buttons interfaced to the GUI. Not only is the interface more natural and appropriate for controlling the action in the games, but it also has the added benefit of getting players off the couch and into the action, an interface feature that is appreciated by parents worldwide!

As can be seen in Figure 1.3, the Wii bowling game enables the player to interact with the game in a manner that is similar to that of real-world bowling. These new interfaces have their own problems, however. For instance, shortly after the Wii was released there were reports of users accidentally letting go of the remote

FIGURE

1.3

Gesture-based interface in action.
The Nintendo Wii, a gesture-based interface, is being played in a bowling simulation. These kinds of interfaces bring their own unique issues to the designer.

(particularly in bowling, since that is what people do when they bowl) and having it crash into, or through, the television [Burnette, 2006]).

Gesture interfaces range from systems that employ hand motion for language input to those that use gestures to navigate (e.g., "I want to go that way") and issue commands (e.g., "Pick that up") in a virtual-reality environment. Issues surrounding the limitations of discriminating gestures and how those limitations guide the design of these interfaces are explored. In addition, the potential for undesirable artifacts when using these kinds of interfaces (fatigue, misinterpretation, etc.) will be discussed along with methods that have been developed to mitigate these potential deficiencies.

1.2.3 Locomotion Interfaces

Locomotion interfaces, although sharing attributes with both haptic interfaces and gesture interfaces, differ because they require gross motor movement and typically deal with large-scale movement or navigation through an interface. Interfaces in research labs not only include treadmill-type interfaces, but have moved in other interesting directions as well, including swimming and hang-gliding applications. These kinds of interfaces are frequently associated with high-end simulators, but the technology has recently moved out of the laboratory and into the commercial realm in the form of body motion arcade games such as dancing and skiing. As with gesture interfaces, several of the most current generations of home gaming boxes have body motion controllers available as well. Issues surrounding the physical limitations of the human body and how those limitations guide the design of these interfaces will be explored. In addition, the potential for undesirable artifacts when using these kinds of interfaces (fatigue, vertigo, etc.) are considered.

1.2.4   Auditory Interfaces

Auditory interfaces have long been used to send simple coded messages across wide areas (e.g., tolling of church bells, wailing of the civil defense sirens). Auditory interfaces have also been used extensively to augment complex interfaces and to spread the cognitive load in highly visual interfaces, particularly in the conveyance of warnings. These kinds of auditory interfaces are relatively simple to implement, but require that the user be able to interpret the meaning of the coded message. Recently, auditory interfaces have been employed as a substitute for more complex visual interfaces, and the term "sonification" has been coined to describe these kinds of auditory interfaces. In a sonified interface, representations that are typically visual, such as graphs and icons, are turned into sound, that is, sonified, so that they can be interpreted in the auditory rather than the visual domain.

The chapter on auditory interfaces will detail the fundamental psychophysics of the human auditory system, and then relate that to the design and implementation of auditory interfaces. Issues of overload, human limitations, and appropriate selection of auditory frequency space for various kinds of auditory and sonified interfaces will be discussed.

1.2.5   Speech User Interfaces

Since Ali Baba proclaimed "Open Sesame!" to magically gain entrance to the Den of the 40 Thieves, speech interfaces have captivated our imagination. With this imagination further fueled by science fiction television series such as *Star Trek*, we have been generally disappointed with the real-world implementations of speech interfaces because they seem to lag so far behind our idea of how well they should work. However, recent advances in computing power have brought the possibility of robust speech interfaces into reality, and commercial systems are now readily available in multiple realms.

Early implementations of single-word speech command interfaces have led to continuous–speech dictation systems and state-of-the-art systems that employ powerful semantic analysis to interpret a user's intent with unstructured speech.

The chapter on speech interfaces will discuss the technology behind these interfaces for both speaking and speech recognition systems and chart the progress of both. It will discuss the implementation of both limited and unbounded vocabulary interfaces, review the advantages of speaker-dependent and speaker-independent systems, and detail the appropriate design of speech prompts and navigation structures. Since speech interfaces have been implemented widely and successfully in telephone-based systems, extensive examples and case studies of successful interfaces in this realm will be used to highlight the human factors of speech user interfaces.

1.2.6   Interactive Voice Response Interfaces

Interactive voice response systems (IVRs) are in widespread use in the commercial world today, yet receive scant attention in traditional human factors. The interface has been embraced by the business community because of its huge potential for cost savings and because when implemented correctly it can result in high customer satisfaction ratings from the user as well. Because of the ubiquity of the interface, however, poorly designed IVR interfaces abound, and users are left to suffer. The chapter on IVRs will discuss the specification of appropriate navigation structures, prompt construction, and transaction management, including user inputs and time-out issues. The selection and use of voice persona and the impact on both user preference and performance will also be considered.

Although IVRs are typically used for routing customers to the right service agent, or to convey limited amounts of information (like a bill balance), some current-generation interfaces provide for more interaction and allow significantly more information to be delivered to the user. These highly interactive IVRs and the special problems associated with them will be considered as well.

1.2.7   Olfactory Interfaces

Olfactory interfaces have typically been used in situations where widespread communication of a message is required, but where traditional interfaces are hampered by the environment. Stench systems to warn miners of danger (where traditional auditory and visual warnings do not function well) and the use of wintergreen as a fire alarm signal in factories where there is significant auditory and visual noise (e.g., a welding shop) are two prime examples of successful olfactory interface implementations. As with many of the interfaces in this book, the advent of computers and simulated environments has pushed secondary interfaces, like olfaction, to the fore.

First introduced in the 1960s as a way to add another dimension to theater movie presentations (where it was a resounding failure), research is continuing to be conducted on these so-called "smell-o-vision" interfaces that will allow the user to experience scents. Unlike the early implementations, research is now focused primarily on the Internet as a delivery medium in such diverse applications as shopping (perfume fragrance samples), entertainment (the smell of burning rubber as you drive a video game race car), and ambience (evergreen in an Internet Christmas shop).

Of course, the introduction of smell as an interface also has tremendous potential in training applications in areas such as medicine and equipment maintenance. When coupled with virtual-reality simulators, the addition of smell may be able to significantly enhance the training experience. The psychophysics of olfaction will be discussed in great detail, since the strengths and limitations of the human olfactory system play a significant role in the correct implementation of these interfaces. The appropriate uses and technology implementations of current- and next-generation systems will also be considered.

1.2.8   Taste Interfaces

Without a doubt, interfaces that rely on taste are one of the least explored of the nontraditional interfaces. Taste interfaces are usually discussed in terms of simulation, in which a particular taste is accurately represented to simulate a real taste (e.g., for a food simulator). However, taste can also be used to convey coded information, much like olfactory displays. Because this interface is in its infancy, the chapter on taste interfaces will focus primarily on the current state-of-the-art taste simulator research in this area and the potential for future uses.

1.2.9   Small-Screen Interfaces

Miniature interfaces have been envisioned since Dick Tracy first used his wrist picture phone, and this vision has become a reality with the successful miniaturization of electronic components. While devices such as mobile telephones and MP3 players have continued to shrink, the problems with controlling and using these miniature devices have grown. From the physical ergonomics associated with using the systems to the navigation of tiny menus, the new systems have proven to be substantially different, and more difficult, to use than their bigger brethren.

The chapter on small-screen interfaces will discuss how these miniature GUIs are designed and tested, and how special interface methods (predictive typing, rapid serial presentation of text, etc.) can be employed to make these interfaces significantly more usable. The chapter will also discuss how to implement common menu, navigation, and information display structures on both monochrome and color screens that contain extremely limited real estate, from tiny cell phone screens to microwave dot matrix displays. Special emphasis will be placed on how to design these interfaces so that the needs of older user populations, who may have reduced visual and motor capabilities, are accommodated.

1.2.10 Multimode Interfaces: Two or More Interfaces to Accomplish the Same Task

In many instances, a task can be accomplished using one or more interfaces that can be used in a mutually exclusive fashion. For example, you can perform simple banking tasks using the Internet or an interactive voice response system. You can choose to use either interface method for any given transaction, but they are mutually exclusive in the sense that the task does not allow or require both to be used simultaneously. Providing multiple interfaces for single systems means that the seemingly independent interfaces must be designed and tested together as a system, to ensure that users who move back and forth between the two interfaces can do so seamlessly.

This chapter will explore the more common mutually exclusive multimode interfaces (MEMM), including IVR/GUI, speech/GUI, small screen/GUI, small screen/IVR, and small screen/speech, and discuss the human factors associated with the design and implementation of these multimode interfaces. Appropriate selection of the different modes will be discussed, as well as consideration of implementing unequal capabilities in these types of system interfaces.

1.2.11 Multimode Interfaces: Combining Interfaces to Accomplish a Single Task

In direct contrast to the MEMM interfaces described in Chapter 11, multimode interfaces that either require or allow the user to interact with the system with more than one interface at the same time (i.e., a mutually inclusive multimode [MIMM] interface) are much more common and typically fall into two distinct classes. The first class of MIMM interfaces are those in which the user can choose among multiple interfaces during a task, and can move back and forth among them at any time. For example, systems that combine speech and interactive voice response interfaces have become more common (e.g., "Press or say one"), and the way these systems must be designed and implemented is decidedly different than if each interface were to be implemented alone.

The second class of systems, and by far the most common of any described in this book, are those interfaces that employ multiple interfaces into a single "system" interface. Auditory and visual interfaces are frequently combined to create effective interfaces. Virtual-reality systems are a prime example of this kind of interface, where vision, audition, speech, haptic, and gesture interfaces are combined in a single integrated experience. This chapter will focus on MIMM interfaces that use one or more of the other nontraditional interfaces described in the book. While the chapter on MIMM interfaces is not meant to be a primer on system human factors, it will discuss how to determine which interface modes to use (interface allocation) and how and when to overcode the interfaces (interface redundancy), and it will deal with issues surrounding the presentation of information using non-native interfaces (i.e., using tactile displays to represent sound).

1.3 DESIGN PRINCIPLES FOR NONTRADITIONAL INTERFACES

Each chapter describes certain human factors principles and design considerations that should be taken into account when working with a particular interface modality. However, the fundamental principles for designing nontraditional interfaces are the same as those for designing traditional GUIs. Schneiderman and Plaisant (2004), Nielsen (1999), Raskin (2000), Norman (2002), Mayhew (1997), and others have described these principles in great detail. The key is to remember that the traditional, fundamental principles still apply, even if the interface you are designing is anything but traditional.

Most users do not care about the interface technology—they simply have a task they want to accomplish and they want the interface to support them in completing that task. Too often, however, the designer is led by other goals (corporate needs, fascination with technology, "cool" factor, etc.), and the human factors of the design suffer. The bottom line is this: Good designs do not just happen! They are the result of careful application of numerous design methodologies that enable you, as the designer, to understand the user, the environment, and how they interact. ISO 9241:11 (ISO, 1998) specifies that usable designs should have three important attributes.

First, they should be *effective*, meaning that the user can successfully use the interface to accomplish a given goal. Second, the interface should be *efficient*. This means that the user not only can accomplish the goal, but can do so quickly and easily with a minimum of error or inconvenience. Finally, the interface should leave the user *satisfied* with the experience. This does not mean the user has to be happy, but it does mean that the user should have high confidence that the task was accomplished according to his intentions. For example, using a bank's automated telephone system to transfer money should be easy to do (effective), quick (efficient), and leave users with the certainty that they really accomplished the task and that the bank has their correct instructions (satisfaction).

The next section summarizes the fundamental human factors design guidelines to keep in mind when pursuing general usability goals, regardless of the interface mode(s) you eventually decide to use in your interface.

1.3.1   Design to Support the User's Goals

Whenever you start a design, you certainly have the user in mind. Unfortunately, somewhere along the way, designers often forget that they should be trying to serve the user rather than another master, such as the corporation, vendor, or technology. Think about what the user wants or needs to do and focus on that. A great example of a violation of this principle can be found in the butterfly ballot debacle in the 2000 Florida elections. Although not maliciously conceived (unless you are a conspiracy buff), the ballot was designed to support a more modest goal: space efficiency. The results, as we all know, were significant questions about voters' intent in the seemingly simple task of selecting their preferred candidate. Each time you are tempted to add, delete, change, prettify, or otherwise alter a good design, ask yourself if this helps the user accomplish the goal. If it does not, think twice!

1.3.2   Design to Support the User's Conceptual Model

Whenever people use a system, they bring with them an idea of how that system works. If your system's design does not match this idea (the user's conceptual model), then the way that the person will use the system can be unpredictable. The classic example of this is how many people interact with the thermostat in their home. For instance, a person comes home from a long day at work, and the house is *hot*. He goes to the thermostat and turns the dial all the way to the left to quickly cool the house down. Why does he turn it all the way to the left, rather than carefully select the exact temperature that he wants the house to be? Does turning the dial all the way to the left make it cooler faster? No! The cooling system in the home is a two-state device—it is either on or off. It cools until it reaches the temperature set on the thermostat. However, the model that many of us have is that the farther you turn the dial to the left, the cooler it gets. In one respect, that is true—left is cooler—but the rate of change is independent of the setting. Why is this a problem?

Invariably, after making this "make-it-as-cold-as-fast-as-possible" adjustment, one gets called away from the house and returns to meat-locker–like temperatures. The interface failed to support what the user thought the system would do (and what the user wanted it to do). Newer electronic thermostats seem to have prevented this problem, since a specific temperature setting is required and so a quick "cold-as-you-can-make-it" command is harder and slower to execute. Figuring out what the user's model is can be difficult (and models may vary from person to person). However, mapping the way the system *really* works to the way people *think* it works makes for superior usability.

1.3.3   Design for the User's Knowledge

Each user of your interface will bring a specific set of knowledge and experience to the table. If your interface requires knowledge that the user does not have, then it will likely fail (unless you can train the user on the spot). For instance, as illustrated in Figure 1.4, the designer took great care to ensure that the interface would support users with multiple language skills (good for the designer!). However, he failed to follow through in his execution of the design. What is wrong with this selection menu? The user must read English in order to select the desired language! Determine what your users know, and then design the interface to require no more knowledge than that.

1.3.4   Design for the User's Skills and Capabilities

In addition to specific knowledge sets, users of your interface will also have specific sets of skills and capabilities that limit how they can use your interface. For example, humans have specific limits on what they can hear, so when you are designing an auditory warning, you want to make sure that you select a frequency that can be easily heard by the human ear. Think of the dog whistle as a dog's basic auditory display—it says "Attention!" to the dog. The interface is worthless for use with humans (like your children) because the whistle operates outside of the human ear's capability to detect sound. It is, however, a perfect interface for the dog. By thinking of what your users can and cannot do, you can design the interface in the most effective fashion.

Figure 1.5 shows an example of a poorly designed interface in that it ignored the specific skills and capabilities of its intended users. While the intent of the designers was noble, the execution of the design resulted in an interface that did not serve its intended purpose.

Language	Russian
	Finnish
	Russian
	French (Standard)
	Saudi Arabia
	French Canadian
	Serbian
	German (Standard)
	Slovak

FIGURE

1.4

Example of a clear violation of designing for the user's knowledge. Users must read English in order to select the language in which they would like to use the interface. (Courtesy of the Interface Hall of Shame.)

FIGURE

1.5
Example of an interface that does not take into account the skills and capabilities of the users for whom it was intended.
The intercom system is supposed to be used by individuals who are mobility impaired and cannot pump their own gas. The problem here is that its placement precludes its use by this very population. Further, the instructions that explain what the user is supposed to do are too small to be easily read from the most likely location of a user in a wheelchair.

1.3.5   Be Consistent—But Not at the Expense of Usability

Consistency is often the hallmark of a usable design. Once you learn to use one interface, then other similar interfaces are easier to learn and use. This consistency can manifest itself as standardization among interfaces made by different designers or can show up as good internal consistency within a given interface. In both cases the usability of the interface is increased because users have to learn and remember less as they become familiar with the interface. Unfortunately, blind adherence to the consistency principle can actually lead to less usable interfaces. Sometimes the existing standard does not lend itself well to a new problem or situation. If the designer makes the decision that consistency is of the utmost importance, then the entire interaction may be "force-fit" into the standard interface. While the end user may enjoy some of the benefits of a consistent interface, the forced fit may actually decrease the overall usability of the system. Small-screen implementations of the Microsoft Windows operating system are a good example in which consistency may not have been the right choice.

Clearly, the users of these devices had the potential advantage of being able to seamlessly transition between desktop and handheld applications. However, many of the GUI elements that Windows uses do not scale well to very small screens and so the consistency begins to degrade. Now, the interface is only partially consistent, so many of the benefits derived from consistency are lost ("I know I can do this in Windows! Where is the icon!?!"). Other manufactures of small-screen devices have determined that there are better interface methods and have designed their own specialized interfaces (Chapter 10 addresses these in much more detail). While the learning consistency has been forfeited, the interface is not being asked to do things that it was not originally designed to do, and so overall usability is enhanced. In the end, you should seriously consider whether consistency enhances or detracts from the overall usability of the interface, and then make the appropriate implementation decision.

1.3.6   Give Useful, Informative Feedback

When I first began giving lectures that included these eight design principles, this guideline simply read "Give Feedback." However, after seeing too many examples of useless feedback messages filled with cryptic error codes or worthless state information (Figure 1.6), the guideline was changed to include the words "useful" and "informative." Users of an interface need to know where they are in the interface, what they are doing, and what state the system is in. By providing feedback, the system aids the user in making correct, efficient decisions about what actions to take next.

The feedback may be explicit and require direct action (e.g., a pop-up dialog box that states, "Clicking yes will erase all the data. Proceed anyway?"), or it

FIGURE

1.6

Interface providing feedback that is not useful or informative. Why can't I do this? What should I do to make it so that I can? Is there a workaround? This is the kind of feedback that gives interface designers a bad name. (Courtesy of the Interface Hall of Shame.)

may be more subtle in nature and provide the user with cues that the state of the system has changed and requires the attention of the user (e.g., antilock-brake feedback systems). As the designer, you should make sure that you provide both kinds of feedback, and test the effectiveness of the feedback with actual users who encounter the cases where it is provided.

1.3.7   Design for Error Recovery

People make mistakes. With that in mind, you should always ensure that the interface is designed to help users by minimizing the mistakes they make (e.g., the auto-spell feature in a word-processing program) and by helping them recover from mistakes that they do make. In the strictest definition of error, no adverse consequence is required for the action (or lack thereof) to be considered as an error. In practical terms, however, an error that is caught and corrected before any adverse impact occurs is of far less concern than an action that has an adverse outcome. Providing users with ways to easily correct their mistakes, even long after they have happened can be very beneficial. The trash can that is employed in many desktop computer interfaces is an excellent example. A document may be thrown away but, just like in the physical world, can be pulled out of the trash and recovered if the user discovers that the deletion was in error. Contrast that with the dialog window shown in Figure 1.7. Even if the user recognizes that she has performed the action in error, the system is offering no recourse. Adding insult to injury, the poor user has to acknowledge and accept the action that is no longer desired!

1.3.8   Design for Simplicity

As Albert Einstein once noted, "Make everything as simple as possible, but not simpler." This axiom is especially true in interface design. Simplicity makes

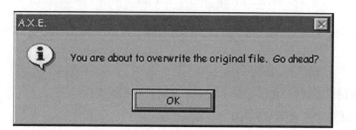

FIGURE

1.7

Poor error recovery for a critical action.
Imagine what you would do (besides scream) if you knew you had just made a mistake, and you did not want to overwrite the original file. (Courtesy of the Interface Hall of Shame.)

interfaces easy to learn and easy to use—complexity does the opposite. Unfortunately, complexity is sometimes a necessary evil that reflects the nature of the task. Interfaces for nuclear power plants and sophisticated fighter jets just cannot be as simple as that for an iPod. In these situations, you must strive to understand the user's task to such a degree as to be able to distill the interface down to its essential components. The result may still be complex, but its functions will represent what the user (the operator or the pilot) actually needs to do.

In many cases, however, the complexity is due to the inclusion of features, usually over time, that "enhance" the product. All of us have seen this feature creep. You only have to look as far as your word processor—features, features, and more features. If only I could just figure out how to use them, they would be great. Design simplicity is where the human factors practitioner and the marketing folks may be at odds, since marketing is often looking for feature superiority ("My toy has more features than your toy!"). The result can be an interface that masquerades as simple, but has significant hidden functionality and complexity behind its beautifully simple exterior. The modern cell phone is a classic example.

In its most basic form, the cell phone is designed to make and receive calls. In most current implementations, it is also a camera, an address book, an MP3 player, a game platform, a pager, and so on and so forth. While each feature on its own may be important, the combination of all these features, particularly on a small-screen device, can lead to usability nightmares. Sometimes the complexity is simply a matter of space—too many features and not enough space to implement them. Figure 1.8 shows an example of a relatively complex design that is driven by space considerations.

As the designer, it will be up to you to try to determine how to maximize both feature inclusion and simplicity in a way that supports all of the user's goals. Anyone can design a complex interface—it takes real skill to design a simple one.

Designing nontraditional interfaces does not release you from the fundamental design principles described here. By understanding what your users want, what they need (which is often different from what they want), what they know, how they do their work, and what their physical and cognitive limitations are, you, the designer, can create superior interfaces that support the user in accomplishing his goal.

1.4   THE FUTURE OF NONTRADITIONAL INTERFACE DESIGN

Undoubtedly, interface design in the future will move toward designs where the right modality for the job is selected, rather than the modality at hand being forced to fit the job. There's an old expression that says that if all you have is a hammer, then everything looks like a nail, and the same can be said of GUIs. The GUI

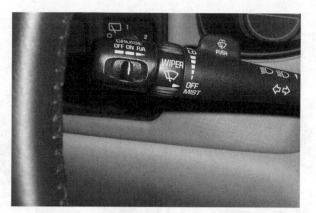

FIGURE

1.8

Complex interface for multiple features on the turn signal stalk in a Chevrolet Suburban.

The stalk used to be reserved for a single function—activating the turn signal. Now it supports the blinker, cruise control, variable front wipers, high-beam headlights, rear wiper, and window washer system. The result is a system that is hard to use and contributes to inadvertent activation of undesired functions.

is a powerful and adaptable interface form, and designers who are trained in the art and science of creating GUIs have tended to approach any given interface implementation with a GUI solution. However, careful analysis of the needs of the user in a given interface environment may suggest that another (nontraditional) form of user interface is required. More likely than not, the interface will be multimodal, as described in Chapters 11 and 12, and the goal of the designer will be to ensure that the correct interface modes are assigned to the information inputs and outputs that provide for the best, most effective experience for the user.

Much of this multimodal interface evolution is bound to move toward more "natural" interaction techniques. In natural interactions, the user does not need to make any translation of coded data or determine how to perform a function— the user simply interacts with the interface as she would in the physical world. Analogs (like the Windows desktop) will disappear, and the interface will become indistinguishable from its representative model. Imagine if the Windows desktop were replaced with an immersive virtual desktop that was indistinguishable from the real-word analog in form and basic function but had all the benefits that can be derived in a computer-driven model.

This vision is most certainly *not* a reversion to the much maligned virtual-world interface proposed by Microsoft in the mid-1990s that was supposed to take the desktop metaphor to the next level. Microsoft Bob (Figure 1.9) tried to replicate the office and home environment through a rich metaphor that had all the

FIGURE

1.9

Home page for Microsoft Bob.
MS Bob was an extension of the desktop metaphor that was supposed to be extremely easy to learn and use. Unfortunately, it suffered from many deficiencies, and Microsoft pulled it from production shortly after its release. *Source:* Mid-1990s screenshot of Microsoft's Bob Operating System.

objects that one would expect in these environments. However, the metaphor was not quite perfect, since there are many things that can (and should) be done on a computer that just do not fit into the physical metaphor. The implementations of these nonconforming functions (not the least of which was the talking dog that was needed to help novice users get their bearings) caused users to have difficulty in learning and using the interface. This complexity, along with a plethora of hidden functions, led the interface to an early demise despite its promise as a research platform into high-performance metaphorical interfaces.

One of the most exciting developments in nontraditional interfaces is the recent advances that are currently being made in brain–computer interfaces. This class of interface completely bypasses the human musculoskeletal body as a mechanism for input/output, and instead interfaces directly with the brain. The interface that the user is trying to use interprets these raw brain waves and then performs the appropriate action. These kinds of interfaces represent a very specialized, newly emerging interface domain, and so they are not addressed in a separate chapter in this book (just wait for the second edition!), but they hold significant promise for the future.

Much of the current brain–computer research is focused on using brain waves to control screen cursor movements (Friedrich, 2004), although recent advances have led to limited success in the operation of robotic arms (Donoghue et al., 2007). Of course, as with many futuristic interfaces, Hollywood has been thinking about this for some time—recall Clint Eastwood flying the super-secret Russian plane in the 1980s thriller *Firefox*, in which the aircraft is controlled by the thoughts of the pilot—but of course those thoughts must be in Russian "Помогите! Я не вижу в русский!"—"Help! I can't think in Russian!".

Although the technology is nowhere near this level of sophistication, Tanaka and his colleagues (Tanaka, Matsunaga, & Wang, 2005) have had tremendous success with research in which patients could control the navigation of their powered wheelchairs using brain interfaces, eliminating the need for joysticks or other input devices. Direct brain interfaces may prove to be a boon for the physically disabled, allowing them to control computers and other assistive devices without physical movement.

Holographic interfaces are another technology that may become important in future applications. One of the biggest problems with current GUIs is that they demand physical space. If the physical space is restricted, as in a mobile phone for instance, then the interface must conform to the reduced space. Holography has the potential to overcome this limitation by using the air as the interface medium. By using holography, no physical interface would be required. Simple commercial holographic interfaces are just now becoming available, and research into more complex holographic interfaces continues (e.g., Bettio et al., 2006; Kurtenbach, Balakrishnan, & Fitzmaurice, 2007).

When mentioning holographic displays, the ones envisioned by George Lucas and his team of special effects wizards immediately come to mind. He and his associates showed us how future holographic interfaces might include games, personal communications devices, and battlefield tactical displays. Reality has been less forthcoming—a holographic-like game was introduced in the early 1980s by Cinematronics that had the illusion of being a projected holographic display (it was an illusion based on, literally, mirrors). Dragon's Lair was immensely popular at the time, but the limitations of the game and its display interface made it the last commercially available game of its type since then. A more realistic depiction of what the future might hold is shown in Figure 1.10.

Although not holographic, the new projected keyboards are similar in concept—keyboards take up space, and so for small devices, a keyboard that could appear out of thin air might be a useful interface mode (although speech might be a viable alternative as well). Figure 1.11 shows one of these virtual keyboards being projected on a flat surface.

FIGURE

1.10

Artist's depiction of what a real desktop holographic interface might look like. Note that the interface has fewer restrictions on space than physical devices, and you could easily imagine that it might be shared with a gesture-based interface to allow manipulation of space with the hands. (Courtesy of Apimac.)

With any luck, these future interfaces will be significantly more intuitive for the average user and become easier to use and hence more powerful. Early computer designers originally envisioned that computers and automation would reduce or eliminate the difficulties of performing tasks, and free people to become creative problem solvers; mundane, repetitive, and insignificant jobs would be relegated to the machines. Anyone who has tried to set a VCR clock or figure out how to format a paper in a word-processing program can easily attest to the fact that the interface often complicates the problem rather than simplify it. These new technologies and the interfaces they may lead to hold the promise of allowing the user to finally focus on the task, not the interface.

While the nontraditional interfaces presented in this book are a start, there is still a tremendous amount of work that must be done, in both the research and application realms, if we as interface designers are going to help our users move beyond the GUI.

FIGURE

1.11

Projected keyboard interface.

While not holographic in the strict technical sense of the term, these kinds of interfaces allow larger interfaces to be created where there are space limitations, as might be the case in a PDA. *Source:* Retrieved from *www.virtuallaserkeyboard. com.*

REFERENCES

Bettio, F., Frexia, F., Giachetti, A., Gobbetti, E., Pintore, G., Zanetti, G., Balogh, T., Forgács, T., Agocs, T., & Bouvier, E. (2006). A holographic collaborative medical visualization system. *Studies in Health Technology and Informatics* 119:52–54.

Burnette, E. (2006). Nintendo to Gamers: "Do Not Let Go of Wii Remote." Ed Burnette's Dev Connection, ZDNet Blog—*http://blogs.zdnet.com/Burnette/?p = 210.*

Donoghue, J. P., Nurmikko, A., Black, M., & Hochberg, L. R. (2007). Assistive technology and robotic control using motor cortex ensemble-based neural interface systems in humans with tetraplegia. *Journal of Physiology* 579(3):603–11.

Friedrich, M. J. (2004). Harnessing brain signals shows promise for helping patients with paralysis. *Journal of the American Medical Association* 291(18):2179–81.

International Standards Organization. (1998). Ergonomic Requirements for Office Work with Visual Display Terminal (VDT's)—Part 11: Guidance on Usability (ISO 9241-11(E)). Geneva, Switzerland.

Kurtenbach, G. P., Balakrishnan, R., & Fitzmaurice, G. W. (2007). Graphical user interface widgets viewable and readable from multiple viewpoints in a volumetric display. US Patent 720599. Washington, DC: U.S. Patent and Trademark Office.

Mayhew, D. J. (1997). *Principles and Guidelines in Software User Interface Design*. Englewood Cliffs, NJ: Prentice Hall.

Nielsen, J. (1999). *Designing Web Usability: The Practice of Simplicity*. Berkeley, CA: Peachpit Press.

Norman, D.A. (2002). *The Design of Everyday Things*. New York: Basic Books.

Raskin, J. (2000). *The Humane Interface: New Directions for Designing Interactive Systems*. Boston: Addison-Wesley.

Schneiderman, B., & Plaisant, C. (2004). *Designing the User Interface: Strategies for Effective Human–Computer Interaction*. Boston: Addison-Wesley.

Tanaka, K., Matsunaga, K., & Wang, H.O. (2005). Electroencephalogram-based control of an electric wheelchair. *IEEE Transactions on Robotics* 21:762–66.

CHAPTER

2 Haptic Interfaces

Marcia K. O'Malley, Abhishek Gupta

In general the word "haptic" refers to the sense of touch. This sense is essentially twofold, including both cutaneous touch and kinesthetic touch. Cutaneous touch refers to the sensation of surface features and tactile perception and is usually conveyed through the skin. Kinesthetic touch sensations, which arise within the muscles and tendons, allow us to interpret where our limbs are in space and in relation to ourselves. Haptic sensation combines both tactile and kinesthetic sensations.

The sense of touch is one of the most informative senses that humans possess. Mechanical interaction with a given environment is vital when a sense of presence is desired, or when a user wishes to manipulate objects within a remote or virtual environment with manual dexterity. The haptic display, or force-reflecting interface, is the robotic device that allows the user to interact with a virtual environment or teleoperated remote system. The haptic interface consists of a real-time display of a virtual or remote environment and a manipulator, which serves as the interface between the human operator and the simulation. The user moves within the virtual or remote environment by moving the robotic device. Haptic feedback, which is essentially force or touch feedback in a man–machine interface, allows computer simulations of various tasks to relay realistic, tangible sensations to a user. Haptic feedback allows objects typically simulated visually to take on actual physical properties, such as mass, hardness, and texture. It is also possible to realistically simulate gravitational fields as well as any other physical sensation that can be mathematically represented. With the incorporation of haptic feedback into virtual or remote environments, users have the ability to push, pull, feel, and manipulate objects in virtual space rather than just see a representation on a video screen.

The application of haptic interfaces in areas such as computer-aided design and manufacturing (CAD/CAM), design prototyping, and allowing users to manipulate virtual objects before manufacturing them enhances production evaluation. Along the same lines, the users of simulators for training in surgical procedures, control panel operations, and hostile work environments benefit from such a capability (Meech & Solomonides, 1996). Haptic interfaces can also be employed to provide force feedback during execution of remote tasks (known as teleoperation) such as telesurgery or hazardous waste removal. With such a wide range of applications, the benefits of haptic feedback are easily recognizable.

2.1   NATURE OF THE INTERFACE

This section describes the fundamental nature of haptic interfaces, introducing the basic components of a haptic display system and describing in detail the capabilities of the human haptic sensing system.

2.1.1   Fundamentals of Haptic Interfaces

A haptic interface comprises a robotic mechanism along with sensors to determine the human operator's motion and actuators to apply forces to the operator. A controller ensures the effective display of impedances, as governed by the operator's interaction with a virtual or remote environment. Impedance should be understood to represent a dynamic (history-dependent) relationship between velocity and force. For instance, if the haptic interface is intended to represent manipulation of a point mass, it must exert on the user's hand a force proportional to acceleration; if it is to represent squeezing of a spring, it must generate a force proportional to displacement (Colgate & Brown, 1994). Finally, the haptic virtual environment is rendered so as to implement the desired representation.

Haptic Interface Hardware

Haptic interface hardware consists of the physical mechanism that is used to couple the human operator to the virtual or remote environment. This hardware may be a common computer gaming joystick, a multiple-degree-of-freedom (DOF) stylus, a wearable exoskeleton device, or an array of tactors that directly stimulate the skin surface. The basic components of the hardware system include the mechanism, which defines the motion capabilities of the human operator when interacting with the device; the sensors, which track operator motion in the virtual environment; and the actuators (motors), which display the desired forces or textures to the operator as defined by the environment model. The final selection of a particular mechanism, sensor, or actuator is typically governed by the target application. Tactile and kinesthetic interfaces provide tactile and kinesthetic feedback to the operator, respectively, and will be treated separately throughout the chapter. Applications where both tactile and kinesthetic feedback is desired can employ tactile displays mounted on a kinesthetic display.

Haptic Interface Control

Haptic devices are typically controlled in one of two manners—impedance or admittance. Impedance control of a robot involves using motion input from the manipulator and calculating the corresponding forces specific to a given system model. For example, when simulating a virtual spring, when the user compresses a spring in the virtual environment, the interface applies forces to the user's hand that oppose hand motion and are proportional to spring displacement. Motion data are available from sensors on the robotic device and are sent to signal-conditioning boards—typically within a desktop personal computer—for processing. The processing calculations involve two differentiations of the position data in order to find velocity and acceleration, or one differentiation to get acceleration if velocity signals are available directly (e.g., from a tachometer). Most simple simulated environments consist only of springs that produce a force proportional to displacement, and dampers that generate forces proportional to velocity. Thus, if position and velocity signals can be obtained directly without any differentiation, impedance control of the robot is the desired approach.

Admittance control of a robot is the opposite operation. Forces are measured, usually with a load cell, and are then sent to the computer. Calculations are performed to find the corresponding motion of the end point according to the simulation's equations of motion, and position control approaches are used to move the robot accordingly. Solving for the output position involves one or two integration steps, depending on the environment model. Typically, integration is a much cleaner operation than differentiation, but problems with offsets and integrator windup are common and detract from this method of robot control. In practice, impedance-controlled interfaces are better at simulating soft, spongy environments, whereas admittance-controlled devices perform better when displaying hard surfaces.

Creating a Haptic Environment

A haptic environment is defined via a mathematical model. For the simple case of a virtual wall, the model of a spring and a damper in parallel is typically used. The higher the stiffness of the spring, the stiffer the virtual wall appears to the user. Using the impedance control mode, where end point motion is measured and force is displayed, the position of the end point is tracked to determine if the user is pushing on the virtual wall. When the plane of the wall is crossed, the corresponding force that the user should feel is calculated according to the model equation, using position sensor data and velocity data to calculate the model unknowns. This virtual wall model is illustrated in Figure 2.1, and serves as the building block for many virtual environments. Haptic rendering will not be a

1

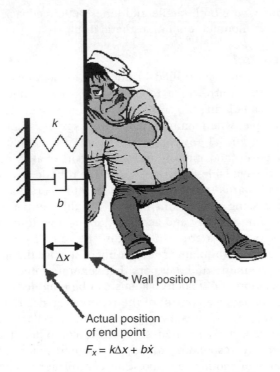

$$F_x = k\Delta x + b\dot{x}$$

FIGURE

2.1

Graphical representation of a virtual wall model.
The virtual wall is a fundamental building block of a haptic virtual environment.
It is typically implemented as a spring and damper in parallel.

focus of this chapter. However, thorough reviews and introductions to the basic concepts of haptic rendering are available, such as the work by Salisbury et al. (2004).

2.1.2   Human Haptic Sensing

Touch can be defined as the sensation evoked when the skin is subjected to mechanical, thermal, chemical, or electrical stimuli (Cholewiak & Collins, 1991). Touch is unlike any other human sense in that sensory receptors related to touch are not associated to form a single organ. Haptic receptors are of three independent modalities: pressure/touch (mechanoreception), heat and cold (thermoreception), and pain (nociception) (Schmidt, 1977). As the mechanoreceptors are responsible for tactile sensation of pressure/touch, and are the primary targets of tactile haptic devices, this section will focus on the pressure/touch modality. Kinesthetic haptic feedback is sensed through receptors in muscles and tendons, and is discussed in Section 2.1.3.

Mechanoreception comprises four sensations: pressure, touch, vibration, and tickle. The distribution of these receptors is not uniform over the body. Hairless (glabrous) skin has five kinds of receptors: free receptors (or nerve endings), Meissner's corpuscles, Merkel's disks, Pacinian corpuscles, and Ruffini endings. In addition to these receptors, hairy skin has the hair root plexus (or follicle) that detects movement on the surface of the skin. Figure 2.2 depicts the location of these receptors in the skin. Each of these mechanoreceptors responds differently to applied pressure/touch stimuli and their combined behavior determines human perception of pressure and vibrations. The study of these properties is critical to successful design of haptic interfaces for temporal as well as spatial detection and/or discrimination by the user.

Sensory adaptation is the tendency of a sensory system to adjust as a result of repeated exposure to a specific type of stimulus. Based on the rate of sensory adaptation, the receptors are classified as slow-adapting (SA) or rapid-adapting

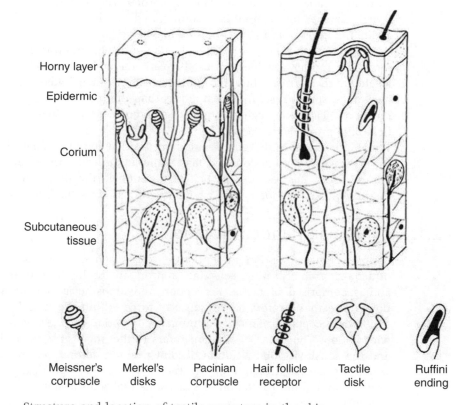

| Meissner's corpuscle | Merkel's disks | Pacinian corpuscle | Hair follicle receptor | Tactile disk | Ruffini ending |

FIGURE

2.2

Structure and location of tactile receptors in the skin.
Source: From Schmidt (1977), with kind permission of Springer Science & Business Media.

(RA) receptors. Merkel's disks produce a long but irregular discharge rate in response to forces on the skin, whereas Ruffini endings produce a regular discharge for a steady load. Meissner's corpuscles discharge mostly at the onset of stimulus, and hence best respond to velocity. Finally, Pacinian corpuscles respond once for every stimulus and are good only as vibration detectors, responding best to frequencies of 200 Hz, which is their lowest stimulus amplitude threshold (Schmidt, 1977). Hence, when designing high-frequency vibrotactile feedback, for example, the behavior of Pacinian corpuscles must be considered to ensure proper detection and discrimination of the stimuli by the user, whereas at lower frequencies, the behavior of other receptors needs to be considered as well.

Mechanoreceptors can also be characterized based on their receptive field size. This is the area in which a stimulus can excite the receptor, and varies from 1 to 2 mm^2 to up to 45 mm^2 depending on the receptor and its location on the body. Pacinian and Ruffini corpuscles have large field size and hence low spatial resolution. On the other hand, Merkel disks and Meissner's endings provide more accurate spatial localization. This is particularly important in tactile display design, as cues that cannot be discriminated by the user will fail to convey any additional information about the simulated environment.

The skin's thermoreceptors are divided into cold- and warmth-sensitive receptors. The former are located just beneath the epidermis, while the latter are located in the dermis. These receptors have a receptive field of 1 to 2 mm in diameter and a spatial resolution that is less than that of pain receptors or mechanoreceptors. Tissue-damaging stimuli trigger nociceptors. These have a receptive field of approximately 25 mm^2.

Haptic interface designers should ensure that the force feedback is sufficient for satisfactory completion of a task while at the same time being comfortable for the user. This requires particular attention to the perceptual capabilities of the human sensorimotor loop, which are discussed in the following sections.

2.1.3   Human Haptic Perception

Human haptic perception (rather than sensing, discussed previously) is the process of acquiring, interpreting, selecting, and organizing haptic sensory information, and is comprised of tactile perception and kinesthesia (including proprioception). Kinesthesia refers to the sense of force within the muscles and tendons, and proprioception refers to the human perception of one's own body position and motion. The sense of position refers to the angle of various skeletal joints, and the sensitivity or resolution of joint position determines the accuracy with which we can control our limbs. Tactile perception specifically concerns the acquisition and interpretation of sensations realized through the mechanoreceptors of the skin.

Many scientists have studied human perception thresholds in order to understand the limits of our abilities. Since the human sense of touch inherently takes place through two separate pathways, namely kinesthetic and tactile information pathways, perception studies in the human sense of touch can also be categorized with respect to the fundamental information contained within the stimuli. Irrespective of whether the dominant pathway is kinesthetic or tactile, existing studies have looked at discrimination or identification of surface properties (e.g., shape and surface texture) and volumetric properties (e.g., mass and sponginess) of objects.

Current studies of the just noticeable differences (JNDs) for kinesthetic and tactile senses have focused on discrimination of geometries, textures, and volumetric properties of objects held by the human, or have focused on discrimination of the subject's own limb movements; see Durlach and Mavor (1995) for a comprehensive review. The JND is the smallest difference in a specified modality of sensory input that is detectable by a human. It is also referred to as the difference limen or the differential threshold.

Early kinesthetic studies by Clark and colleague and Jones and Hunter (Clark & Horch, 1986; Clark, 1992; Jones & Hunter, 1992) investigated human perception of limb positions and concluded that humans are capable of detecting joint rotations of a fraction of a degree performed over a second of time interval. Jones and Hunter (1992) also reported the differential threshold for limb movement as 8 percent. Further psychophysical experiments conducted by Tan and colleagues (1994) determined the JND for the finger joints as 2.5 percent, for the wrist and elbow as 2 percent, and for the shoulder as 0.8 percent.

Durlach and colleagues (1989) investigated the length resolution for rigid objects held in a pinch grasp between the thumb and the forefinger (Durlach et al., 1989). Commonly accepted perception thresholds for length resolution are given as about 1 mm for a reference length of 10 mm, increasing to 2 to 4 mm for a reference length of 80 mm. For purposes of comparison, the thickness of a penny is approximately 1.57 mm, whereas its diameter is about 19 mm.

Later experiments focusing on object size characterized the effect of varying levels of force output and virtual surface stiffness on the ability of human subjects to perform size identification and size discrimination tasks in a simulated environment (O'Malley & Goldfarb, 2002, 2004; Upperman et al., 2004; O'Malley & Upperman, 2006). In an application where haptic cues are provided for navigation, detection of the stimuli is important and not their discrimination from each other. In such a scenario, low forces and virtual surface stiffness may suffice. Note that these cues will feel soft or squishy due to low force and stiffness levels. On the other hand, tasks that require size discrimination, such as palpation in a medical trainer, require larger force and stiffness values and, consequently, a haptic interface capable of larger force output and of higher quality. Recently, McKnight and colleagues (2004) extended these psychophysical size discrimination experiments to include two- and three-finger grasps.

The bandwidth of the kinesthetic sensing system has been estimated at 20 to 30 Hz (Brooks, 1990). In other words, the kinesthetic sensing system cannot sense movements that happen more frequently than 30 times in a second. Hence, in studies on perception of high-frequency vibrations and surface texture, the tactile pathway serves as the primary information channel, whereas the kinesthetic information is supplementary. Early tactile perception studies concluded that the spatial resolution on the finger pad is about 0.15 mm for localization of a point stimulus (Loomis, 1979) and about 1 mm for the two-point limen (Johnson & Phillips, 1981). Other parts of the body have much less spatial resolution. For example, the palm cannot discriminate between two points that are less than 11 mm apart (Shimoga, 1993).

A related measure, the successiveness limen (SL), is the time threshold for which subjects are able to detect two successive stimuli. An approximate SL value for the mechanoreceptors is 5 msec, with a required interval of 20 msec to perceive the order of the stimuli. The human threshold for the detection of vibration of a single probe is reported to be about 28 dB for the 0.4- to 3-Hz range. An increase in level of 6 dB represents a doubling of amplitude, regardless of the initial level. A change of 20 dB represents a change in amplitude by a factor of 10. This threshold is shown to decrease at the rate of −5 dB per octave in the 3- to 30-Hz range, and at the rate of −12 dB per octave in the 30- to about 250-Hz range, with an increase for higher frequencies (Rabinowitz et al., 1987; Bolanowski et al., 1988).

2.1.4   Sensory Motor Control

In addition to tactile and kinesthetic sensing, the human haptic system includes a motor subsystem. Exploratory tasks are dominated by the sensorial part of the sensory motor loop, whereas manipulation tasks are dominated by the motor part (Jandura & Srinivasan, 1994). The key aspects of human sensorimotor control are maximum force exertion; force-tracking resolution; compliance, force, and mass resolution; finger and hand mechanical impedance; and force control bandwidth.

Maximum Force Exertion

A maximum grasping force of 400 N for males and 228 N for females was measured in a study by An and coworkers (An et al., 1986). In a study on maximum force exertion by the pointer, index, and ring fingers, it was found that the maximum force exerted by the pointer and index fingers was about 50 N, whereas the ring finger exerted a maximum force of 40 N (Sutter et al., 1989). These forces were found to be constant over 0 to 80 degrees of the metacarpal (MCP) joint angle. This work was later extended to include the proximal-interphalangeal (PIP) joints and MCP joints, as well as the wrist, elbow, and shoulder (with arm extended to the side and in front) (Tan et al., 1994). Note that the maximum force exertion capability is dependent on the user's posture. It was found that maximum force exertion grows from the most distal joint in the palm to the most proximal joint (shoulder). In addition, it was found that controllability over the maximum force decreased from the shoulder to the PIP joint. In order to ensure user safety, a haptic interface should never apply forces that the user cannot successfully counter.

Sustained Force Exertion   Prolonged exertion of maximum force leads to fatigue. Fatigue is an important consideration when designing feedback for applications like data visualization where force feedback may be present for extended periods of time. Wiker and colleagues (1989) performed a study of the relationship between fatigue during grasping as a function of force magnitude, rest duration, and progression of the task. The tests showed a direct correlation between magnitude of discomfort and magnitude of pinch force. The work-versus-rest ratio was not found to be important for low forces but was effective in reducing fatigue for high pinch forces.

Force-Tracking Resolution

Force-tracking resolution represents the human ability to control contact forces in following a target force profile. Srinivasan and Chen (1993) studied fingertip force tracking in subjects using both constant and time-varying (ramp and sinusoid) forces. For some participants, a computer monitor provided a display of both the actual and target forces. Subjects also performed the tests under local cutaneous anesthesia. It was found that when no visual feedback was available, the absolute error rate increased with target magnitude. When visual feedback was present, the error rate did not depend on target magnitude.

Compliance Resolution

Compliance, or softness, resolution is critical in certain applications such as training for palpation tasks or telesurgery, since many medical procedures require accurate discrimination of tissue properties. The following discussion presents a short summary of the literature on compliance resolution both with and without the presence of additional visual or auditory clues. If a haptic interface is to be used for exploratory tasks that require discrimination among objects based on their compliance, then designers should ensure that the simulated virtual objects appear sufficiently different to the human operator.

Human perception of compliance involves both the kinesthetic and tactile channels since spatial pressure distribution within the contact region sensed through the tactile receptors plays a fundamental role in compliance perception. However, for deformable objects with rigid surfaces, the information available through the tactile sense is limited and kinesthetic information again becomes the dominant information channel. In such cases, human perceptual resolution is much lower than the cases for compliant objects with deformable surfaces (Srinivasan & LaMotte, 1995). In studies involving deformable objects with rigid surfaces, Jones and Hunter (1990, 1992) reported the differential thresholds for stiffness as 23 percent. Tan and colleagues (1992, 1993) observed that for such objects held in a pinch grasp, the JND for compliance is about 5 to 15 percent when the displacement range is fixed, and 22 percent when the displacement range is randomly varied. Moreover, they reported a minimum stiffness value of 25 N/m (Newtons per meter) for an object to be perceived as rigid (Tan et al., 1994).

In further studies, Tan and coworkers (1995) investigated the effect of force work cues on stiffness perception and concluded that JND can become as high as 99 percent when these cues are eliminated. Investigating the effect of other cues on compliance perception, DiFranco and colleagues (1997) observed the importance of auditory cues associated with tapping harder surfaces and concluded that the objects are perceived to be stiffer when such auditory cues are present.

In a similar study, Srinivasan and coworkers (1996) reported dominance vision in human stiffness perception. In related studies, Durfee et al. (1997) investigated the influence of haptic and visual displays on stiffness perception, while O'Malley and Goldfarb (2004) studied the implications of surface stiffness for size identification and perceived surface hardness in haptic interfaces. Observing the importance of the initial force rate of change in stiffness perception, Lawrence and colleagues (2000) proposed a new metric for human perception of stiffness, called rate hardness.

Force Resolution

In related experiments, human perceptual limitations of contact force perception—when the kinesthetic sense acts as the primary information channel—have been studied. The JND for contact force is shown to be 5 to 15 percent of the reference force over a wide range of conditions (Jones, 1989; Pang et al., 1991; Tan et al., 1992). Accompanying experiments revealed a JND value of about 10 percent for manual resolution of mass (Tan et al., 1994), while a JND value of about 34 percent has been observed for manual resolution of viscosity (Jones & Hunter, 1993). Recently, Barbagli and colleagues (2006) studied the discrimination thresholds of force direction and reported values of 25.6 percent and 18.4 percent for force feedback only and visually augmented force feedback conditions, respectively. Special attention is required when designing feedback for applications like grasping and manipulation, where subtle changes in force can be important, such as in making the difference between holding and dropping an object in the virtual environment.

Mechanical Impedance

The impedance of the human operator's arm or finger plays an important role in determining how well the interface performs in replicating the desired contact force at the human–machine contact point. Hajian and Howe (1997) studied the fingertip impedance of humans toward building a finger haptic interface. Over all subjects and forces, they estimated the equivalent mass to vary from 3.5 to 8.7 g, the equivalent damping from 4.02 to 7.4 Ns/m, and stiffness from 255 to

1,255 N/m. It was noted that the damping and stiffness increased linearly with force. In similar work, Speich and colleagues (2005) built and compared two- and five-DOF models of a human arm toward design of a teleoperation interface.

Sensing and Control Bandwidth

Sensing bandwidth refers to the frequency with which tactile and/or kinesthetic stimuli are sensed, and control bandwidth refers to the frequencies at which the human can respond and voluntarily initiate motion of the limbs. In humans, the input (sensory) bandwidth is much larger than the output bandwidth. As noted earlier, it is critical to ensure that the level of haptic feedback is sufficient for task completion while being comfortable for the user. In a review paper, Shimoga showed that the hands and fingers have a force exertion bandwidth of 5 to 10 Hz, compared to a kinesthetic sensing bandwidth of 20 to 30 Hz (Shimoga, 1992). Tactile sensing has a bandwidth of 0 to 400 Hz. Keeping this in mind, if we design an application that requires repetitive force exertion by the user, to guarantee user comfort the required rate should not be more than 5 to 10 times a second. Similarly, any kinesthetic feedback to the user should be limited to 20 to 30 Hz.

2.2   TECHNOLOGY OF THE INTERFACE

Haptic interfaces are a relatively new technology, with increased use for human interaction with virtual environments since the early 1990s. A primary indicator of the proliferation of haptic devices is the number of companies that now market devices, including Sensable Technologies, Immersion, Force Dimension, Quanser, and Novint, among others. The commercialization of haptic devices is due primarily to technological advances that have reduced the cost of necessary components in haptic systems, including materials, actuation, sensing, and computer control platforms.

Novel materials, such as carbon fiber tubing, have enabled the design and fabrication of light-weight yet stiff kinematic mechanisms that are well suited to the kinesthetic type of haptic display. Power-dense actuators, such as brushless DC motors, have allowed for increased magnitude force output from haptic devices with minimal trade-offs in terms of weight. However, it should be noted that actuation technology is still a key limitation in haptic device design, since large forces and torques obtained via direct drive actuation are often desired, while still achieving minimal inertia (mass) in the mechanism. Improved sensor technology has also enabled an increase in the availability of high-quality haptic interface hardware. The key requirement of sensors for haptic applications is high resolution, and many solutions such as optical encoders and noncontact potentiometers are providing increased resolution without compromising the back-drivability of haptic devices due to their noncontact nature.

The final set of technological advances is in the area of computational platforms. First, data acquisition systems, which enable transformation from analog and digital signals common to the sensors and actuators to the digital computation carried out by the control computer, are achieving higher and higher resolutions. Second, real-time computation platforms and increasing processor speeds are enabling haptic displays (typically rendered at a rate of 1,000 Hz) to exhibit increasingly greater complexity in terms of computation and model realism. This in turn broadens the range of applications for which haptic feedback implementation is now feasible. Finally, embedded processors and embedded computing are enabling haptic devices to be more portable.

2.3 CURRENT INTERFACE IMPLEMENTATIONS

Over the last several years, a variety of haptic interfaces have been developed for various applications. They range from simple single-DOF devices for research (Lawrence & Chapel, 1994) to complex, multi-DOF wearable devices (Frisoli et al., 2005; Kim et al., 2005; Gupta & O'Malley, 2006). DOF refers to the number of variables required to completely define the pose of a robot. A higher-DOF device has a larger workspace—the physical space within which the robot end point moves—as compared to a low-DOF device of similar size. Haptic devices are also used in various applications (Hayward et al., 2004). For instance, haptic interfaces have been employed for augmentation of graphical user interfaces (GUIs) (Smyth & Kirkpatrick, 2006), scientific data visualization (Brooks et al., 1990), enhancement of nanomanipulation systems (Falvo et al., 1996), visual arts (O'Modhrain, 2000), CAD/CAM (Nahvi et al., 1998; McNeely et al., 1999), education and training, particularly surgical training (Delp et al., 1997), master interfaces in teleoperation (Kim et al., 2005), rehabilitation (Bergamasco & Avizzano, 1997; Krebs et al., 1998), and the scientific study of touch (Hogan et al., 1990; Weisenberger et al., 2000).

The PHANToM desktop haptic interface (Sensable Technologies), shown in Figure 2.3, is probably the most commonly used haptic interface. It is a pen- or stylus-type haptic interface, where the operator grips the stylus at the end of the robot during haptic exploration. The PHANToM desktop device has a workspace of about 160 W × 120 H × 120 D mm. The device provides feedback to the operator in three dimensions with a maximum exertable force capability of 1.8 foot-pounds (lb_f) (7.9 N) and a continuous exertable force capability (over 24 hours) of 0.4 lb_f (1.75 N). A number of device models are available that vary in workspace and force output specifications. Several other haptic interfaces are commercially available, such as the six-DOF Delta haptic device (Force Dimension), the three-DOF planar pantograph (Quanser), and the force feedback hand controller (MPB Technologies) (Figure 2.3).

The common feature of most commercially available haptic devices is that they are point contact devices, in that the end point of the robot is mapped to a

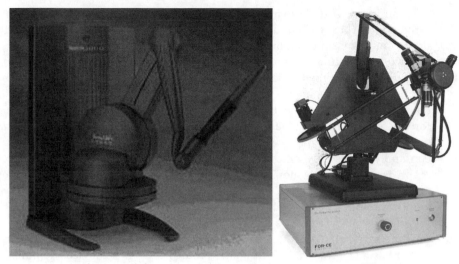

(a)   (b)

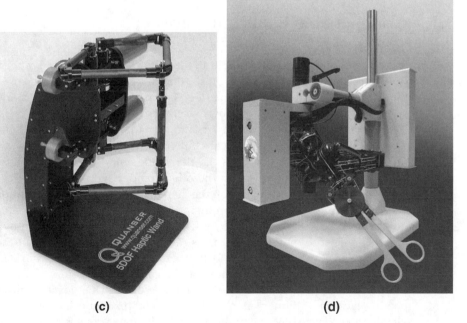

(c)   (d)

FIGURE   Selected commercially available haptic interfaces.

2.3   (a) PHANToM desktop, (b) six-DOF Delta haptic interface, (c) three-DOF
              planar pantograph, and (d) force feedback hand controller. (Courtesy of
              (a) SensAble Technologies, Inc.; (b) Francois Conti; (c) Quanser Inc.;
              and (d) MPB Technologies Inc.

position in the virtual environment and forces are applied back to the user at the same point. Thus, within the virtual environment, the user can interact with only one point. One can think of this as being similar to interacting with objects in the real world with the aid of tools like a pen, screwdriver, or scalpel. Even with this limitation, these types of devices have been employed in applications such as scientific visualization, augmentation of GUIs, CAD, and psychophysical studies. The choice of a specific device depends on the desired workspace and DOF, the type of force feedback desired, and the magnitude of forces to be displayed. For example, the PHANToM can move within the three-dimensional (3D) physical space, but can apply forces only on the user's hands. In comparison, the three-DOF pantograph is restricted to moving in a plane, but can apply a torque or wrench to the user's hand in addition to forces in the two planar directions.

In the following subsections, we take a closer look at selected current implementations of haptic interfaces. The examples presented have been chosen to demonstrate essential features, and do not necessarily represent the state of the art but rather basic features of their respective categories. These devices demonstrate the wide range of technologies involved in haptics.

2.3.1   Nonportable Haptic Interfaces

Haptic Joysticks

Joysticks are widely used as simple input devices for computer graphics, industrial control, and entertainment. Most general-purpose joysticks have two DOF with a handle that the user can operate. The handle is supported at one end by a spherical joint and at the other by two sliding contacts (Figure 2.4). Haptic joysticks vary in both mechanical design and actuation mechanisms. Adelstein and Rosen (1992) developed a spherical configuration haptic joystick for a study of hand tremors. Spherical joysticks, as the name implies, have a sphere-shaped workspace. Cartesian joysticks, on the other hand, have two or three orthogonal axes that allow the entire base of the handle to translate. An example is the three-DOF Cartesian joystick comprising a moving platform that slides using guiding blocks and rails proposed by Ellis and colleagues (1996). The moving block supports an electric motor that actuates the third DOF (rotation about the z-axis). This Cartesian joystick (Figure 2.5) has a square-shaped planar workspace, and each axis is actuated using DC motors and a cable transmission.

Other examples of haptic joysticks include a four-DOF joystick based on the Stewart platform (Millman et al., 1993) and a magnetically levitated joystick (Salcudean & Vlaar, 1994). The magnetically levitated joystick has no friction at all and is particularly suited for display of small forces and stiff contact.

These haptic joysticks are point contact devices, and each type varies according to workspace shape and size. While the spherical joystick can be used with just wrist movements, Cartesian joysticks require the user to employ other joints of

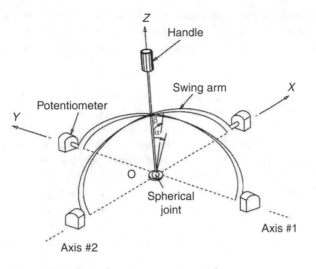

FIGURE   Two-DOF slotted swing arm joystick.
           Source: From Adelstein and Rosen (1992).
2.4

FIGURE   Three-DOF Cartesian joystick.
           The joystick comprises a central stage that moves using a guiding block and rails.
2.5   *Source*: From Ellis et al. (1996).

the arm, like the elbow or the shoulder. Consequently, the workspace and force
output of Cartesian joysticks can be greater than that of similarly sized spherical
models. Note that most commercially available force feedback joysticks, typically
marketed for computer gaming applications, lack the quality necessary to achieve

high-quality feedback for exploratory or manipulative tasks. While they may suffice to provide haptic cues that can be detected, high-quality hardware and a fast computer platform are necessary to ensure proper discrimination of cues.

Pen-Based Masters

Pen-based haptic devices allow interaction with the virtual environment through tools such as a pen (or pointer) or scalpel (in surgical simulations). These devices are compact with a workspace larger than that of spherical and magnetically levitated joysticks and have three to six DOF. The best-known example of a pen-based haptic interface is the PHANToM, mentioned earlier in this section. Originally developed by Massie and Salisbury (1994), the PHANToM is an electrically actuated serial-feedback robotic arm that ends with a finger gimbal support that can be replaced with a stylus (Figure 2.6). The gimbal orientation is passive and the serial arm applies translational forces to the operator's fingertip or hand. A six-DOF interface that can apply forces as well as torques to the operator is presented

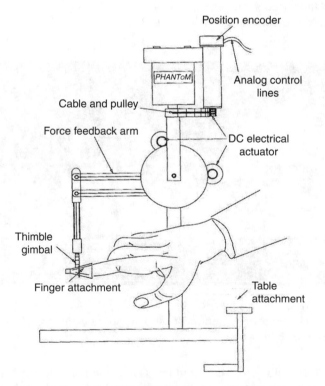

FIGURE

2.6

Schematic of the PHANToM desktop haptic interface.
Source: From Massie and Salisbury (1994).

in Iwata (1993), extending the complexity of force and torque interactions between a human operator and the remote or virtual environment.

Floor- and Ceiling-Mounted Interfaces

Generally, floor- and ceiling-mounted interfaces are larger and more complex and expensive than desktop devices. They have a large force output, and as a result user safety becomes critical. This is especially true for exoskeletons where the operator is inside the device workspace at all times. Figure 2.7 shows one of the first generalized master arms that was developed at the National Aeronautical and Space Administration (NASA) Jet Propulsion Laboratory (JPL) (Bejczy & Salisbury, 1980). It is a six-DOF interface with a three-axis hand grip that slides and rotates about a fixed support attached to the floor. The hand grip can apply forces up to 9.8 N and torques up to 0.5 N/m. The JPL device is another example of a point contact haptic interface where the forces are applied at the user's hands. As compared to joysticks or desktop devices, though, it provides a much larger work volume with greater force output capabilities, coupled with greater freedom of arm movement. These larger devices are useful for remotely manipulating large robotic manipulators like those used in space.

An example of a grounded exoskeletal haptic interface is the MAHI arm exoskeleton (Figure 2.8) built at Rice University (Gupta & O'Malley, 2006; Sledd & O'Malley, 2006). This five-DOF exoskeleton was designed primarily for rehabilitation and training in virtual environments. The device encompasses most of the human arm workspace and can independently apply forces to the elbow, forearm, or wrist joints. Note that this is no longer a point contact device, but can provide independently controlled feedback to various human joints. This feature makes it extremely suitable as a rehabilitation interface that allows the therapist to focus treatment on isolated joints.

FIGURE

2.7

Six-DOF JPL arm master.
Two hand controllers used by a human operator. *Source*: From O'Malley and Ambrose (2003).

FIGURE

2.8

MAHI haptic arm exoskeleton.
The five-DOF arm exoskeleton applies forces to the operator's elbow, forearm, and wrist joints. *Source*: From Sledd and O'Malley (2006); © 2006 IEEE.

2.3.2   Portable Haptic Interfaces

All elements of portable haptic interfaces are worn by the user. Based on their mechanical grounding, they can be classified as arm exoskeletons or hand masters. Arm exoskeletons are typically attached to a back plate and to the forearm. Hand masters, on the other hand, are attached to the user's wrist or palm. As compared to point contact devices, exoskeletal devices are capable of measuring the location of various human joints and can provide feedback at multiple locations. Thus, with an exoskeleton-type interface the user is no longer restricted to interacting with a single point in the workspace, but can use the whole arm, as with an arm exoskeleton, or grasp and manipulate multidimensional objects using a hand master. In addition, wearable devices have a workspace that is comparable to the natural human workspace.

One of the earliest modern haptic arm exoskeletons was developed by Bergamasco and colleagues (Bergamasco et al., 1994). The five-DOF arm provides feedback to the shoulder, elbow, and forearm joints using DC motors and a complex cable transmission. The user controls the exoskeleton through a handle attached to the last rigid link. The device weighs 10 kg and can apply torque up to 20 N/m at the shoulder, 10 N/m at the elbow, and 2 N/m at the wrist joint. Recently, Bergamasco and colleagues (Frisoli et al., 2005) developed a newer version of the device, the Light Exoskeleton (Figure 2.9), which has improved weight and torque output properties.

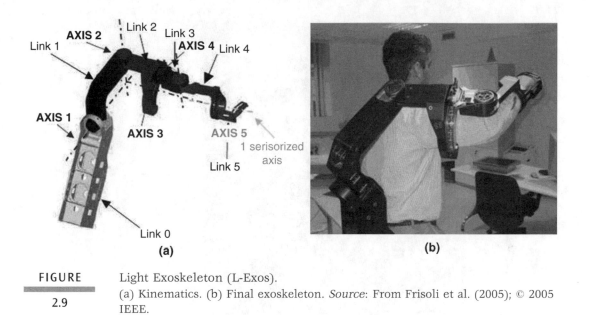

Link 1
AXIS 2
Link 2
Link 3
AXIS 4
Link 4
AXIS 1
AXIS 3
AXIS 5
1 serisorized axis
Link 5
Link 0
(a)
(b)

FIGURE
2.9
Light Exoskeleton (L-Exos).
(a) Kinematics. (b) Final exoskeleton. *Source*: From Frisoli et al. (2005); © 2005 IEEE.

An example of a portable hand master is the Rutgers Master II built at Rutgers University (Figure 2.10). The Rutgers Master II incorporates the robot's actuators into the palm of the user, thereby eliminating the need for a transmission or bulky cable routing over the backs of the fingertips. The total weight of the interface is about 100 g, and it can apply forces of up to 4 N at the fingertip. With this hand master, the positions of each of the four fingers can be separately mapped in the virtual or remote environment and respective forces displayed back to the user. This makes it an ideal interface for tasks where grasp or manipulation of objects is desirable. Examples of such an application include palpation, virtual tours of homes and museums, and remote manipulation of robotic grippers. One drawback of the design is a limitation on the tightness of the grip that can be achieved due to location of the actuators within the palm.

2.3.3   Tactile Interfaces

Tactile interfaces convey tactual information, that is, information related to heat, pressure, vibration, and pain. Just like the wide range of stimuli displayed by tactile interfaces, the interfaces themselves come in various designs with a variety of sensing and actuation technologies. Hence, classification of tactile interfaces is nearly impossible, and no single interface is representative of the state of art in tactile interface design. Most tactile interfaces provide feedback to the fingertips of the operator, although some interfaces intended for other parts of the body, like

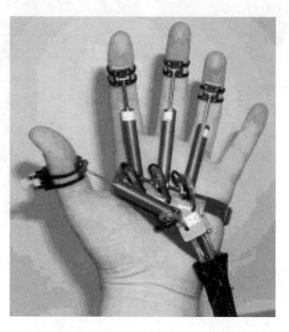

Rutgers Master II.
This hand master employs pneumatic actuators in the palm, thereby eliminating the need for a transmission. *Source*: From Bouzit et al. (2002); © 2002 IEEE.

the back, have also been implemented. In this section, we take a look at vibrotactile displays, which are one of the most common forms of tactile interfaces, and tactile interfaces for the torso.

Vibrotactile Interfaces

These are tactile interfaces for conveying vibratory information to an operator. Applications where vibrotactile displays can be particularly useful include inspection tasks, texture perception, scientific data visualization, and navigational aids. Kontarinis and Howe (1995) were the first to present design guidelines for implementation of vibration displays. Based on the properties of the human tactile system, they noted that a vibration display device should produce mechanical vibrations in the range of 60 to 1,000 Hz with variable amplitude and frequency. In order to achieve this goal, they employed modified 0.2-Watt loudspeakers. The complete setup is shown in Figure 2.11. The range of motion of the device is 3 mm, and it can produce up to 0.25 N peak force at 250 Hz. The user grasps the master manipulator as shown in the figure. Another similar robot manipulator, known as the slave, that has acceleration sensors mounted on its ends, sends

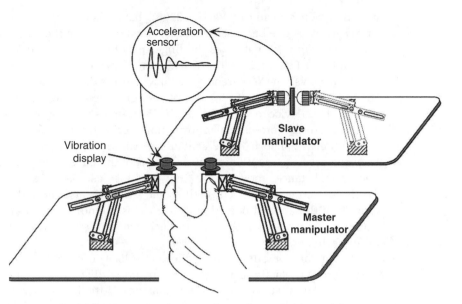

FIGURE

2.11

Loudspeaker-based vibrotactile display.
Vibrations are conveyed to a human operator through brackets mounted at the end of the master manipulator. Subjects perform various tasks in a teleoperation setting. *Source*: From Kontarinis and Howe (1995).

back data related to the vibrations felt while grasping an object. These data are then displayed to the human operator.

Okamura and colleagues (2001) developed decaying sinusoidal waveform-based models for vibration feedback during haptic interaction. Through experiments performed with real materials, they recorded amplitude, frequency, and decay rates of vibrations during impact events. They noted that for some materials the parameters were beyond the bandwidth of their haptic display and hence the interface was not capable of displaying those vibrations. These authors reported that incorporation of vibration feedback along with force feedback led to improved performance during material discrimination tasks. Experiments were conducted using the 3GM haptic interface by Immersion. Similar results were presented in Okamura et al. (1998) using the IE 2000 joystick, also by Immersion.

Wearable Tactile Interfaces

Most tactile interfaces are made for the fingertip, given its high perceptual sensitivity. However, given the small size of the fingertip, tactile interfaces for other parts of the body, including the torso (Ertan et al., 1998; Traylor & Tan, 2002) and the mouth (Tang & Beebe, 2006), have also been explored. The torso is primarily attractive because the surface area of skin on the back can convey twice the amount of information as the fingertips (Jones et al., 2004).

Tan and Pentland (1997) provide an overview of technologies for wearable tactile displays. The researchers also developed a directional haptic display, called the "rabbit," composed of a 3×3 array of nine vibrotactile actuators for the back of a user. The device makes use of the "sensory saltation" phenomenon (Gerald, 1975) to provide directional information to the human user. If tactile cues are sequentially applied to three spatially separated points on the arm of a subject, then the actual perception of the subject is of the cues to be uniformly distributed over the distance between the first and the last tactile actuator. This spatial resolution of the discrete cues felt by the subject is also more than the spatial resolution of the applied cues themselves. This phenomenon is known as "sensory saltation," and allows researchers to achieve high spatial resolution with the use of few actuators.

In similar work, Jones and colleagues (2004) presented the design of a tactile vest using vibrotactile actuators for directional display of spatial information to the blind. The vest provided directional cues to the blind user through the tactile actuators mounted on the back. They evaluated and compared four different electrical actuators for the vibrotactile vest, and chose a pancake motor after considering peak frequency, power requirements, and size of the actuators. They found that the participants identified the directions 85 percent of the time, with most errors being in the diagonal direction. These results indicate that wearable tactile interfaces are promising candidates to serve as navigational aids for the disabled.

Jones and colleagues (2004) also built and tested a shape memory alloy (SMA)–based tactor unit for tactile feedback to the torso (Figure 2.12). The unit had an overall height of 17 mm, length of 42 mm, and width of 22 mm. In experiments, the tactor unit produced a peak force in the range of 5 to 9 N, with an average of 7 N with a displacement of 3 mm. The bandwidth of the tactors was less than 0.3 Hz. In experimental studies, tactors were arranged in 1×4 and 2×2 arrays, and were activated sequentially as well as together. They noted that although the users perceived the stimulation, it was not well localized and felt like firm pressure such as a finger prodding the skin. Furthermore, stimulations on fleshier areas of the back were found to be more easily detectable than near the spinal cord. These experiments suggest that SMA tactors may be used for tactile feedback to the torso, which can lead to lighter, more compact vests due to better power-to-weight characteristics of the SMA as compared to electrical actuators.

2.3.4   Applications of Interface to Accessibility

The modalities of both haptics and vision are capable of encoding and decoding important structural information on the way object parts relate to each other in a 3D world (Ballestero & Heller, 2006). Due to this similarity in the role of the haptic and visual modalities, engineers and researchers have been interested in the

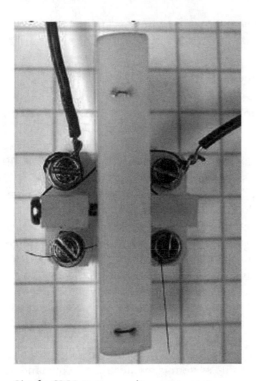

FIGURE

2.12

Single SMA tactor unit.
SMA-based vests have the potential to be lighter and more compact than vibrotactile ones. *Source*: From Jones et al. (2004); © 2004 IEEE.

use of touch in the absence of vision. The Tactile Vision Substitution System (TVSS) was, perhaps, one of the most dramatic early examples of this interest (Bach-y-Rita, 1972). The original TVVS employed a camera connected to a computer and an array of vibrating stimulators on the skin of the back. The basic idea was to allow people to "see" with their skin. A derivative of the TVVS was the Optacon, a portable tactile display to permit blind people to read printed material. The main unit of the Optacon contained a template or "array" with 144 tiny pins. The pins of the array vibrated to create a tactile image of alphabets and letters as a camera lens was moved over them.

Recent advances in haptic interfaces have led to renewed research efforts to build haptic interfaces for the blind or visually impaired. There are three types of haptic interfaces for accessibility: devices like the Optacon that tactually display material to be read, haptic navigational aids for navigation without sight, and haptic interfaces for web or computer access. Christian (2000) provides a broad overview of haptic display design for blind users. He notes that even though little research has focused on the design of tactile displays for the blind, already the tactile displays outperform speech interfaces both in terms of speed and performance. Also, tactile displays, as compared to auditory ones, speak a universal language. They can be understood by any blind user, regardless of language.

Haptic Braille Displays

A Braille display is a device, typically attachable to a computer keyboard, that allows a blind person to read the textual information from a computer monitor one line at a time. Each Braille character consists of six or eight movable pins in a rectangular array. The pins can rise and fall depending on the electrical signals they receive. This simulates the effect of the raised dots of Braille impressed on paper. Several Braille displays are commercially available.

Pantobraille is a single-cell bidirectional Braille display developed at the Centre for Information Technology Innovation, Canada (Ramstein, 1996). A Braille module is coupled with the pantograph, a planar haptic display. The device provides the user with a combination of tactile stimulation and strong feedback. It has a workspace of 10 cm × 16 cm and, unlike traditional Braille displays, allows the reader to move over the material in a bidirectional fashion. In a pilot study with two users, Ramstein found that the users preferred the interface over the Optacon, even though no significant improvement in performance over large Braille displays was realized.

HyperBraille is another text screen–oriented application that integrates tools for creating, retrieving, and sharing printed and electronic documents (Kieninger, 1996). As compared to other similar tools that provide blind users access to specific applications, the HyperBraille system promotes the use of standard document formats and communication protocols. For example, various parts of a letter such as the sender, recipient, and body are prelabeled, and HyperBraille automatically generates links to take the blind reader to those parts of the documents.

Haptic Access to Graphical Information

Unlike the Braille displays that provide textual information to the blind, some haptic interfaces provide blind users access to the elements of a regular GUI. An example of such a device is the Moose, a haptic mouse (O'Modhrain & Gillespie, 1997). The Moose, shown in Figure 2.13, is effectively a powered mouse that displays elements of a GUI using haptic cues. For example, window edges are represented by grooves, and checkboxes use attractive and repulsive force fields. Yu and colleagues (2000) have investigated the use of haptic graphs for data visualization in blind users. Based on experiments on blind and sighted users, they recommend the use of engravings and textures to model curved lines in haptic graphs. Furthermore, they propose the integration of surface properties and auditory cues to aid the blind user.

The previously mentioned interfaces allow exploration of a GUI in two dimensions. The Haptic and Audio Virtual Environment (HAVE) developed as part of the European Union GRAB project seeks to provide blind users access to 3D virtual environments. This is achieved by the means of a dual-finger haptic interface, as shown in Figure 2.14. The haptic display is further augmented with the use of audio input and output. Wood and colleagues (2003) evaluated the interface through a simple computer game for blind users and concluded that users can easily find and identify objects within

FIGURE

2.13

Moose haptic interface.
The Moose reinterprets the Microsoft Windows screen for blind users.
Source: From O'Modhrain and Gillespie (1997).

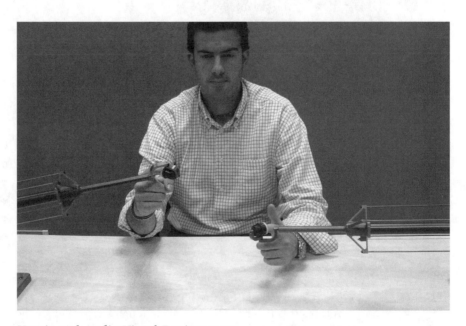

FIGURE

2.14

Haptic and Audio Visual Environment.
HAVE provides a multimodal display for 3D exploration for blind users. The haptic
interface is pictured. (Courtesy of Scuola Superiore Sant'Anna, Lab. PERCRO.)

the game, and can cause changes in the game environment and perceive them. In addition, all users improved quickly and reported an immersive experience.

Haptic Navigational Aids

Haptic navigational aids, as the name implies, attempt to provide navigational information to blind users. As compared to auditory aids, the haptic signals provided by the haptic aids cannot be confused with environmental signals by the blind user. The two can also be used together to augment each other. Ertan and colleagues (1998) presented a wearable haptic guidance system that uses a 4 × 4 grid of micromotors to tactually provide navigational information to a user's back. In addition to the tactile display, the proposed interface comprises an infra-red-based system to locate the user in the environment and a computer for route planning. In similar work, a tactile vest (see Section 2.3.3) has been developed at the Massachusetts Institute of Technology (MIT) that provides navigational information to the user (Jones et al., 2004). The authors note that if the total area of the skin is considered, the torso can convey twice the amount of information of the fingertips. In more recent work, Tang and colleagues (2006) presented an oral tactile navigational aid for the blind (Figure 2.15).

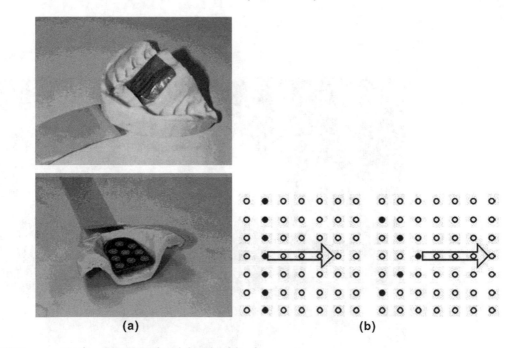

(a)   (b)

FIGURE

2.15

Oral navigational aid for the blind.
(a) Aid fits the upper teeth and tongue touch keypad for the bottom (*left*).
(b) Moving lines and arrows are used to provide directional cues as shown on the right. *Source*: From Tang and Beebe (2006); © 2006 IEEE.

The device is a mouthpiece with a microfabricated electrotactile display to provide tactile information to the roof of the mouth. A tongue touch keypad is provided for simultaneous operation. The device provides directional cues to the operator in four different directions—left, right, forward, and backward—using lines and arrows. In preliminary experiments, the researchers found that small electrical signals were sufficient for stimulating the roof of the mouth. In a preliminary experimental study, user performance was found to be good for discrimination of left and right signals, and mixed for the forward and backward ones.

2.4   HUMAN FACTORS DESIGN OF THE INTERFACE

There are two basic functions of haptic interfaces. First, the device is used to measure the motion (position, velocity, and possibly acceleration) and the contact forces of the user's entire body or arm, foot, or hand. Second, the device is used to display contact forces and motion along with spatial and temporal distributions to the user (Tan et al., 1994). While current technology in haptic interfaces is limited, allowing the display of only approximate interactions with a comparable real environment, the feedback experienced via a haptic device can feel very realistic, and can indeed improve human performance and sensations of realism when interacting with a virtual environment. These experiences are primarily attributed to the device's ability to exploit limitations of the human tactile and kinesthetic sensory channels. To specify a haptic interface for a given application, it is therefore necessary to understand the biomechanical, sensorimotor, and cognitive abilities of the human system (Tan et al., 1994). Sections 2.4.1 and 2.4.2 discuss the possible uses of haptic interfaces and the factors to be considered when selecting interfaces for a particular application, respectively. Section 2.4.3 is aimed at readers interested in designing and building their own interfaces.

2.4.1   When to Select a Haptic Interface

Haptic interfaces have a number of beneficial characteristics, such as enabling perception of limb movement and position, improving skilled performance of tasks (typically in terms of increased precision and speed of execution of the task), and enabling virtual training in a safe and repeatable environment. Force feedback has been shown, specifically for teleoperator systems, to improve performance of the operator in terms of reduced completion times, decreased peak forces and torques, and decreased cumulative forces and torques (Hill, 1979; Draper et al., 1987; Hannaford et al., 1991; Kim, 1991; Massimino & Sheridan, 1994; Murray et al., 1997; Williams et al., 2002; O'Malley & Ambrose, 2003). For training, virtual environments can provide a setting for safe, repeatable practice, and the inclusion of haptic feedback in such environments improves feelings of realism in the task, increasing the likelihood for skill transfer from the virtual to the real environment.

Haptic feedback is also shown to support hand–eye coordination tasks, specifically improving performance in dexterous manipulation (Hale & Stanney, 2004). Broadly speaking, haptic feedback can effectively alert people to critical tasks, provide a spatial frame of reference for the operator, and improve performance of tasks requiring hand–eye coordination (Hale & Stanney, 2004). Specifically, tactile cues, such as vibrations or varying pressures applied to the hand or body, are effective as simple alerts, while kinesthetic feedback is key for the more dexterous tasks that humans carry out (Biggs & Srinivasan, 2002; Hale & Stanney, 2004).

Hale and Stanney (2004) provide an excellent summary of the benefits of adding tactile and kinesthetic feedback via diverse interface types, which are summarized here. First, they discuss texture perception. The addition of tactile feedback via a tactile display, in addition to visual feedback, results in more accurate judgment of softness and roughness compared to human performance of such tasks with visual feedback alone. If the tactile feedback is added to the visual display by means of a probe-based device rather than a tactile display, research shows that it is possible for the operator to judge softness and roughness with the same accuracy as when using the fingertip directly. Finally, if the tactile feedback is displayed to the operator via an exoskeleton device with tactile actuators in the fingertips, it is possible for the person to judge texture. Texture perception could be important in applications that involve exploration or object manipulation.

Tactile feedback can also be used to assist in two-dimensional (2D) form perception. For such tasks, visual feedback alone enables perception of the form's relative depth within the field of view. The addition of tactile feedback via a tactile display, either directly to the fingertip or through a probe device, does not do much to improve 2D form perception, and can be ignored when irrelevant. For example, vibrotactile (binary) feedback added to visual feedback during a pick-and-place task does not significantly improve performance because the information is not rich enough for the operator (Murray et al., 1997). However, there is some benefit to having cross-modal cueing—for example, if tactile actuators are used within the fingers of an exoskeleton, it is possible to judge 2D form perception.

These researchers also summarize the benefits of adding kinesthetic feedback to visual displays for various tasks (Hale & Stanney, 2004). For the purpose of spatial awareness in terms of position (of objects in the environment or of self), visual displays alone enable comprehension of the relative depth of objects and visual proprioception within the field of view. The addition of kinesthetic feedback via a positional actuator further allows for an egocentric frame of reference within the operator's personal space, gestures for navigation of the environment, and target location (with less decay than visual target location). When using a probe-based device or an exoskeleton to provide kinesthetic feedback, the operator will experience enhanced distance judgments within the workspace. For example, perception of contact can be greatly improved with the use of haptic feedback over visual-only displays in situations where the viewing angle does not permit a clear view of the contacting points.

When the task is 3D form perception, visual feedback alone will result in identification and discrimination performance that is dependent on viewing angle, and the user will have no indication of the weight of objects in the environment. By including kinesthetic feedback via a probe-based system, the operator will experience deformability of objects in the environment through the force feedback, thereby aiding discrimination and identification. With an exoskeleton system, the user will experience improved weight discrimination of objects along with improved object interaction. This kind of information can be critical in applications like surgical training and telesurgery, where high manual dexterity is desired. For both probe-based and exoskeleton devices, the inclusion of haptic feedback to a visual virtual environment results in increased sensations of "presence" or embodiment within the virtual world.

Finally, haptic interfaces could be included to augment visual information. Examples of such applications include augmentation of a GUI (see Section 2.3.4), scientific data visualization, or CAD.

2.4.2   Data Needed to Build the Interface

Upon determining that the inclusion of haptic feedback is beneficial to a virtual or remote environment display, a number of decisions must be made in order to build a haptic interface system, even if commercial hardware is to be selected. First, the designer must determine if tactile or kinesthetic feedback is preferred. These decisions are dependent on the type of feedback that the designer wishes to provide. For example, if the desire is to provide a simple alert to the user or to display textures or surface roughness, then a tactile device is most appropriate. In contrast, if 2D or 3D shape perception, discrimination, or presence in the virtual or remote environment is the goal, then kinesthetic devices are preferred. Refer to Section 2.4.1 for a discussion of situations where haptic feedback might be beneficial.

If kinesthetic, then the designer must select a probe- or joystick-type device that is grasped by the user, or an exoskeleton device that is worn by the user. When selecting a desktop device versus a wearable exoskeleton device for kinesthetic force display, the designer must decide on the importance of mobility when using the interface and the nature of the feedback to be displayed. Often, we choose to simulate interactions with an environment through use of a tool wielded in the hand, in which case the desktop devices are entirely suitable. Exoskeleton devices, on the other hand, enable joint-based feedback to simulate grasping of objects for hand exoskeletons, or manual object manipulation for the more complex exoskeletons.

For all applications, the designer must consider the perceptual elements of the application for which haptic feedback is to be designed. For example, does the user need to discriminate between two cues or just detect them? Is the feedback within human perceptual capabilities? Will the user be fatigued after prolonged use? Can the user discriminate among the different vibratory cues provided? These factors are highly task dependent, and answering such questions for an array of possible applications is beyond the scope of this chapter. Designers are encouraged to consult the literature for numerous examples of applications of haptic interfaces and to refer to Section 2.1.3 for a summary of human perceptual capabilities. Due to the wide range of applications for which haptic feedback may be desired, variations introduced by different robotic devices, parts of the body with which the device may interface, the nature of the feedback, the precise role of a cue in a particular application, and users, the designers may need to conduct their own experiments and user studies to fully answer some of these questions.

Kinesthetic Feedback Devices

For both probe-based and exoskeleton devices, the design decisions that must be made in order to implement the device are similar. Again, as with the nature of the feedback (tactile versus kinesthetic), decisions are often task dependent. The designer must determine an appropriate DOF number for the device—too few and the flexibility of the device will be compromised, while too many will overcomplicate the implementation. For example, is planar motion in the virtual environment sufficient, or does it require motion in three-dimensional space? It is also important to consider how many DOF the user requires, and how many DOF of force feedback are appropriate. For example, the popular PHANToM haptic devices (Sensable Technologies) allow for six DOF of motion control (three Cartesian motions and three rotations), while force feedback is provided only for the translational DOF, with limited detrimental effects on sensation of presence in the virtual environment by the user.

The size of the workspace is another significant design decision for any haptic device, and should be considered carefully. Scaling of operator motion from the device to the virtual environment—either to amplify operator motion or scale down motions if working in a microscopic environment, for example—is a feasible solution in many cases. It is not always necessary that motions with the haptic interface exactly mimic real-world tasks. However, if the application of the system is for training, then careful consideration of the workspace size and scaling should be practiced. Exoskeleton devices typically do not involve scaling of the user motion, except in some applications for upper extremity rehabilitation (Brewer et al., 2005). However, decisions about DOF and required workspace in terms of joint range of motion should be determined.

Decisions regarding the range of forces to be displayed by the haptic device are often directly coupled to workspace size and to the size and cost of the haptic interface hardware. Designers should consider the typical range of interaction forces that are required for the desired tasks to be rendered in the environment, but consideration of maximum human force output capabilities (to ensure user safety and reduce fatigue) along with knowledge of the limits of human force-sensing capabilities for small forces (as presented in the section on human haptic sensing) will influence decisions. Designers should be careful to ensure that the selected device is capable of providing a quality of feedback that allows for successful task execution. For example, is a vibratory cue like that of a cell phone ringer all that is required, or should the user be able to discriminate between two separate cues?

Finally, the designer must implement the virtual environment via computer control of the haptic interface and (typically) a coupled visual and haptic display. Basic haptic rendering is not a focus of this chapter; however, the reader is referred to an excellent introduction to haptic rendering concepts presented by Salisbury and colleagues (2004).

Designers are strongly encouraged to review the commercially available haptic devices, as their implementation will be more straightforward than building haptic devices for custom applications. As noted previously, there are a wide variety of devices available on the market with varying DOF and workspace dimensions. The remainder of this section focuses on additional decisions that must be made when fabricating a custom haptic interface.

When building a custom haptic interface, the designer must select the basic control approach for the system—impedance or admittance. Impedance devices will require sensors to record operator motions, and will be controlled by specifying the forces and torques that the environment will apply to the user based on his or her interactions. Such devices require mechanical designs that are very lightweight, stiff, and easily back-drivable. Admittance devices will require sensors to record operator forces, in addition to position sensors to allow for closed-loop position control of the device. Often these systems are non–backdrivable and exhibit properties of typical industrial robots due to the position control approach of display of the virtual environment. Most commercial haptic interfaces are of the impedance display type.

Tactile Feedback Devices

The specifications required for tactile feedback devices are fewer in number than those required for kinesthetic haptic devices. The designer must determine the body location for the tactile display. Typically, this is the finger pad of the operator, although tactile displays have also been developed for the torso (Tan & Pentland, 1997; Traylor & Tan, 2002). Most tactile feedback devices are pin arrays; therefore, the designer must specify the density of the pin array and the method of actuation for the pins. For example, if the pins are too close, the user may not be able to discriminate simultaneous cues from two adjacent pins, whereas if they are too far apart the cues may appear disjointed. Tactile arrays for the finger pad can be static, or can be coupled to mouse-like devices that allow translation in a plane. The designer must determine if this capability is required for the application. Finally, lateral skin stretch (in addition to, or in place of, finger pad deflection normal to the skin surface) can generate very different sensations for the human operator (Hayward & Cruz-Hernandez, 2000). Again, the appropriateness of the method of feedback and the selection of actuators and mechanical design will be task-dependent, and designers are strongly encouraged to review commercially available solutions along with other tactile display devices that have been presented in the literature.

2.4.3 Detailed Description of What Is Needed to Specify Such Interfaces for Use

The features of kinesthetic and tactile haptic interfaces are described separately in this section. Sensing and actuation are discussed specifically for kinesthetic haptic interfaces, since selection of components for these tasks is closely coupled to the mechanical design of such devices.

Mechanical Design of Kinesthetic Haptic Interfaces

The mechanical design of a haptic interface includes specification of the device DOF, the kinematic mechanism, and portability. The most prominent feature of a haptic interface is the number and the nature of the DOF at the active end or ends. The active end refers to the part of the robot that is connected to the body of the operator. At the active end, the hand holds the device or the device braces the body; otherwise, the interaction is unilateral (Hayward & Astley, 1996). The DOF that are actuated or active and others that are passive are also critical. For example, the PHANToM is a pen-based mechanism that has six DOF at the end point, but only three of these are actuated (Massie & Salisbury, 1994). Through the PHANToM haptic interface, an operator can explore a virtual environment in six DOF (three in translation and three in rotation), but receives force feedback only in the three translational DOF. The choice of DOF of a particular haptic interface depends primarily on the intended application. Haptic interfaces range from simple single-DOF devices built for research (Lawrence & Chapel, 1994) to a 13-DOF exoskeleton master arm for force-reflective teleoperation (Kim et al., 2005).

The choice of mechanism for a haptic interface is influenced by both the application and the part of the body interfaced with. Robotic mechanisms can be serial or parallel. A serial mechanism is composed of a sequence of links connected end to end, with one end of the resulting linkage connected to the ground (base) and the other being free. Serial mechanisms provide simplicity of design and control, but typically require larger actuators than parallel mechanisms. In addition, errors in the motion of links near the base of the robot are propagated to the end effector, resulting in loss of precision. A parallel mechanism, on the other hand, contains closed loops of these linkages with two or more connections to ground. Parallel mechanisms offer high structural stiffness, rigidity, precision, and low apparent inertia, which are desirable for the display of high-fidelity virtual environments, but these mechanisms tend to have singularities, limited workspaces, and more complex control schemes than their serial counterparts.

Haptic interfaces apply forces to the human operator. As a result, equal and opposite forces act on the interface and need to be distributed in order to maintain force equilibrium. Based on the grounding of these feedback forces, haptic devices can be classified as nonportable (grounded) or portable (ungrounded). A grounded haptic device is affixed to a rigid base, transferring reaction forces to ground. An ungrounded haptic device is attached only to the operator's body, exerting reaction forces on the user at the point(s) of attachment. Most of today's haptic interfaces, like pen-based haptic devices and joysticks, are grounded.

Typically, ungrounded haptic interfaces are good at providing feedback such as grasping forces during object manipulation, and have workspaces that permit natural movement during haptic interactions—but at the expense of design simplicity. Alternatively, grounded devices perform better when displaying kinesthetic forces to the user, like forces that arise when simulating static surfaces (Burdea, 1996). The workspace of a grounded device is limited by the manipulator's link lengths and joint limits, such as in common desktop interfaces like the PHANToM Desktop by Sensable Technologies (workspace: 6.4 in W $\times$ 4.8 in H $\times$ 4.8 in D) or the Impulse Engine 2000 by Immersion Corporation (workspace: 6 in $\times$ 6 in).

Some haptic interfaces, mostly exoskeleton-type interfaces, can be wearable. Examples of such interfaces include the Rutgers Master II force feedback glove (Bouzit et al., 2002), the Salford arm exoskeleton (Tsagarakis & Caldwell, 2003), the L-Exos force feedback exoskeleton (Frisoli et al., 2005), and the MAHI arm exoskeleton (Gupta & O'Malley, 2006).

Sensing and Actuation

Sensing and actuation are critical components of a haptic interface. Section 2.1 presented the human sensory and sensory motor capabilities. An effective haptic interface needs to match these requirements through its sensors and actuators. For high-quality haptic display, the actuators of a haptic interface should have a high power-to-weight ratio, high force/torque output, and high bandwidth. The bandwidth of an actuator refers to the range of frequency of forces that can be applied with the actuator. In addition, the actuator should have low friction and inertia as these can mask small feedback forces, thereby destroying the sense of realism. Sensors for haptic interfaces should have high resolution. Due to the difference in human tactile and kinesthetic sensing, tactile and kinesthetic displays typically employ different sets of sensors and actuators.

Kinesthetic interfaces may use electrical actuators, hydraulic actuators, or pneumatic actuators. Electrical actuators are currently the most used haptic actuators. These include DC motors (both brushed and brushless), magnetic-particle brakes, and SMA. Specific trade-offs of these types of actuators are discussed in detail in the online case studies mentioned in Section 2.7.

High-resolution sensors for kinesthetic interfaces are readily available, and generally noncontact optical encoders that measure position of a motor shaft are used. These encoders typically have a resolution of less than half a degree, which is sufficient for most applications. Another option for position sensing is noncontact rotary potentiometers that, like the rotary encoders, are placed in line with the actuator shafts.

Mechanical Design of Tactile Haptic Interfaces

Tactile feedback can be provided using pneumatic stimulation by employing compressed air to press against the skin (Sato et al., 1991), vibrotactile stimulation by applying a vibration stimulus locally or spatially over the user's fingertips using voice coils (Patrick, 1990) or micropin arrays (Hasser & Weisenberger, 1993), or electrotactile stimulation through specially designed electrodes placed on the skin to excite the receptors (Zhu, 1988). In addition, single-stage Peltier pumps have been adapted as thermal/tactile feedback actuators for haptic simulations, such as in Zerkus et al. (1993). A tactile display using lateral skin stretch has also been developed (Hayward & Cruz-Hernandez, 2000). Sensing in tactile interfaces can be achieved via force-sensitive resistors (Stone, 1991), miniature pressure transducers (MPT) (Burdea et al., 1995), the ultrasonic force sensor (Burdea et al., 1995), and the piezoelectric stress rate sensor (Son et al., 1994). For a thorough survey of current technologies related to tactile interface design, see Pasquero (2006).

2.5 TECHNIQUES FOR TESTING THE INTERFACE

As with all interfaces, haptic interfaces need to be evaluated to ensure optimal performance. Sections 2.5.1 and 2.5.2 describe some of the special considerations that must be taken into account when testing and evaluating haptic interfaces.

2.5.1 Testing Considerations for Haptic Devices

The evaluation can focus on the hardware alone, or performance can be measured by testing human interaction with the environment. Hardware testing is required to determine the quality of force feedback achievable by the interface. This involves examination of the range of frequencies of forces the device can display and the accuracy with which those forces can be displayed. In testing of the machine, which is application independent, the capabilities of the devices themselves are measured and comparison among various devices is allowed. User-based testing is task dependent and carried out to study the perceptual effectiveness of the interface, which is important from a human factors point of view.

When testing the hardware, the procedures are fairly straightforward and should simply follow best practices for experiment design, data collection, and data analysis, including statistical considerations. For tests involving human subjects, first it is important to follow governmental regulations for human-subject testing, often overseen by an Institutional Review Board. Second, a sufficient number of subjects should be enrolled and trials conducted for the results to have statistical significance. Third, and specifically for haptic devices, it is necessary to isolate only those sensory feedback modalities that are of interest in a given study. For example, often the haptic interface hardware can provide unwanted auditory cues due to the amplifiers and actuators that provide the force sensations for the operator. In such studies, it is often beneficial to use noise-isolating headphones or some other method of masking the unwanted auditory feedback. Similarly, if focusing on the haptic feedback capabilities of a particular device, it may be necessary to remove visual cues from the environment display, since in many cases of human perception the visual channel dominates the haptic channel.

2.5.2   Evaluating a Haptic Interface

Several approaches can be taken when evaluating a haptic interface. First, performance of the hardware can be assessed using human-subject testing, usually via methods that assess human performance of tasks in the haptic virtual environment or that measure the individual's perception of the qualities of the virtual environment. To this end, researchers have studied the effects of software on the haptic perception of virtual environments (Rosenberg & Adelstein, 1993; Millman & Colgate, 1995; Morgenbesser & Srinivasan, 1996). Morgenbesser and Srinivasan (1996), for example, looked at the effects of force-shading algorithms on the perception of shapes.

It is also common to compare performance of tasks in a simulated environment with a particular device to performance in an equivalent real-world environment (Buttolo et al., 1995; West & Cutkosky, 1997; Richard et al., 1999; Shimojo et al., 1999; Unger et al., 2001). Work by O'Malley and Goldfarb (2001, 2005) and O'Malley and Upperman (2006) extended these comparisons to include performance in high- and low-fidelity virtual environments versus those in real environments, demonstrating that although performance of some perceptual tasks may not be degraded with lower-fidelity haptic devices, human operators can still perceive differences in quality of the rendered virtual environments in terms of the perceived hardness of surfaces. Such studies can give an indication of the extent to which a particular device and its accompanying rendered environment mimic real-world scenarios and enable humans to perceive the virtual environment with the same accuracy as is possible in the natural world.

Finally, performance can be assessed using typical measures and characteristics of quality robotic hardware. Primary requirements for a haptic system are the ability to convey commands to the remote or virtual plant and to reflect relevant sensory information, specifically forces in the remote or virtual environment, back to the operator. In essence, the dynamics of the device must not interfere with the interaction between the operator and the environment. Jex (1988) describes four tests that a haptic interface should be able to pass. First, it should be able to simulate a piece of light balsa wood, with negligible inertia, friction, or perceived friction by the operator. Second, the device should be able to simulate a crisp hard stop. It should simulate coulomb friction, that is, the device should drop to zero velocity when the operator lets go of the handle. Finally, the device should be able to simulate mechanical detents with crisp transition and no lag.

In practice, performance of a haptic interface is limited by physical factors, such as actuator and sensor quality, device stiffness, friction, device workspace, force isotropy across the workspace, backlash, computational speed, and user's actions (hand grip, muscle tone). From the previous discussion, it is clear that there is a wide range of haptic devices both in terms of their DOF and applications, making the task of generating a common performance function particularly challenging.

Various researchers have attempted to design a set of performance measures to compare haptic devices independent of design and application (Ellis et al., 1996; Hayward & Astley, 1996), including kinematic performance measures, dynamic performance measures (Colgate & Brown, 1994) (Lawrence et al., 2000), and application-specific performance measures (Kammermeier & Schmidt, 2002; Kirkpatrick & Douglas, 2002; Chun et al., 2004). Additionally, when developing custom hardware, sensor resolution should be maximized, structural response should be measured to ensure that display distortion is minimized, and closed-loop performance of the haptic interface device should be studied to understand device stability margins. Detailed discussion of the techniques necessary for measuring these quantities is beyond the scope of this chapter.

2.6 DESIGN GUIDELINES

This section provides guidance on how to effectively design a haptic interface that is both safe and effective in its operation.

2.6.1 Base Your Mechanical Design on the Inherent Capabilities of the Human Operator

Because the haptic device will be mechanically coupled to the human operator, it is important to ensure that the characteristics of the system, such as workspace size, position bandwidth, force magnitude, force bandwidth, velocity, acceleration, effective mass, accuracy, and other factors, are well matched to the human operator (Stocco & Salcudean, 1996). The design goals for the system, if based on the inherent capabilities of the human hand (or other body part using the display), will ensure a safe and well-designed system that is not overqualified for the job.

2.6.2 Consider Human Sensitivity to Tactile Stimuli

Sensitivity to tactile stimuli is dependent on a number of factors that must be considered. For example, the location of application of the stimuli or even the gender of the user can affect detection thresholds (Sherrick & Cholewiak, 1986). Stimuli must be at least 5.5 msec apart, and pressure must be greater than 0.06 to 0.2 N/cm^2 (Hale & Stanney, 2004). Additionally, vibrations must exceed 28 dB relative to a 1-microsecond peak for 0.4- to 3-Hz frequencies for humans to be able to perceive their presence (Biggs & Srinivasan, 2002).

2.6.3 Use Active Rather Than Passive Movement

To ensure more accurate limb positioning, use active movement rather than passive movement of the human operator. Additionally, avoid minute, precise joint rotations, particularly at the distal segments, and minimize fatigue by avoiding static positions at or near the end range of motion (Hale & Stanney, 2004).

2.6.4 Achieve Minimum Force and Stiffness Display for Effective Information Transfer from the Virtual Environment

When implementing the virtual environment and selecting actuator force output and simulation update rates, ensure that the minimum virtual surface stiffness is 400 N/m (O'Malley & Goldfarb, 2004) and minimum end point forces are 3 to 4 N (O'Malley & Goldfarb, 2002) to effectively promote haptic information transfer.

2.6.5 Do Not Visually Display Penetration of Virtual Rigid Objects

A virtual environment simulation with both visual and haptic feedback should not show the operator's finger penetrating a rigid object, even when the stiffness of the virtual object is limited such that penetration can indeed occur before significant forces are perceived by the operator (Tan et al., 1994). This is because when no visual feedback is available, people tend to fail to differentiate the deformation of the soft finger pad from movements of the finger joints.

2.6.6 Minimize Confusion and Control Instabilities

In multimodal systems, it is important to minimize confusion of the operator and limit control instabilities by avoiding time lags among haptic/visual loops (Hale & Stanney, 2004).

2.6.7 Ensure Accuracy of Position Sensing in Distal Joints

Serial linkages require that the distal joints have better accuracy in sensing angular position than proximal joints, if the accuracy of all joints is constrained by cost or component availability (Tan et al., 1994). This is because joint angle resolution of humans is better at proximal joints than at distal ones.

2.6.8 For Exoskeleton Devices, Minimize the Contact Area at Attachment Points for Mechanical Ground

It is important to minimize the contact area for ground attachment points because humans are less sensitive to pressure changes when the contact area is decreased (Tan et al., 1994).

2.6.9 Ensure Realistic Display of Environments with Tactile Devices

Note that a human operator must maintain active pressure to feel a hard surface after contact, and maintaining the sensation of textured surfaces requires relative motion between the surface and the skin (Hale & Stanney, 2004).

2.6.10 Keep Tactile Features Fixed Relative to the Object's Coordinate Frame

It is important to maintain this fixed relative position for realistic perception of objects with a tactile display. This requirement translates to a need for high temporal bandwidth of the pins (imagine fast finger scanning) (Peine et al., 1997). Matching maximum finger speeds during natural exploration is a useful goal.

2.6.11   Maximize Range of Achievable Impedances

Because it is just as important for the haptic device to be light and back-drivable as it is for the device to be stiff and unyielding, it can be difficult to optimize design parameters for a specific hardware system. Therefore, it is recommended to generally achieve a broad range of impedances with the device (Colgate & Schenkel, 1997). Such a range can be achieved by carefully selecting robot configuration, defining geometric parameters, using transmission ratios, incorporating external dampers, or enabling actuator redundancy (Stocco et al., 2001). Specifically, techniques include lowering the effective mass of the device (Lawrence & Chapel, 1994), reducing variations in mass (Ma & Angeles, 1993; Hayward et al., 1994; Massie & Salisbury, 1994), designing a device such that it exhibits an isotropic Jacobian (Kurtz & Hayward, 1992; Zanganeh & Angeles, 1997), or adding physical damping (mechanical or electrical) to the system (Colgate & Schenkel, 1997; Mehling et al., 2005).

2.6.12   Limit Friction in Mechanisms

To reduce nonlinearities in the haptic device, limiting friction is important. If using impedance control techniques, minimal friction is key to back-drivability of the device as well. Friction can be limited through the use of noncontacting supports like air bearings or magnetic levitation, by incorporating direct drive actuation techniques, or, if transmissions are needed to achieve desired forces and torques, by selecting cable drives over gears or other transmission methods.

2.6.13   Avoid Singularities in the Workspace

In certain parts of the workspace, the robot end point may lose (inverse kinematics singularity) or gain (forward kinematics singularity) a degree of freedom. For example, if a robot arm with revolute joints is fully stretched, then the end point of the robot loses a degree of freedom as it cannot be moved along the line connecting the joints. Similarly, for parallel mechanisms in certain configurations, it is possible that the end point gains a degree of freedom, that is, it can be instantaneously moved without affecting the actuated joints. Near workspace locations that exhibit inverse kinematics singularities, significantly high torques are required to move the robot in the singular direction. Near these points, even during free movement, the operator of a haptic interface would need to exert considerable forces to move, thereby reducing the realism of display. Conversely, at a forward kinematics singularity, it is possible to initiate end point motion with little force, which is especially detrimental for haptic interfaces as it is not possible to display any force to the operator at these locations. Hence, singularities in the robot workspace should be avoided.

2.6.14   Maximize Pin Density of Tactile Displays

Objects feel more realistic as the spatial density of the pins is increased, although this will be limited by the size of the actuators selected. Vertical displacement of pins should be 2 to 3 mm while providing 1 to 2 N of force to impose skin deflections during large loads (Peine et al., 1997).

2.7   CASE STUDIES

Case studies for several haptic interfaces can be found at *www.beyondthegui. com.*

2.8   FUTURE TRENDS

The commercial applications of haptic and tactile displays have been simple and inexpensive devices, such as the vibrations of a cellular telephone or pager, or force feedback joysticks common to video games. Haptic interfaces, both kines-thetic and tactile displays, which have greater capability and fidelity than these examples, have seen limited application beyond the research lab. The primary barrier has been cost, since high-fidelity devices typically exhibit higher numbers of DOF, power-dense actuation, and high-resolution sensing. Wearable haptic devices, specifically wearable tactile displays, will also likely see increased demand from defense to consumer applications for the purpose of situational awareness, with developments in flexible materials that can be woven into fabric. Therefore, a prediction for the next 10 to 20 years is much greater accessibility to haptic devices in commercial applications as the price of improved sensor and actuator technology comes down. Such widespread applicability of haptic inter-face technology, especially in gaming, will be catalyzed by the recent increase in video games that encourage and even require active human intervention (e.g., Dance Dance Revolution, Wii).

The second driver of haptic device proliferation will be the sheer number of applications where haptic feedback will prove itself beneficial. Improved data visualization by use of haptic (including kinesthetic and tactile) displays that enable increased channels of information conveyance to the user will be realized. Haptic devices are already under development in geoscience and pharmaceutical research and testing via haptic-enriched protein-docking displays (Salisbury, 1999; Fritz & Barner, 1999). The most likely discipline for widespread adoption of haptic technologies will be medicine. From robot-assisted rehabilitation in virtual environments with visual and haptic feedback, to surgical robotics that enable realistic touch interactions displayed to the remote surgeon, to hardware plat-forms that, due to their reconfigurability and flexibility, will enable new discoveries in cognitive neuroscience, to the incorporation of haptic sensory feedback in prosthetic limbs for the increasing population of amputees, haptic devices will enable a new dimension of interaction with our world.

REFERENCES

Adelstein, B., & Rosen, M. (1992). Design and implementation of a force reflecting manipulandum for manual control research. *Proceedings of ASME*, 1–12.

An, K.-N., Askew, L., & Chao, E. Y. (1986). Biomechanics and functional assessment of upper extremities. In Karwowski, W., ed., *Trends in Ergonomics/Human Factors III*. Amsterdam: Elsevier Science Publishers, 573–80.

Bach-y-Rita, P. (1972). *Brain Mechanisms in Sensory Substitution*. New York: Academic Press.

Ballestero, S., & Heller, M. A. (2006). Conclusions: Touch and blindness. In Heller, M. A., & Ballestero, S., eds., *Touch and Blindness*. Mahwah, NJ: Erlbaum, 197–219.

Barbagli, F., Salisbury, K., et al. (2006). Haptic discrimination of force direction and the influence of visual information. *ACM Transactions on Applied Perception* 3(2):125–35.

Bejczy, A., & Salisbury, K. (1980). Kinematic coupling between operator and remote manipulator. *Advances in Computer Technology* 1:197–211.

Bergamasco, M., Allotta, B., et al. (1994). An arm exoskeleton system for teleoperation and virtual environments applications. *Proceedings of IEEE International Conference on Robotics and Automation*, 1449–54.

Bergamasco, M., & C. A. Avizzano, C. A. (1997). Virtual environment technologies in rehabilitation. *Proceedings of IEEE-RSJ-SICE Symposium on Robot and Human Interactive Communication*.

Biggs, J., & Srinivasan, M. (2002). Haptic interfaces. In Stanney, K., ed., *Handbook of Virtual Environments*. London: Erlbaum, 93–116.

Bolanowski, S. J., Gescheider, G. A., et al. (1988). Four channels mediate the mechanical aspects of touch. *Journal of the Acoustical Society of America* 84(5):1680–94.

Bouzit, M., Burdea, G., et al. (2002). The Rutgers Master II—new design force-feedback glove. *IEEE/ASME Transactions on Mechatronics* 7(2):256–63.

Brewer, B. R., Fagan, M., et al. (2005). Perceptual limits for a robotic rehabilitation environment using visual feedback distortion. *IEEE Transactions on Neural Systems and Rehabilitation Engineering* 13(1):1–11.

Brooks, F. P., Ouh-Young, M., et al. (1990). Project GROPE: Haptic displays for scientific visualization. *Proceedings of the Association for Computing Machinery Special Interest Group on Graphics and Interactive Techniques (SIGGRAPH) Conference*, 177–95.

Brooks, T. L. (1990). Telerobotic response requirements. *Proceedings of IEEE International Conference on Systems, Man and Cybernetics*, 113–20.

Burdea, G., Goratowski, R., et al. (1995). A tactile sensing glove for computerized hand diagnosis. *The Journal of Medicine and Virtual Reality* 1(1):40–44.

Burdea, G. C. (1996). *Force and Touch Feedback for Virtual Reality*. New York: John Wiley & Sons.

Buttolo, P., Kung, D., et al. (1995). Manipulation in real, virtual and remote environments. *Proceedings of IEEE International Conference on Systems, Man and Cybernetics*, 4656–61.

Cholewiak, R., & Collins, A. (1991). Sensory and physiological bases of touch. In Heller, M., & Schiff, W., eds., *The Psychology of Touch*. Hillsdale, NJ: Erlbaum, 23–60.

Christian, K. (2000). Design of haptic and tactile interfaces for blind users. College Park, MD: Department of Computer Science, University of Maryland.

Chun, K., Verplank, B., et al. (2004). Evaluating haptics and 3D stereo displays using Fitts' law. *Proceedings of Third IEEE International Workshop on Haptic, Audio and Visual Environments and Their Applications*, Ottawa, 53–58.

Clark, F. J. (1992). How accurately can we perceive the position of our limbs? *Behavioral Brain Science* 15:725–26.

Clark, F. J., & Horch, K. (1986). Kinesthesia. In Boff, K., Kauffman, L., & Thomas, J., eds., *Handbook of Perception and Human Performance*. New York: John Wiley, 1–62.

Colgate, J. E., & Brown, J. M. (1994). Factors affecting the Z-width of a haptic display. *Proceedings of IEEE International Conference on Robotics and Automation*, 3205–10.

Colgate, J. E., & Schenkel, G. G. (1997). Passivity of a class of sampled-data systems: Application to haptic interfaces. *Journal of Robotic Systems* 14(1):37–47.

Delp, S. L., Loan, J. P., et al. (1997). Surgical simulation: An emerging technology for training in emergency medicine. *Presence-Teleoperators and Virtual Environments* 6:147–59.

DiFranco, D. E., Beauregard, G. L., et al. (1997). The effect of auditory cues on the haptic perception of stiffness in virtual environments. *Proceedings of ASME Dynamic Systems and Control Division*, 17–22.

Draper, J., Herndon, J., et al. (1987). The implications of force reflection for teleoperation in space. *Proceedings of Goddard Conference on Space Applications of Artificial Intelligence and Robotics*.

Durfee, W. K., Hendrix, C. M., et al. (1997). Influence of haptic and visual displays on the estimation of virtual environment stiffness. *Proceedings of ASME Dynamic Systems and Control Division*, 139–44.

Durlach, N. I., Delhorne, L. A., et al. (1989). Manual discrimination and identification of length by the finger-span method. *Perception & Psychophysics* 46(1):29–38.

Durlach, N. I., & Mavor, A. S., eds. (1995). *Virtual Reality: Scientific and Technological Challenges*. Washington, DC: National Academy Press.

Ellis, R. E., Ismaeil, O. M., et al. (1996). Design and evaluation of a high-performance haptic interface. *Robotica* 14:321–27.

Ertan, S., Lee, C., et al. (1998). A wearable haptic navigation guidance system. *Proceedings of Second International Symposium on Wearable Computers*, 164–65.

Falvo, M. R., Finch, M., Superfine, R., et al. (1996). The Nanomanipulator: A teleoperator for manipulating materials at nanometer scale. *Proceedings of International Symposium on the Science and Engineering of Atomically Engineered Materials*, 579–86.

Frisoli, A., Rocchi, F., et al. (2005). A new force-feedback arm exoskeleton for haptic interaction in virtual environments. *Proceedings of First Joint Eurohaptics Conference and Symposium on Haptic Interfaces for Virtual Environment and Teleoperator Systems*, 195–201.

Fritz, J. P., & Barner, K. E. (1999). Design of a haptic visualization system for people with visual impairments. *Proceedings of IEEE Transactions on Neural Systems and Rehabilitation Engineering* 7(3):372–84.

Gerald, F. A. (1975). *Sensory Saltation: Metastability in the Perceptual World*. Hillsdale, NJ: Erlbaum, 133.

Gillespie, B., O'Modhrain, S., et al. (1998). The virtual teacher. *Proceedings of ASME Dynamic Systems and Control Division*, 171–78.

Gupta, A., & O'Malley, M. K. (2006). Design of a haptic arm exoskeleton for training and rehabilitation. *IEEE/ASME Transactions on Mechatronics* 11(3):280–89.

Hajian, A., & Howe, R. (1997). Identification of the mechanical impedance of the human fingertips. *ASME Journal of Biomechanical Engineering* 119(1):109–14.

Hale, K. S., & Stanney, K. M. (2004). Deriving haptic design guidelines from human physiological, psychophysical, and neurological foundations. *IEEE Computer Graphics and Applications* 24(2):33–39.

Hannaford, B., Wood, L., et al. (1991). Performance evaluation of a six-axis generalized force-reflecting teleoperator. *IEEE Transactions on Systems, Man and Cybernetics* 21(3): 620–33.

Hasser, C., & Weisenberger, J. (1993). Preliminary evaluation of a shape-memory alloy tactile feedback display. *Proceedings of Symposium on Haptic Interfaces for Virtual Environments and Teleoperator Systems, American Society of Mechanical Engineers*, Winter Annual Meeting, 73–80.

Hayward, V., & Astley, O. R. (1996). Performance measures for haptic interfaces. In Giralt, G., & Hirzinger, G., eds., *Robotics Research: The Seventh International Symposium*. New York: Springer, 195–207.

Hayward, V., Astley, O. R., et al. (2004). Haptic interfaces and devices. *Sensor Review* 24(1):16–29.

Hayward, V., Choksi, J., et al. (1994). Design and multi-objective optimization of a linkage for a haptic interface. In Lenarcic, J., & Ravani, B., eds., *Advances in Robot Kinematics*. New York: Kluwer Academic, 359–68.

Hayward, V., & Cruz-Hernandez, M. (2000). Tactile display device using distributed lateral skin stretch. *Proceedings of Symposium on Haptic Interfaces for Virtual Environments and Teleoperator Systems*, Orlando, 1309–16.

Hill, J. W. (1979). Study of modeling and evaluation of remote manipulation tasks with force feedback. Pasadena, CA: SRI International for Jet Propulsion Laboratory.

Hogan, N., Kay, B. A., et al. (1990). Haptic illusions: Experiments on human manipulation and perception of virtual objects. *Proceedings of Cold Spring Harbor Symposia on Quantitative Biology*, 925–31.

Iwata, H. (1993). Pen-based haptic virtual environment. *Proceedings of IEEE Virtual Reality Annual International Symposium*, 287–92.

Jandura, L., & Srinivasan, M. (1994). Experiments on human performance in torque discrimination and control. *Proceedings of ASME Dynamic Systems and Control Division*, New York, 369–75.

Jeong, Y., Lee, Y., et al. (2001). A 7 DOF wearable robotic arm using pneumatic actuators. *Proceedings of International Symposium on Robotics*.

Jex, H. (1988). Four critical tests for control-feel simulators. *Proceedings of 23rd Annual Conference on Manual Control*.

Johnson, K. O., & Phillips, J. R. (1981). Tactile spatial resolution—I. two point discrimination, gap detection, grating resolution and letter recognition. *Journal of Neurophysiology* 46(6):1171–92.

Jones, L. A., & Hunter, I. W. (1990). A perceptual analysis of stiffness. *Experimental Brain Research* 79(1):150–56.

Jones, L. A., Hunter, I. W., & Irwin, R. J. (1992). Differential thresholds for limb movement measured using adaptive techniques. *Perception & Psychophysics* 52:529–35.

Jones, L. A. & Hunter, I. W. (1993). A perceptual analysis of viscosity. *Experimental Brain Research* 94(2):343–51.

Jones, L. A., et al. (2004). Development of a tactile vest. *Proceedings of Twelfth International Symposium on Haptic Interfaces for Virtual Environment and Teleoperator Systems*, 82–89.

Jones, L. A. (1989). Matching forces: Constant errors and differential thresholds. *Perception* 18(5):681–87.

Kammermeier, P., & Schmidt, G. (2002). Application-specific evaluation of tactile array displays for the human fingertip. *Proceedings of IEEE/RSJ International Conference on Intelligent Robots and Systems*, Lausanne, Switzerland, 3:2937–42.

Kieninger, T. G. (1996). The growing up of HyperBraille—an office workspace for blind people. *Proceedings of Association for Computing Machinery Symposium on User Interface Software and Technology*, 67–73.

Kim, W. S. (1991). A new scheme of force reflecting control. *Proceedings of Fifth Annual Workshop of Space Operations Applications and Research*, NASA, Lyndon B. Johnson Space Center, 254–61.

Kim, Y. S., Lee, J., et al. (2005). A force reflected exoskeleton-type master arm for human-robot interaction. *IEEE Transactions on Systems, Man and Cybernetics, Part A* 35(2):198–212.

Kirkpatrick, A. E., & Douglas, S. A. (2002). Application-based evaluation of haptic interfaces. *Proceedings of Tenth Symposium on Haptic Interfaces for Virtual Environment and Teleoperator Systems*, IEEE Computing Society, Orlando, 32–39.

Kontarinis, D. A., & Howe, R. D. (1995). Tactile display of vibratory information in teleoperation and virtual environments. *Presence-Teleoperators and Virtual Environments* 4(4):387–402.

Krebs, H. I., Hogan, N., et al. (1998). Robot-aided neurorehabilitation. *IEEE Transactions on Rehabilitation Engineering* 6(1):75–87.

Kurtz, R., & Hayward, V. (1992). Multiple-goal kinematic optimization of a parallel spherical mechanism with actuator redundancy. *IEEE Transactions on Robotics and Automation* 8(5):644–51.

Lawrence, D. A., & Chapel, J. D. (1994). Performance trade-offs for hand controller design. *Proceedings of IEEE International Conference on Robotics and Automation*, 4:3211–16.

Lawrence, D. A., Pao, L. Y., et al. (2000). Rate-hardness: A new performance metric for haptic interfaces. *IEEE Transactions on Robotics and Automation* 16(4):357–71.

Lee, S., Park, S., et al. (1998). Design of a force reflecting master arm and master hand using pneumatic actuators. *Proceedings of IEEE International Conference on Robotics and Automation*, 3:2574–79.

Loomis, J. M. (1979). An investigation of tactile hyperacuity. *Sensory Processes* 3(4):289–302.

Ma, O., & Angeles, J. (1993). Optimum design of manipulators under dynamic isotropy conditions. *Proceedings of IEEE International Conference on Robotics and Automation*, 1:470–75.

Massie, T., & Salisbury, J. K. (1994). The PHANToM haptic interface: A device for probing virtual objects. *Proceedings of ASME*, Winter Annual Meeting, New York, 55(1):295–300.

Massimino, M. J., & Sheridan, T. B. (1994). Teleoperator performance with varying force and visual feedback. *Human Factors* 36(1):145–57.

McKnight, S., Melder, N., et al. (2004). Psychophysical size discrimination using multifingered haptic interfaces. *Proceedings of EuroHaptics Conference.*

McNeely, W., Puterbaugh, K., et al. (1999). Six degree-of-freedom haptic rendering using voxel sampling. *Proceedings of Association for Computing Machinery Special Interest Group on Graphics and Interactive Techniques Conference.*

Meech, J. F., & Solomonides, A. E. (1996). User requirements when interacting with virtual objects. *Proceedings of IEEE Colloquium on Virtual Reality—User Issues* (Digest No. 1996/068), 3/1–3/3.

Mehling, J. S., Colgate, J. E., et al. (2005). Increasing the impedance range of a haptic display by adding electrical damping. *Proceedings of First Joint Eurohaptics Conference and Symposium on Haptic Interfaces for Virtual Environment and Teleoperator Systems,* 257–62.

Millman, P., Stanley, M., et al. (1993). Design of a high performance haptic interface to virtual environments. *Proceedings of IEEE Virtual Reality Annual International Symposium.*

Millman, P. A., & Colgate, J. E. (1995). Effects of non-uniform environment damping on haptic perception and performance of aimed movements. *Proceedings of ASME Dynamic Systems and Control Division,* 57(2):703–11.

Morgenbesser, H. B., & Srinivasan, M. A. (1996). Force shading for haptic shape perception. *Proceedings of ASME Dynamic Systems and Control Division,* 407–12.

Moy, G., Singh, U., et al. (2000). Human psychophysics for teletaction system design. *Haptics-e, The Electronic Journal of Haptics Research,* 1(3).

Murray, A. M., Shimoga, K. B., et al. (1997). Touch feedback using binary tactor displays: Unexpected results and insights. *Proceedings of ASME Dynamic Systems and Control Division,* 3–9.

Nahvi, A., Nelson, D., et al. (1998). Haptic manipulation of virtual mechanisms from mechanical CAD designs. *Proceedings of IEEE International Conference on Robotics and Automation,* 375–80.

Nakai, A., Ohashi, T., et al. (1998). 7 DOF arm–type haptic interface for teleoperation and virtual reality systems. *Proceedings of IEEE/RSJ International Conference on Intelligent Robots and Systems.*

Okamura, A. M., Cutkosky, M. R., et al. (2001). Reality-based models for vibration feedback in virtual environments. *IEEE/ASME Transactions on Mechatronics* 6(3):245–52.

Okamura, A. M., Dennerlein, J. T., et al. (1998). Vibration feedback models for virtual environments. *Proceedings of IEEE International Conference on Robotics and Automation,* 674–79.

O'Malley, M. K., & Ambrose, R. O. (2003). Haptic feedback applications for Robonaut. *Industrial Robot: An International Journal* 30(6):531–42.

O'Malley, M. K., & Goldfarb, M. (2001). Force saturation, system bandwidth, information transfer, and surface quality in haptic interfaces. *Proceedings of IEEE International Conference on Robotics and Automation,* 1382–87.

O'Malley, M. K., & Goldfarb, M. (2002). The effect of force saturation on the haptic perception of detail. *IEEE/ASME Transactions on Mechatronics* 7(3):280–88.

O'Malley, M. K., & Goldfarb, M. (2004). The effect of virtual surface stiffness on the haptic perception of detail. *IEEE/ASME Transactions on Mechatronics* 9(2):448–54.

O'Malley, M. K., & Goldfarb, M. (2005). On the ability of humans to haptically identify and discriminate real and simulated objects. *Presence: Teleoperators and Virtual Environments* 14(3): 366–76.

O'Malley, M. K., & Gupta, A. (2003). Passive and active assistance for human performance of a simulated underactuated dynamic task. *Proceedings of Eleventh Symposium on Haptic Interfaces for Virtual Environment and Teleoperator Systems*, 348–55.

O'Malley, M. K., Gupta, A., et al. (2006). Shared control in haptic systems for performance enhancement and training. *ASME Journal of Dynamic Systems, Measurement, and Control* 128(1):75–85.

O'Malley, M. K., & Upperman, G. (2006). A study of perceptual performance in haptic virtual environments. *Journal of Robotics and Mechatronics*, Special Issue on Haptics: Interfaces, Applications, and Perception, 18(4):467–75.

O'Modhrain, M. S. (2000). Playing by Feel: Incorporating Haptic Feedback into Computer-Based Musical Instruments. PhD diss., Stanford University.

O'Modhrain, M. S., & Gillespie, B. (1997). The Moose: A haptic user interface for blind persons. *Proceedings of Sixth World Wide Web Conference*, Santa Clara, CA.

Pang, X. D., Tan, H. Z., et al. (1991). Manual discrimination of force using active finger motion. *Perception & Psychophysics* 49:531–40.

Pasquero, J. (2006). *Survey on Communication through Touch*. Montreal: Center for Intelligent Machines, McGill University.

Patrick, N. J. M. (1990). *Design, Construction and Testing of a Fingertip Tactile Display for Interaction with Virtual and Remote Environments*. Master's thesis, Department of Mechanical Engineering, Massachusetts Institute of Technology.

Peine, W. J., Wellman, P. S., et al. (1997). Temporal bandwidth requirements for tactile shape displays. *Proceedings of ASME Dynamic Systems and Control Division*, 61:107–14.

Rabinowitz, W. M., Houtsma, A. J. M., et al. (1987). Multidimensional tactile displays: Identification of vibratory intensity, frequency, and contactor area. *Journal of the Acoustical Society of America* 82(4):1243–52.

Ramstein, C. (1996). Combining haptic and Braille technologies: Design issues and pilot study. *Proceedings of Association for Computing Machinery Conference on Assistive Technologies*, 37–44.

Richard, P., Coiffet, Ph., et al. (1999). Human performance evaluation of two handle haptic devices in a dextrous virtual telemanipulation task. *Proceedings of IEEE/RSJ International Conference on Intelligent Robots and Systems*, 1543–48.

Rosenberg, L. B. (1993). Virtual fixtures: Perceptual tools for telerobotic manipulation. *Proceedings of IEEE Virtual Reality Annual International Symposium*, 76–82.

Rosenberg, L. B., & Adelstein, B. D. (1993). Perceptual decomposition of virtual haptic surfaces. *Proceedings of IEEE Symposium on Research Frontiers in Virtual Reality*, 46–53.

Salcudean, S., & Vlaar, T. (1994). On the emulation of stiff walls and static friction with a magnetically levitated input/output device. *Proceedings of ASME Dynamic Systems and Control Division*, 303–309.

Salisbury, J. K. (1999). Making graphics physically tangible. *Communications of the ACM* 42(8):74–81.

Salisbury, K., Conti, F., et al. (2004). Haptic rendering: Introductory concepts. *IEEE Computer Graphics and Applications* 24(2):24–32.

Sato, K., Igarashi, E. et al. (1991). Development of non-constrained master arm with tactile feedback device. *Proceedings of IEEE International Conference on Advanced Robotics* 1:334–38.

Schmidt, R. (1977). Somatovisceral sensibility. In Schmidt, R., ed., *Fundamentals of Sensory Physiology*. New York: Springer-Verlag, 81–125.

Sherrick, C. E., & Cholewiak, R. W. (1986). Cutaneous sensitivity. In Boff, K., Kaufman, L., & Thomas, J. L., eds., *Handbook of Perception and Human Performance*. New York: Wiley, 1–58.

Shimoga, K. (1992). Finger force and touch feedback issues in dexterous telemanipulation. *Proceedings of NASA-CIRSSE International Conference on Intelligent Robotic Systems for Space Exploration*, 159–78.

Shimoga, K. (1993). A survey of perceptual feedback issues in dexterous telemanipulation. Part II: Finger touch feedback. *Proceedings of IEEE Virtual Reality Annual International Symposium*, New York.

Shimojo, M., Shinohara, M., et al. (1999). Human shape recognition performance for 3D tactile display. *IEEE Transactions on Systems, Man and Cybernetics, Part A* 29(6):637–44.

Sledd, A., & O'Malley, M. K. (2006). Performance enhancement of a haptic arm exoskeleton. *Proceedings of Fourteenth Symposium on Haptic Interfaces for Virtual Environment and Teleoperator Systems*, 375–81.

Smyth, T. N., & Kirkpatrick, A. E. (2006). A new approach to haptic augmentation of the GUI. *Proceedings of International Conference on Multimodal Interfaces*, 372–79.

Son, J., Monteverde, E., et al. (1994). A tactile sensor for localizing transient events in manipulation. *Proceedings of IEEE International Conference on Robotics and Automation*, 471–76.

Speich, J. E., Shao, L., et al. (2005). Modeling the human hand as it interacts with a telemanipulation system. *Mechatronics* 15(9):1127–42.

Srinivasan, M., & Chen, J. (1993). Human performance in controlling normal forces of contact with rigid objects. *Proceedings of American Society of Mechanical Engineers Conference on Advances in Robotics, Mechatronics and Haptic Interfaces*, New York, 119–25.

Srinivasan, M. A., Beauregard, G. L., et al. (1996). The impact of visual information on the haptic perception of stiffness in virtual environments. *Proceedings of ASME Dynamic Systems and Control Division*, 555–59.

Srinivasan, M. A., & LaMotte, R. H. (1995). Tactual discrimination of softness. *Journal of Neurophysiology* 73(11):88–101.

Stocco, L., & Salcudean, S. E. (1996). A coarse-fine approach to force-reflecting hand controller design. *Proceedings of IEEE International Conference on Robotics and Automation*, 404–10.

Stocco, L. J., Salcudean, S. E., et al. (2001). Optimal kinematic design of a haptic pen. *IEEE/ASME Transactions on Mechatronics* 6(3):210–20.

Stone, R. (1991). Advanced human-system interfaces for telerobotics using virtual-reality & telepresence technologies. *Proceedings of Fifth International Conference on Advanced Robotics*, 168–73.

Sutter, P., Iatridis, J., et al. (1989). Response to reflected-force feedback to fingers in teleoperation. *Proceedings of NASA Conference on Space Telerobotics*, 157–66.

Tan, H. Z., Durlach, N. I., et al. (1995). Manual discrimination of compliance using active pinch grasp: The roles of force and work cues. *Perception & Psychophysics* 57(4):495–510.

Tan, H. Z., Durlach, N. I., et al. (1994). Tactual performance with motional stimulation of the index finger. *Journal of the Acoustical Society of America* 95(5):2986–87.

Tan, H. Z., Durlach, N. I., et al. (1993). Manual resolution of compliance when work and force cues are minimized. *Proceedings of ASME Dynamic Systems and Control Division*, 99–104.

Tan, H. Z., Eberman, B., et al. (1994). Human factors for the design of force-reflecting haptic interfaces. *Proceedings of International Mechanical Engineers Congress and Exposition*, Chicago, 353–59.

Tan, H. Z., Pang, X. D., et al. (1992). Manual resolution of length, force, and compliance. *Proceedings of American Society of Mechanical Engineers*, Winter Annual Meeting, Anaheim, CA, 13–18.

Tan, H. Z., & Pentland, A. (1997). Tactual displays for wearable computing. *Digest of the First International Symposium on Wearable Computers*, 84–89.

Tang, H., & Beebe, D. J. (2006). An oral tactile interface for blind navigation. *IEEE Transactions on Neural Systems and Rehabilitation Engineering* 14(1):116–23.

Traylor, R., & Tan, H. Z. (2002). Development of a wearable haptic display for situation awareness in altered-gravity environment: Some initial findings. *Proceedings of Tenth Symposium on Haptic Interfaces for Virtual Environment and Teleoperator Systems*, 159–64.

Tsagarakis, N., Caldwell, D. G., et al. (1999). A 7 DOF pneumatic muscle actuator (pMA) powered exoskeleton. *Proceedings of Eighth IEEE International Workshop on Robot and Human Interaction*, 327–33.

Tsagarakis, N. G., & Caldwell, D. G. (2003). Development and control of a "soft-actuated" exoskeleton for use in physiotherapy and training. *Autonomous Robots* 15(1):21–33.

Unger, B. J., Nicolaidis, A., et al. (2001). Comparison of 3-D haptic peg-in-hole tasks in real and virtual environments. *Proceedings of IEEE/RSJ International Conference on Intelligent Robots and Systems*, 1751–56.

Upperman, G., Suzuki, A., et al. (2004). Comparison of human haptic size discrimination performance in simulated environments with varying levels of force and stiffness. *Proceedings of International Symposium on Haptic Interfaces for Virtual Environment and Teleoperator Systems*, 169–75.

Verillo, R. T. (1966). Vibrotactile sensitivity and the frequency response of the Pacinian corpuscle. *Psychonomic Science* 4:135–36.

Weisenberger, J. M., Krier, M. J., et al. (2000). Judging the orientation of sinusoidal and square wave virtual gratings presented via 2-DOF and 3-DOF haptic interfaces. *Haptics-e* 1(4).

West, A. M., & Cutkosky, M. R. (1997). Detection of real and virtual fine surface features with a haptic interface and stylus. *Proceedings of ASME International Mechanical Engineers Congress Dynamic Systems and Control Division*, Dallas, 159–65.

Wiker, S., Hershkowitz, E., et al. (1989). Teleoperator comfort and psychometric stability: Criteria for limiting master-controller forces of operation and feedback during telemanipulation. *Proceedings of NASA Conference on Space Telerobotics*, Greenbelt, MD, 99–107.

Williams, L. E. P., Loftin, R. B., et al. (2002). Kinesthetic and visual force display for telerobotics. *Proceedings of IEEE International Conference on Robotics and Automation*, 1249–54.

Williams II, R. L., Murphy, M. A., et al. (1998). Kinesthetic force/moment feedback via active exoskeleton. *Proceedings of Image Society Conference*.

Wood, J., Magennis, M., et al. (2003). The design and evaluation of a computer game for the blind in the GRAB haptic audio virtual environment. *Proceedings of Eurohaptics Conference.*

Yu, W., Ramloll, R., et al. (2000). Haptic graphs for blind computer users. *Lecture Notes in Computer Science* 2058:41–51.

Zanganeh, K. E., & Angeles, J. (1997). Kinematic isotropy and the optimum design of parallel manipulators. *International Journal of Robotics Research* 16(2):185–97.

Zerkus, M., Becker, B., et al. (1993). Temperature sensing in virtual reality and telerobotics. *Virtual Reality Systems* 1(2):88–90.

Zhu, H. (1988). Electrotactile stimulation. In Webster, J., ed., *Tactile Sensors for Robotics and Medicine*. New York: John Wiley & Sons, 341–53.

3

CHAPTER

Gesture Interfaces

Michael Nielsen, Thomas B. Moeslund, Moritz Störring, Erik Granum

This chapter provides an introduction to the domain of gesture interfaces. We begin with the foundation of gesture interfaces and how and where they are commonly used, followed by basic theory that will help you select and design a gesture interface, and related advice and warnings. Next, we present a procedure to identify and test good gestures for a design and a practical example of how to use this procedure. Finally, we provide a short perspective on our vision of the future as the interface evolves.

3.1 GESTURES

Gestures originate from natural interaction between people. They consist of movements of the body and face as nonverbal communication that complements verbal communication. This is the inspiration behind using gesture interfaces between man and machine. Figure 3.1 shows an interaction between two people that relies on nonverbal communication. The camera is mounted on the ear of one person. As seen here, interpretation is context dependent. They cannot hear each other because of loud music, and the person with the camera wants to order a drink. The gesture by the person at the left means "I cannot hear you." The person (at the right) with the camera on his head is using an iconic gesture to show what he wants. The other person imitates his gesture.

A gesture interface can be seen as an alternative or complement to existing interface techniques, such as the old desktop paradigm. Good examples are the new alternatives to the mouse, such as ergonomic track balls, mouse tablet pens (e.g., from Palm and a TabletPC), and the iGesturePad (Figure 3.2).

FIGURE
3.1

Interaction based on nonverbal communication.
Gestures between two people who cannot hear each other.

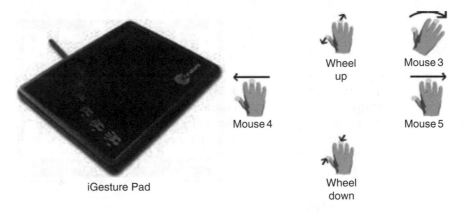

iGesture Pad

FIGURE
3.2

The iGesturePad reacts to the relative positions and movements of the user's
fingertips as they touch the pad. (Courtesy of Fingerworks.)

Gesture interfaces can navigate a Windows interface just as well or better than
the mouse cursor, while they may be more or less useless when it comes to fast
computer games, such as three-dimensional (3D) shooters and airplane simulators. For these applications, specialized interfaces have been developed that are
based on wheels and flight sticks from the real world.

When developing a gesture interface, the objective should not be "to make a
gesture interface." A gesture interface is not universally the best interface for
any particular application. The objective is "to develop a more efficient interface"
to a given application.

This can be illustrated by an example of an interface for artistic modeling of a
sculpture. The artist may be given a mouse and a keyboard for a CAD program.
The result is perfect to the smallest detail regarding accuracy of the lines, because

it is possible to set coordinates explicitly. If the artist is provided with a gesture interface in which a virtual clay model can be altered by touching and squeezing it, it will not be accurate in terms of coordinates and straight lines, but it might be aesthetically closer to the artist's vision. Thus, selecting an interface is a matter of which outcome of the application is desired.

Consequently, the first step is analysis of the kind of interface that is most suitable for this task. Such analysis may lead to the conclusion that a gesture interface is the most suitable type of interface.

3.2 TECHNOLOGY AND APPLICABILITY

Much work has been done in the investigation and development of natural interaction interfaces, including gesture interfaces (Cassell, 1998; Freemann & Weissman, 1995; Hummels & Stapiers, 1998; Storring et al., 2001; Paggio & Music, 2000; Steininger et al., 2002; Streitz et al., 2001). Gesture interfaces have also been developed in science fiction literature and movies, such as the movies *Johnny Mnemonic* (1995), *Final Fantasy* (2001), and *Minority Report* (2002). Furthermore, gesture interfaces are applied to solve problems for people with physical disabilities (Keates & Robinson, 1998; Jaimes & Sebe, 2005).

The most interesting potential in this field of research is to make accessory-free and wireless gesture interfaces, such as in virtual-reality and intelligent rooms, because the use of physical and wired gadgets makes the interface and gesturing tedious and less natural. The first solutions required expensive data gloves or other such intrusive equipment with wires that made the user feel uncomfortable. Greater success came with pen-based gestures (e.g., Palm handheld devices), where trajectories were recognized as gestures.

A common motivation behind the analogy between nonverbal communication and human–computer communication is that it allows for better, more natural and intuitive interaction. However, Cassell (1998) claims that this cannot be liberally assumed to always be the case. For example, a command language is as natural as writing a to-do list. Another idea involved accessibility studies, where people with physical disabilities are not capable of using tactile input modalities.

While natural gestures are often subtle, gesture interfaces rely on emphasized gestures. Interfaces have mostly revolved around simple hand gestures for pointing out objects or controlling a mouse-like pointer and a few gestures that are linked to specific functions in the application. Simple usage of these motion-tracking interfaces is found in game consoles (e.g., Playstation Eye Toy).

Recent work has also focused on facial gestures (face expressions and poses)—detecting reactions and emotions. This information can be used for automatic annotation in human behavior studies (Emotion Tool from iMotions, *http://www.imotions.dk*), accessibility for paralyzed people (Betke et al., 2002), and feedback to an intelligent learning system (Pantic & Patras, 2006).

3.2.1 Mechanical and Tactile Interfaces

Early gesture interfaces relied on mechanical or magnetic input devices. Examples include the data glove (Figure 3.3), the body suit (Suguru Goto, 2006), and Nintendo Wii.

Single-point touch interfaces are well known as pen gestures (Long et al., 2000), most commonly seen in Palm handheld devices. But recent research has developed multipoint touches directly onto the screen, used in the iGesturePad (Jefferson Han, 2005; Tse et al., 2006), which open up a new and more efficient interface potential.

There are examples of research in making the computer aware of human emotions shown in body language. De Silva et al. (2006) detected emotion intensity from gestures using sensors that read galvanic skin response. However, people in general do not like the idea of having such sensors or attachments on their bodies.

3.2.2 Computer Vision Interfaces

When the aim is to make gesture interfaces invisible to the user (Jaimes & Sebe, 2005), computer vision is a nice way to detect gestures. Computer vision is inherently wireless, and people have become accustomed to surveillance cameras in retail and airport environments.

Computer vision algorithms often consist of three parts: *segmentation* that spots relevant parts in the field of view, *tracking* that follows the movements, and *classification* that finds meaningful information.

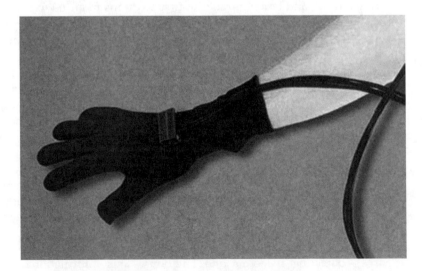

FIGURE   Data glove from 5DT. (Courtesy of *www.5DT.com.*)

3.3

Segmentation detects the body parts in the images, such as relying on shape or skin color, or motion detection in video sequences for dynamic gestures (Stoerring et al., 2004). In order to move past this step, many solutions have simplified the segmentation by using infrared reflectors or other such markings that are easy to segment and track.

Tracking follows a set of features when it moves with, for example, condensation or Kalman filtering (Rasmussen et al., 2006).

Classification detects the actual gestures using hidden Markov models (Sage et al., 2003; De Silva et al., 2006; Min et al., 1999), template or model matching (Moeslund et al., 2003; Riviera & Guitton, 2004; Fihl et al., 2006), or fuzzy or Bayesian logic (Aveles-Arriga et al., 2006; Wachs et al., 2006; Moustakas et al., 2006).

A new approach (Wang et al., 2006) tracks the movement of a camera phone by analyzing its images of surroundings. These phone movements are used as gestures.

3.2.3 Face Gaze and Expression

Face gaze and expression are a subdomain of gesture research (Jaimes & Sebe, 2005). Face gaze tracking is traditionally used for viewpoint estimation of virtual-reality and stereo displays, but recent research aims to extract all visual information from the face.

Facial expressions may be used as input modalities in accessibility research (Betke et al., 2002), such as for disabled people who cannot make any movement other than facial, such as blinking or smiling. Furthermore, facial movements convey information about the emotions of the user (Pantic & Rothkrantz, 2003). These can be used to detect confusion, user reactions, and intentions, and for purposes of surveillance and automatic video feed annotation such as in human behavior research. A computer without empathy is much like an apathetic person, and interacting with it can cause frustration (Klein et al., 2002).

Computer vision is also used for detection of facial expressions, which allows for an impressive perceptive interface and efficient video annotation tools, and can give emotional awareness to virtual agents (Betke et al., 2002; Bowyer et al., 2006; Pantic & Patras, 2006). However, the technology for doing this is still under development and results are usually given as technical detection rates. Testing user experience with such systems is going to be very interesting.

3.2.4 Applicability

Gesture interfaces are popular wherever the interface requires some freedom of movement or immersive feeling such as in virtual-reality environments (Stoerring et al., 2001), intelligent rooms (Streitz et al., 2001), medical tools during surgery (Wachs et al., 2006), and medical simulators (Tsagarakis et al., 2006). Other typical applications are advanced electronic whiteboards and handling of 3D objects (Müller-Tomfelde & Steiner, 2001; Moustakas et al., 2006).

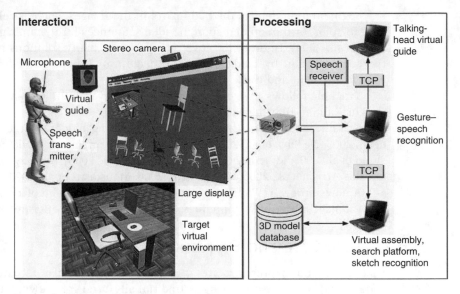

FIGURE

3.4

Example of a gesture interface system.
Object search using speech and gestures and a virtual agent in Masterpiece.
Source: Adapted from Moustakas et al. (2006); © 2006 IEEE.

An example of a gesture interface system (Moustakas et al., 2006) is shown in Figure 3.4. The user can retrieve 3D objects from a database and manipulate them. Commands such as retrieve, rotate, and so on, are accessed through speech. The user manipulates the 3D space with gestures such as showing how to rotate, scale, move, and draw primitives. A virtual agent acts as a help wizard if the user forgets the commands and gestures.

Gesture interface systems are also very applicable in accessibility research because they are naturally hands-off and hands-free interactions that can be done with any functional part of the body. Interfaces for accessibility focus on creating alternative input modes to WIMP (windows, icons, mouse, pointers) applications, such as controlling the mouse pointer with the face (Figure 3.5) or any single limb that the disabled person can use (Betke et al., 2002).

3.3   FUNDAMENTAL NATURE OF THE INTERFACE

The simplest gesture interface is the well-known motion detector that turns on a light in a room or triggers the water in a water fountain at a public lavatory. We will focus on the more explicit gesture interfaces in which specific gestures such as postures or movements are to be interpreted by a computer to motivate something in an application. This section describes general theory related to such gesture interfaces.

FIGURE

3.5

Controlling a mouse pointer with the face.
The camera mouse is controlled by a 30-month-old user. *Source:* From Betke et al. (2002); © 2002 IEEE.

3.3.1 How Gestures Relate to Other Modes

It is not always feasible to design an interface that relies completely on gestures because there may be issues relating to user or computer system precision and accuracy or the complexity of the input required. Hence, it is often beneficial to make gestures part of a multimodal system. To know when to use one mode or the other, we need to be aware of perceptive abilities in relation to other senses. Table 3.1 shows the senses that contain information to be interpreted in human perception.

Audible, visual, and tactile senses are the most important in human–computer interaction, and these are the commonly used senses thus far to make use of in interface design. Scent and taste are less relevant in a computer interaction context, but they may become more important in the future as technology for detection and synthesis emerges.

The senses can obtain conflicting, complementary, irrelevant, inadequate, or redundant information. These data must be filtered in the cognitive system and processed in order to be understood (Figure 3.6). This is done using expectations based on a known or assumed context regarding the world, history, and personal attributes. Sometimes these expectations are wrong, which leads to prejudice, among other things. Figure 3.6 also lists what those expectations could consist of if a similar interpretation system were developed on a computer.

If the computer would be able to interpret the human behavior and provide natural sensory feedback, it would enhance the immersive feeling of the user.

TABLE 3.1   The Five Senses and Relevant Information about
              Surroundings

Sense	Relevant Information
Audio	Speech
	Identity
	Intonation
	Precision (timing)
Vision	Identity
	Facial expression
	Body language
	Gesture
	Accuracy (spatial)
Tactile/somesthetic	Tabs, pads, devices
	Texture
	Precision (timing)
	Accuracy (spatial)
Scent	Atmosphere, likability
Taste	Clarification, enjoyment

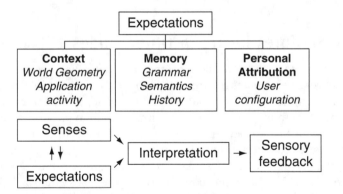

FIGURE   Understanding sensory data.

3.6   Interpretation of integrated senses relies on expectations, which also direct the
              focus of attention of the senses. Possible expectations of interpretation system
              on a computer.

An immersive interface would be like driving a car, where the driver does not
focus on the interaction of the car but becomes as one with it.

It is challenging to use device-free interaction, as the tactile sensory feedback
is difficult to provide. It is as if the somesthetic sense is lacking, such as in a
disabled patient, as follows:

- ✦ Loss of the capability to sense limb movement and position.
- ✦ Major impairment in skilled performance, even with full vision and hearing. This is worsened as visual information degrades.
- ✦ Abnormal movements and the inability to walk following the loss of somesthesis.
- ✦ Patients must exert immense effort to relearn how to walk.
- ✦ Major loss of precision and speed of movement, particularly in the hands.
- ✦ Major difficulty performing tasks that combine significant cognitive loads and fine motor skills such as writing minutes during meetings.
- ✦ Major difficulty learning new motor tasks, relearning lost ones, or using previous experience to guide these processes (Tsagarakis et al. 2006).

Figure 3.7 shows results from Tsagarakis et al. (2006), who designed a virtual surgical knife through a physical handheld device posing as the knife. They found

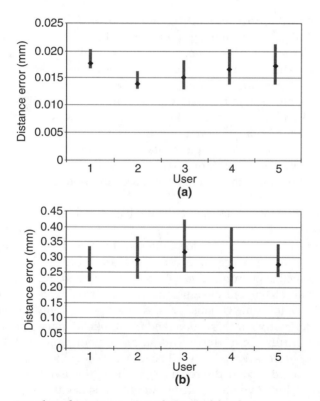

FIGURE

3.7

Results of testing a virtual surgical knife.
The accuracy of the surgical knife is measured in millimeters for five test users.
(a) Test performed using haptic and visual feedback. (b) Test using only visual feedback. *Source:* From Tsagarakis et al. (2006); © 2006 IEEE.

that the gesture accuracy improved 15-fold by adding tactile feedback when the virtual knife hit the virtual flesh.

Applying such feedback to accessory-free gesture interfaces such as a virtual reality CAVE (an environment in which the user is surrounded by back-projected screens on all four sides, ceiling, and floor) with camera-based tracking of body parts will be challenging. Tactile (or haptic) feedback is often given through physical contact with accessory devices via vibrations (Caldwell et al., 1999; Langdon et al., 2000). However, device-free tactile feedback could also be possible through subsonic sound waves (Müller-Tomfelde & Steiner, 2001). Using this approach could enable usage in applications where physical and wired devices would be unnatural.

A common solution is to use visual and audio feedback, but these are not tactile. Another solution could be to research the usage of directed low-frequency audio waves, as this would feel like tactile feedback.

Spatial versus Temporal Perceptive Relation and Precision

A popular solution in multimodal interface studies is the complementary usage of speech and gesture. These modalities complement each other well because vision relates mainly to spatial perception, while sound relates mainly to temporal perception. For example, an experiment was conducted where test subjects saw a dot blinking once on the monitor while hearing two clicks within a certain time frame (Vroomen & Gelder, 2004). The result was that the test subjects perceived two blinks and even three blinks when the sound clicked thrice. This demonstrated a clear complementary merging of senses with a dominant audio cue for temporal cognition. But humans are much better at establishing distance and direction from visual cues than from auditory cues (Loomis et al., 1998).

When you design a visual- and an audio-based detection and synthesis system synchronization problems can arise because their response times may be different. A physically accurate detection and synthesis model tends to reduce response time performance. Long response time can cause ambiguity and error between the modes, while good synchronization solves ambiguity and minimizes errors.

Consider an elaborated example such as a virtual-reality application where you can pick up items for various tasks. A voice recognition system reacts to the phrase "Pick up" to trigger this action, and the application uses your hand position to identify the virtual item. However, you move your hand slightly fast, because you are going to do a lot of work. The visual recognition system lags behind, which causes a 0.5-second discrepancy between the hand position and where the application thinks you are. Furthermore, the speech recognition might have a 1.5-second lag. If you say "Pick up" when your real hand is on top of the virtual object, your hand has moved on from the item that you wanted and you may pick up another item or none at all. If it were a menu item, you might have chosen Delete everything instead of Save.

3.3.2   Gesture Taxonomies

A number of taxonomies on gestures are found in the literature. The best one for use in a given application depends on the concrete context. Basically, there are three ways of labeling gestures:

Semantic labels describe the meaning of the gesture, that is, what it communicates and its purpose. Commonly used in nonverbal communication studies.

Functional labels describe what the gesture does in an interface. Commonly used in technical human–computer (HCI) descriptions.

Descriptive labels describe how the gesture is performed, such as its movement. Commonly used in technical HCI descriptions.

Semantic Labels

Semantic labels, as described by Cassell (1998), can be conscious or spontaneous. Furthermore, they can be interactional or propositional. Conscious gestures have meaning without speech, while spontaneous gestures only have meaning in the context of speech. Examples of conscious gestures follow:

Emblems are conscious communicative symbols that represent words. These are interactional gestures. An example is a ring formed by the thumb and index finger. In Western culture this means "Okay," and in Japan it means "Money."

Propositional gestures consciously indicate places in the space around the performer and can be used to illustrate size or movement. Examples include "It was this big" (using a propositional gesture to show the size) or "Put that there."

Spontaneous gestures are less controllable and emerge spontaneously as the sender is engaged in an interaction:

- ✦ *Iconic* gestures are illustrations of features in events and actions, or how they are carried out. Examples are depicting how a handle was triggered or looking around a corner.
- ✦ *Metaphoric* gestures are like iconic gestures, but represent abstract depictions of nonphysical form. An example is circling the hand to represent that "The meeting went on and on."
- ✦ *Deictic* gestures refer to the space between the narrator and the listener(s). For instance, these can point to objects or people being discussed or refer to movement or directions. They can also be used in a more abstract way, such as in waving away methods, "We don't use those," and picking desired methods as in "These are what we use." These gestures are mainly spontaneous, but can also occur consciously, such as when pointing at an object.
- ✦ *Beat* gestures are used to emphasize words. They are highly dynamic, as they do not depict the spoken messages with postures. An example is when the speaker makes a mistake and then corrects him- or herself while punching with the hand.

Cassell (1998) states that emblems and metaphoric gestures are culturally dependent, while other types are mostly universal, although some cultures use them more than others. Asian cultures use very little gesturing while Southern European cultures use a great deal.

The gestures that are generally relevant for machine interactions are deictic, iconic, propositional, and emblems. This is because the applications have had common themes such as conveying commands and manipulating and pointing out entities. This observation leads to functional labels.

Functional Labels

Functional labels explain intended usage in an application.

Command gestures access system functions in an application, such as Quit, Undo, and Configure. Typically emblems can be used for these because their appearance signals something specific.

Pointing gestures are commonly deictic or propositional gestures that point out entities in the space around the user. This is a core gesture type that is used for selection. For example, if you want to buy something in a market and you do not speak the local language, you might point to it.

Manipulation gestures relate to functions that manipulate or edit the data space, such as scaling or rotating an image or 3D model. Propositional gestures are closely related to this type of interaction because they resemble the usage of those gestures in nonverbal communication.

Control gestures mimic the control over entities in the application such as avatar or camera movements. Iconic, propositional, and deictic gestures can be used because all are used when one person imitates another.

Descriptive Labels

Descriptive labels refer to the manner in which the gestures are performed in a spatiotemporal sense.

Static gestures are postures, that is, relative hand and finger positions, that do not take movements into account. Any emblem can be seen as an example of a static gesture, as in "Thumbs up."

Dynamic gestures are movements, that is, hand trajectory and/or posture switching over time. In nonverbal communication, dynamics in a gesture can alter its meaning (Hummels & Stapiers, 1998) as it is commonly used in sign language. A dynamic gesture can be defined by a trajectory that resembles a figure (Figure 3.8) or simply a sequence of static gestures.

Spatiotemporal gestures are the subgroup of dynamic gestures that move through the workspace over time.

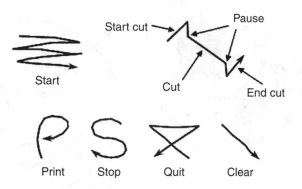

FIGURE

3.8

Dynamic gestures.
Examples of command gestures that are dynamic and spatiotemporal, as they track the position of a hand or handheld device.

3.4   HUMAN FACTORS INVOLVED IN INTERFACE DESIGN

One of the most difficult tasks is to find a feasible gesture vocabulary that is easy for the user to remember and perform (Keates & Robinson, 1998). Sign language is not convenient because the gestures are rather complicated, and sign languages differ according to the underlying vocal language.

Limiting the vocabulary is important, and will benefit both users and designers. Methods that can be used for this purpose follow:

Context dependence: Available options vary with the context of the current selection. This is similar to the context menu in Windows applications, where a right click spawns a menu of items that are relevant only to the selected object. It is important to make the state of the system visible to the user at all times in such interfaces.

Spatial zones: This method is known as the surround user interface (Figure 3.9). The space around the user is divided into spatial zones, each of which has its own context that defines the functions of the gestures. For example, command gestures are close to the user and manipulation gestures are farther away. Special utility functions could be stowed to both sides of the user, much like stowing things away in pockets.

3.4.1   Technology- versus Human-Based Gestures

A typical approach to defining an application's gesture vocabulary is to make it easy for the computer system's recognition algorithm to recognize the gestures. The result of this approach for finding gestures can be called a technology-based

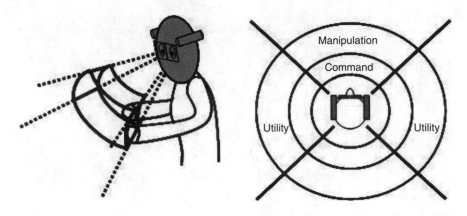

FIGURE

3.9

Example of a surround user interface.
Left: The command gesture shell is close to the user. Gestures are interpreted as command gestures when the hands are inside the bent cube. *Right:* Gestures performed farther away are interpreted as data manipulation gestures. Utility functions are available at the sides of the user.

gesture vocabulary. An example is to define a gesture by how many fingers are stretched (Figure 3.10) because an algorithm has been developed that can count extended fingers. These gestures together create a vocabulary that might be used where there is no particular meaning to the gesture itself; in brief, the association between gesture and meaning is arbitrary. On the other hand, it is easy to implement a computer vision–based recognizer for such gestures.

Table 3.2 shows two example applications in which functionalities have been assigned to gestures. A neutral hand position that is not to be interpreted is called "residue."

Applications containing technology-based gestures have been implemented and tested. Figure 3.11 shows the interface of the application. The tests showed that the gestures were inappropriate for one or more of the following reasons:

+ Stressful/fatigue producing for the user
+ Nearly impossible for some people to perform
+ Illogically imposed functionality

As an alternative, the human-based gesture approach investigates the people who are going to use the interface, and makes use of human factors derived from HCI research, user-centered design, ergonomics, and biomechanics.

3.4.2   HCI Heuristics and Metaphors

Numerous guidelines for user-friendly designs are available. Usability can be described using the following five parameters (Federico, 1999; Bevan & Curson,

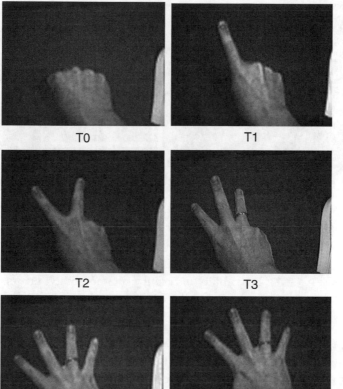

T0   T1

T2   T3

T4   T5

FIGURE

3.10

Technology-based gestures.
These gestures are easy for a computer to recognize using computer vision,
but they are difficult to relate to functions and some of them are difficult to
perform.

TABLE 3.2   Imposed Functionalities in Two Demo Applications

Application Gesture*	Painting	Object Handling
T0	Residue/release	Residue
T1	Paint/select	Select
T2		Copy and paste
T3		Delete
T4–T5	Menu	Release

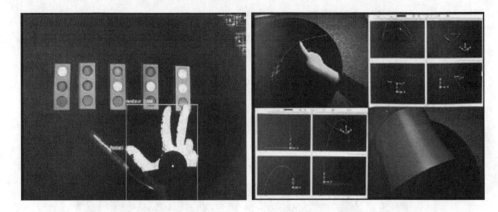

FIGURE

3.11

Technology-based gesture.
A pointing gesture is used for shape creation in a computer-aided design tool.
Source: From Moeslund et al. (2004).

1997; Nielsen, 1992): learnability, efficiency, memorability, errors, and coverage. All parameters are important for the entire interface in which the gestures are used, encompassing the structure, sequence, and feedback from the user interface as well as the gestures themselves. Sequencing refers to the steps required to get from one state to the goal state.

An example is word processing or file management. There are two methods to move a section or file: Copy-Paste-Delete, or Cut-Paste. This is an example of minimizing the steps in the sequencing. However, some people may prefer the three steps because it provides added security by leaving the original in place in case something goes wrong in the transaction.

The core of the human-based approach comprises the following characteristics for the chosen gestures:

Intuitive: Users can use the interface with little or no instruction.

Metaphorically or iconically logical toward functionality: Users can easily see what the gestures are for.

Easy to remember without hesitation: Users focus on their tasks rather than on the interface.

Ergonomic: Not physically stressful when used often.

Because of the technical limitations of gesture recognition technology, it is even more necessary to simplify the dialog and sequencing than it is with conventional input methods. Furthermore, it is wise to keep in mind that the application must be able to distinguish the different gestures. Hence, the principle from the technology-based gestures may have to direct the choice in the development of human-based gestures to some extent.

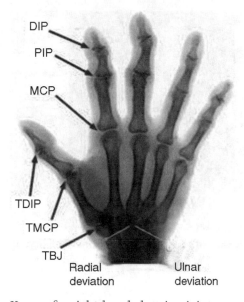

DIP
PIP
MCP
TDIP
TMCP
TBJ
Radial
deviation
Ulnar
deviation

FIGURE

3.12

X-ray of a right hand showing joint name abbreviations and wrist movement.

3.4.3   Ergonomics and Intuitivity

Another way to approach the development of a gesture interface is to look into the ergonomics and biomechanics of gesturing to ensure that a physically stressing gesture is avoided. In this section the biomechanics of the hand (Lin et al., 2001; Lee & Kunjii, 1995) is described. Figures 3.12 and 3.13 show the terms used in this section.

As a handy reference, Table 3.3 lists the ranges of motion for the joints (Eaton, 1997) for the average hand in degrees, where zero degrees in all joint angles is a stretched hand. An example of how to read the numbers follows: The wrist extension/flexion of 70/75 means that the wrist can extend 70 degrees upward and flex 75 degrees downward.

Hyperextension means extending the joint farther than naturally by external force. *Adduction* means moving the body part toward the central axis, which in the hand is between the middle and ring fingers (i.e., gathering the fingers). *Abduction* means moving the body part away from the central axis (i.e., spreading the fingers). *Pronation* and *supination* mean rotating the wrist around the forearm. If the neutral position faces the palm sideways, *pronation* faces the palm downward and *supination* faces it upward.

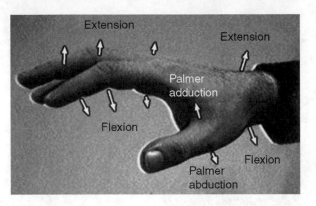

FIGURE

3.13

Side view of a hand.

Finger and wrist motion direction is shown with arrows. For example, bending fingers into a fist is flexion of joints. Stretching fingers is extension of joints.

TABLE 3.3   Range of Motion in Degrees in Hand and Wrist Joints

Joint	Extension/ Flexion	Hyperextension	Adduction/ Abduction
Metacarpophalangeal (MCP)[1]	90	0–45	30
Proximal interphalangeal (PIP)	100	0	0
Distal interphalangeal (DIP)	80	0	0
Trapezio-metacarpophalangeal (TMCP)	90	10	0
Thumb distal interphalangeal (TDIP)	80	15	0
		Palmar Adduction/ abduction	**Radial Adduction/ abduction**
Thumb basal joint (TBJ)		Contact/45	Contact/60
		Pronation/ supination[2]	**Radial/ ulnar**
Wrist[3]	70/75	70/85	20/35

[1]See Figure 3.12.
[2]Rotation is done with the entire forearm.
[3]More than one joint.

The numbers in Table 3.3 are the static constraints for gesture postures. However, there are also dynamic constraints—that is, intra- and interfinger constraints, as in the following list.

✦ Intrafinger constraints are the dependencies between the finger joints.
✦ Interfinger constraints are the dependencies between the postures of neighboring fingers.

Furthermore, finger postures affect pressure in the carpal tunnel (Keir et al., 1998), most severely with metacarpophalangeal joint angles at zero degrees (fingers extended) and at least 45 degrees.

Major principles in ergonomics (see Grant, 2002; Hedge et al., 1999; Shaw & Hedge, 2002) follow:

✦ Avoid outer positions.
✦ Avoid repetition.
✦ Relax muscles.
✦ Relaxed neutral position is in the middle between outer positions.
✦ Avoid staying in a static position.
✦ Avoid internal and external force on joints and stopping body fluids.

Given these guidelines, a gesture interface is potentially ergonomically superior to physical handheld devices, which introduce external force. When you consider and compare biomechanics in gestures, it is important to evaluate internal forces and posture angles. Such evaluation is applied in the following paragraphs to a commonly used gesture.

The TV control gesture in Freeman & Weissman (1995) was a tight fist. Feedback was provided to navigate a pointer on the screen. The dialog was very simple, but tests showed that the gesture was very tiring. The ergonomics of the gesture support this conclusion: The neutral position is for all joints to be in the middle of respective ranges of motion. This means that a fist is a forced position by the muscles in the hand. Furthermore, the hand is raised to head height, or higher, which places the shoulder in or close to an outer position of external rotation (rotated away from the body), with the weight of the arm and hand on it. On the positive side, the wrists are kept straight while operating the TV.

The gestures in Figure 3.8 are stressing because they do not follow the interfinger constraints of the hand. While constraint limitations are different for each person, the user must use more or less force to position the fingers for the system to recognize the correct posture. Furthermore, the recognition is vulnerable to how stretched the fingers are. This means that the users must stretch them to outer positions.

The ergonomics show that making the recognition algorithms tolerant to non-stressing movements is important, which allows the user to avoid remaining in a fixed or static "residue" or "pointing" gesture. Tolerance for deviations in gestures is desirable when implementing gesture interfaces, also because of varying hand shapes and posture performance.

DESIGN GUIDELINES

In this section we discuss some important issues in designing a gesture interface.

3.5.1   The Midas Touch

In Greek mythology, the hedonist King Midas helped Dionysius, the god of wine. As a reward Midas asked that Dionysius make it possible for all that he, Midas, touched to turn into gold. And so it was until he realized that everything turned to gold, including his daughter and his food and drink. Seeing his gift as a curse, he prayed to Dionysius and he was told to pass on the power to a river, Pactolus, which historically has been rich in gold and other metals.

The relevance of this story to gesture interface design is important: The Midas touch refers to the ever-returning problem of when to start and stop interpreting a gesture. As a designer you can expect spontaneous gestures from users all the time if the goal is natural immersive behavior. Therefore, the gesture recognition must be very tolerant. Otherwise, users would suffer rigid constraints to their behavior while in the system. Unfortunately, designers often select rigid constraints as solutions, such as forcing the user not to move between gestures. Alternatively, users may have a manual trigger that tells the system when a gesture starts and stops.

3.5.2   Cultural Issues

It is widely recognized that nonverbal communication is culturally dependent (Cassell, 1998; Agliati et al., 2006) in typology (semantics), rhythm, and frequency. Perhaps there are even gender differences, but this is an infant theory in HCI (Hall, 2006). An important question to ask is whether cultural dependence is a problem. Conventional interfaces that are international are generally in English, but most software is available with a series of national language packages, and people in some nations use different keyboards.

In a gesture interface, this can be translated to selectable gesture vocabularies if it should become a problem that an emblem is illogical to another culture. Furthermore, if a culturally dependent gesture is used, this does not necessarily mean that it is utterly illogical for people of other cultures to learn it.

It is critical to consider cultural aspects when analyzing and developing gesture interfaces/detectors with a focus on natural human conversation and behavior. The system must be able to distinguish (and/or synthesize) those parameters on rhythm, frequency, and typology.

3.5.3   Sequencing

The choice of gestures may depend on your choice of sequencing. Avoiding problems later in the process is easier if you design the sequencing from the start. Sequence design involves deciding each step that a user and the system will go through to accomplish a given task. For example, one way to design an insert task follows:

1. The user makes a command gesture for insert and the gesture is performed at the location of the object's placement.
2. The system feeds back a list of objects to insert.
3. The user uses a pointing gesture to select the type.
4. The system removes the object list and inserts the object.

An alternative sequencing could be the following:

1. The user makes a pointing gesture at an object zone next to the workspace.
2. The system highlights the object.
3. The user uses a pointing gesture in the workspace to place the object.
4. The system inserts the object.

Note that an emblem for insert is only needed in the first example. It is obvious that sequencing can affect interface efficiency and learnability. This is why classical WIMP interfaces usually provide multiple paths to the same objective using the mouse only, the keyboard only, or a combination of the two. In a gesture interface, which is quite limited, this may not be a feasible solution.

In conclusion, choose a sequencing that works well for the intended user group and their preferred gesture vocabulary. It may be a matter of habit, which you can determine from a Wizard of Oz experiment or by observing users' behavior in existing applications.

3.5.4   Use Context Awareness

As discussed in Section 3.3.1, human perception relies on context awareness and expectations related to this context and personal attributes. Furthermore, a WIMP interface has a context-aware menu by right-clicking on an object. This metaphor can be implemented into your gesture interface as well. Context-dependent recognition can be used for limiting the gesture vocabularies and to solve ambiguity.

Spatial Zones (Surround GUI)

One technically tangible way of using context awareness is by using spatial zones (Figure 3.7). The space around the user can be divided into various interaction zones such as a workspace, menu space, template space, and command space. The position where a gesture is performed directs the expectation of the gesture recognition.

Input Modes

Another method is to reuse gestures in different input modes. Metaphorically this is similar to changing input modes for the mouse in the WIMP interface. For example, the mouse in Adobe Photoshop can be used for drawing or for cutting, depending on the input mode.

FIGURE

3.14

ARTHUR project.
Left: Architects investigate the location and size of a new building in London.
Right: Physical placeholder objects are moved in place to select the number of beams in a rooftop. *Source:* From Moeslund et al. (2004).

Use of "Arguments"

A third method is to assign arguments to gestures. An argument can be a physical or a virtual placeholder that alters the meaning of a gesture.

In the project called ARTHUR (Aish et al., 2004), these arguments were physical placeholder objects made out of colored wooden bricks. Architects would use these placeholders to move virtual buildings around in a virtual model of a city. Other placeholders were used for parameter selection or to modify the visual appearance of the virtual buildings. This solution provided robust selection and placement as well as intuitive interaction for the users, who had natural tactile and visual feedback when they moved objects (Figure 3.14).

3.5.5   Feedback

Using a hands-free interface can make the user feel disconnected from the system. It is important to maintain a connection by supplying the user with feedback that tells her that the system is aware of her and what she is doing. This feedback can also tell the user in which context she is acting and how the system interprets the actions.

The feedback can be any sensory input (tactile, auditory, or visual) depending on the nature of the conveyed information. For example, if the user interacts with 3D objects in a workspace, the object under the hand is highlighted in a way other than the selected objects are. This serves two goals: (1) the user knows that the system is aware of the hand, and (2) any discrepancy between the physical world and the virtual environment is calibrated in the consciousness of the user. The outcome is that the user feels freer to move less carefully in the interface.

When the user moves an object close to another object, a sound alerts the user at the time the objects intersect while simultaneously the intersected parts are colored red. The sound tells the user exactly when the intersection occurs and allows the user to react promptly. The color shows the spatial extent of the intersection.

3.5.6   Integration and Disambiguation

Designing a gestures-only interface is difficult. The taxonomy for cooperating with other modalities is called types of cooperation (TYCOON):

Transfer: Serial chronology, that is, first one and then the other mode.

Equivalence: Either or, that is, use one or the other mode.

Complement: Put that there (e.g. Say "Put" while pointing at "that" and moving the hand to the destination and saying "there" to complete the action).

Redundancy: Simultaneous modes to ensure robustness.

Specialization: "Monopoly," that is, only one mode has access to the function.

One way to organize the types of cooperation for an imaginary application is shown in Figure 3.15. Consider having three input modes: speech, vision, and traditional WIMP. Each function in the application should be assigned to one or more modes using TYCOON. For example, "Exit" is a sensitive function, which should not be accessed by accident. Thus, it is determined that all three modes

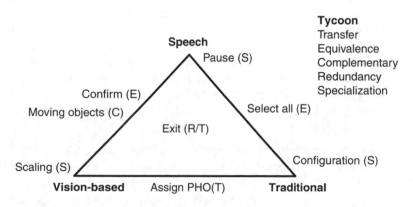

FIGURE

3.15

Organizing types of cooperation.
Assignment of functions to three modalities using the TYCOON taxonomy.

should be used simultaneously to exit the application. The user must say "Exit" and make an exit gesture (redundancy), and then confirm via the keyboard (transfer).

When using multiple modalities, there is a risk of conflicting information. An advance expectation module can prevent such conflicts, or two modes can also disambiguate information. Another solution to conflicting information is to assign *mode priority* and *action priority*. One mode or action can have precedence over the other. For example, when one mode says *Yes* and the other says *No*, mode priority given to the first leads to system execution of the *Yes* command. If the action *No* is given priority, in this instance the *No* command is executed.

HOW TO BUILD AND TEST A GESTURE VOCABULARY

Section 3.4 outlined the importance of choosing a logical and ergonomic gesture vocabulary, and discussed parameters that can be tuned to achieving the same. This section presents a procedure for how to find an appropriate gesture vocabulary for a given application. The description is accompanied by a practical example that shows how it can be used.

To ensure intuitive and logical mapping, an investigation into the interfaces of known applications should be conducted to identify the needed functionalities. It is important to remember that a gesture vocabulary must be tailored for the purpose of the application and for the user group of the application. A gesture vocabulary should not be imposed arbitrarily on any given application.

In the development of interfaces, scenarios have proven valuable (Dzida & Freitag, 1998) to define the context and functionalities and to investigate the user and problem domains. A scenario can be used to examine human–human nonverbal communication. Users would participate in scenarios in which they communicate the same things to a person that they would communicate to the computer application.

There are bottom-up and top-down approaches to this investigation. The bottom-up type finds matching gestures to functions, while the top-down approach presents gestures and identifies logically matched functions. Outlines of variations on these approaches follow:

Subconscious, bottom-up: Create scenarios of communication between people. Record the transactions on video and analyze the subconscious gesturing that occurs.

Conscious, bottom-up: Ask for each functionality for which a gesture would be used, or make a guided drawing test. A testee receives a random drawing project and guides the operator with gestures on what to draw. If an object template–based application is considered, those objects can be provided on the table.

Top-Down: Ask each testee what a given gesture could mean, or do a drawing test. The operator guides the testee with gesturing what to put on paper using a pre-defined gesture vocabulary. This is useful for testing a gesture vocabulary.

Another necessary tool is a benchmark to measure the quality of a gesture vocabulary according to principles in the human-based approach.

3.6.1   The Procedure to Find Gestures

This section describes the proposed procedure and benchmark when developing a gesture interface. In the online case study referred to in Section 3.7, we demonstrate our own experiment using the procedure. It is an iterative approach, where gestures are found and tested and results are used to refine the choices.

Step A: Find the Functions

Find the functions that the gestures will have to access. Keep in mind the user interface in existing similar applications on standard interfaces. (For example, we were designing an architectural design application, so we looked at 3D Studio, among others.)

Step B: Find Logical Gestures

Find the logical gestures that represent the functions found in Step A. This can be done through interviews or experiments with people by taking them through scenarios under camera surveillance where they communicate the same messages that they would communicate to the computer to the "operator" (i.e., the person who conducts the experiment).

It is important that the scenarios take the testees away from normal technical thinking, especially when conducting the tests on technical people. If you tell an engineer that it is a test for the gesture recognition system, the engineer will have a conscious design in mind and act according to the hypothetical restrictions of that design. In brief, you risk that such a person will still think in terms of interfaces and algorithms. If you want to write a scenario with a technical interface aspect, it can be performed as a Wizard of Oz experiment, which tests not only the gesturing but also the design of the entire interface, including the feedback from the system and the sequencing in the interface.

The number of people required for this investigation depends on how broad the user group is and how diverse the results of the test are. Because it is a time-consuming process, using up to 10 participants is typical.

Step C: Process the Data

Note and capture the frames with the commonly used gestures, and note how consistently the various testees use them. Note if they are used only as static postures or if dynamics play an important part in interpreting the gesture.

The theory discussed in Section 3.4.3 should be taken into account in the selection of gestures:

+ Evaluate internal force caused by posture.
  - Deviation from neutral position
  - Outer limits
  - Forces from interjoint relations
+ Evaluate frequency and duration of that gesture.
+ Consider effect on wrist from wrist and finger posture.

See the online case study (Step C) to examine how this is done in praxis.

Step D: Benchmark the Chosen Gesture Vocabulary

The final step is to test the resulting gesture vocabulary. The attributes shown in Table 3.4 are to be tested in the benchmark.

Test 1: Guess the Function   Give the testee a list of functions. Present the gestures and explain the general idea of the application. Ask the person to guess the functions. Gestures that depend on context must be presented in context, but not how the application reacts to them. Explain carefully that it is the gestures that are being tested, not the intelligence of the testee. This takes tension away from the testee.

$$\text{Score} = \text{wrong guesses divided by number of gestures}$$

Test 2: Memory   This test measures how long it takes to become comfortable with the gestures. Only when the gestures are performed without hesitation will the user keep the application in focus rather than the interface.

Give the gesture vocabulary to the testee, who will experiment with the gestures to make sure they are understood. Remind the testee again that it is the gesture vocabulary that is subject to the test, not the testee's ability to perform it.

TABLE 3.4   Overview of What is Actually Tested in Each Test

Attribute	Evaluated in Test ___
Semantic interpretation	1
Generalization	1
Intuitivity	1, 2
Memory, learning rate	2
Stress	3

Present a slide show of functions at a swift pace, two seconds per function. The testee must perform them correctly. Continue until they are all correct. The order should be logical toward sequences in the application, but it should not be the same every time. Then the testee will simply memorize the order of gestures and not the mapping. Restart the slide show after every mistake, and show the gesture vocabulary to the testee between each retry.

$$\text{Score} = \text{number of restarts}$$

The outcome is an indicator of how difficult it will be for a new user to become comfortable with the gesture vocabulary to the extent that the gestures will not steal the focus from the user's tasks.

Test 3: Stress   This is a subjective evaluation of ergonomics. Present the testee with a list of a sequence of gestures. The testee must perform the sequence X times, where X times the size of the gesture vocabulary equals at least 200. Go back to the neutral hand position (residue) between each gesture. Afterward ask the testee how stressful each gesture was and to provide an overall rating for using each gesture over a long time. Meanwhile the operator should note how ergonomically the testee is performing the gestures and how the testee is sitting at the table.

We used the following score list for each gesture and overall for the sequence:

1. No problem
2. Mildly tiring/stressing
3. Tiring/stressing
4. Very annoying/painful
5. Impossible

To place the process in context, as it is very subjective, a computer mouse would be scored at around 2 for normally able people. A more objective stress test would require electrodes on the nerves at the wrist and forearm to measure strain. Such testing may be of interest to pain researchers in particular.

The benchmark can be used to compare multiple gesture vocabularies or user groups. Test 2 is only comparable if the vocabularies are of the same size. If testing a single vocabulary, reasonable success criteria must be ascertained. These aims depend on the gesture vocabulary at hand. See the online case study (Step D) to see how this is done in praxis.

Wizard of Oz experiments have proven valuable in the development of gestures (Beringer, 2002; Carbini et al., 2006). The experiments simulate the response of the system by having a person respond to the user commands. This approach tests a developed interface and can identify problems, with the choice of gestures and sequencing as the necessary feedback.

3.6.2   Technical Tests

Once the interface has been implemented, it is important to conduct a series of technical performance tests. An interface that responds in a delayed fashion or is imprecise will not be successful. Interfaces that assume slow or fairly static movements are less sensitive to this than those in which movement is allowed to be fluent and natural.

Precision of Position, Orientation, and Timing

The important precision parameters that should be tested follow:

Position: Measure any discrepancy between where the system thinks the gesture is performed and where it really is in real-world units.

Orientation: When applicable, measure any discrepancy between the direction that a gesture is pointing and the real direction.

Timing: Measure the response delay in time units. Note start, stroke, and end times, where stroke is the climax of the gesture. For example, the stroke of raising a hand in the air is when the hand reaches its highest position.

Evaluate whether the results are critical to the application and whether it is safe to have the user compensate by using a given feedback. This feedback could be sound to mark the moment that the application triggers an action, or a visual marker that shows where the application thinks the fingers are.

Accuracy of Classifications

This parameter is a measure of gesture detection robustness. Accuracy is the percentage of gestures detected by the computer based on a large series of gesture classifications.

3.7   CASE STUDY

A case study for a gesture interface can be found at *www.beyondthegui.com*.

3.8   SUMMARY

A gesture interface can be a good way to enhance the immersive feeling in a virtual-reality application and a good means of making the user's experience feel more natural. This chapter provided background information on various work and types of gesture interfaces. It also outlined fundamental theory and issues to be addressed when considering and designing such an interface.

Gestures are particularly useful when relating to the space around the user as well as for conveying commands. In particular, we stressed that the gesture vocabulary must be concise, intuitive, and ergonomic to perform, because it can be tiring to use such an interface for a prolonged time.

A procedure for achieving this was proposed and demonstrated, and the resulting gesture vocabulary was tested on various user groups and compared to a purely technical-based gesture vocabulary. The success of the gesture vocabulary depended on the user groups, and the technical gestures were not as easy to use as the intuitive set.

3.9 FUTURE TRENDS

The future will likely bring many interesting breakthroughs in this area. Raw computer vision and other detection systems will evolve into robust classifiers with less obstructive designs. This means that there will be no devices attached to the user, so that the user can feel completely immersed in the virtual environment that blends perfectly with the real world.

Classical computer systems see input modes and output modes as interchangeable input modules to the system. This defines a clear boundary between the user and the system. We should stop thinking of an interface as just … an interface. We do not say that we interface with a car. We say that we drive to some location, that is, the task of driving to the location is in focus. If we can bring the application and the user closer to each other in an invisible interface, the user focuses on the objective, not the computer or interface, which is just a natural part of the environment.

We predict that hands-free gesture interfaces will be as smooth as Jefferson's (Han, 2005) multitouch interface, and will be used to create intelligent rooms with ubiquitous computing and advanced immersive virtual- and mixed-reality environments, which will be aware of people in the real world. These environments will be aware not only of people's presence but of their emotions, actions, and intentions as well. Computers will be able to adapt to users to such a degree that they become a "natural" part of life—almost like a living organism.

REFERENCES

Agliati, A., Vescovo, A., & Anolli, L. (2006). A new methodological approach to nonverbal behavior analysis in cultural perspective. *Behavior Research Methods* 38(3):364–71.

Aish, F., Broll, W., Stoerring, M., Fatah, A., & Mottram, C. (2004). ARTHUR: An augmented reality collaborative design system. *Proceedings of First European Conference on Visual Media Production*, London.

Aviles-Arriaga, H. H., Sucar, L. E., & Mendoza, C. E. (2006). Visual recognition of similar gestures. *Proceedings of Eighteenth International Conference on Pattern Recognition* 1:1100–103.

Beringer, N. (2002). Evoking gestures in SmartKom—Design of the graphical user interface. *Lecture Notes in Artificial Intelligence* 2298, 228–40.

Betke, M., Gips, J., Fleming, P. (2002). The camera mouse: visual tracking of body features to provide computer access for people with severe disabilities. *IEEE Transactions on Neural Systems and Rehabilitation Engineering* 10(1):1–20.

Bevan, N., & Curson, I. (1997). Methods for measuring usability. *Proceedings of Sixth International Federation of Information Processing Conference on Human–Computer Interaction*, Sydney, Australia.

Bowyer, K. W., Chang, K., & Flynn, P. 2006. A survey of approaches and challenges in 3D and multi-modal 3D + 2D face recognition. *Computer Vision and Image Understanding* 101:1–15.

Caldwell, D. G., Tsagarakis, N., & Giesler, C. (1999). An integration tactile/shear feedback array for simulation of finger mechanoreceptor. *Proceedings of 1999 IEEE International Conference on Robotics and Automation*, Detroit.

Carbini, S., Delphin-Poulat, L., Perron, L., & Viallet, J. E. (2006). From a Wizard of Oz experiment to a real-time speech and gesture multimodal interface. *Signal Processing* 86:3559–77.

Cassell, J. (1998). A framework for gesture generation and interpretation. In Cipolla, R., & Pentland, A., eds., *Computer Vision in Human–Machine Interaction*. New York: Cambridge University Press, 191–215.

De Silva, P. R., Marasinghe, C. A., Madurapperuma, A. P., Lambacher, S. G., & Osano, M. (2006). Modeling cognitive structure of emotion for developing a pedagogical agent based interactive game. *IEEE International Conference on Systems, Man and Cybenetics* 1(8–11):339–44.

Dzida, W., & Freitag, R. (1998). Making use of scenarios for validating analysis and design. *IEEE Transactions on Software Engineering* 24:1182–96.

Eaton, C. (1997). *Electronic Textbook on Hand Surgery*. Available at *http://www.eatonhand.com/*.

Federico, M. (1999). *Usability Evaluation of a Spoken Data-Entry Interface*. Povo, Trento, Italy: ITC-Irst Centro per la Ricera Scientifica e Technologica.

Fihl, P., Holte, M. B., Moeslund, T. B, & Reng, L. (2006). Action recognition using motion primitives and probabilistic edit distance. *Proceedings of Fourth International Conference on Articulated Motion and Deformable Objects*. Mallorca, Spain.

Fingerworks. (2002). iGesture Pad. Available at *http://www.fingerworks.com/igesture.html*.

Freeman, W. T., & Weissman, C. D. (1995). Television control by hand gestures. *Proceedings of IEEE International Workshop on Automatic Face and Gesture Recognition*.

Goto, S. (2006). The case study of an application of the system, "BodySuit" and "Robotic-Music": Its introduction and aesthetics. *Proceedings of Conference on New Interfaces for Musical Expression*, Paris, 292–95.

Grant, C. (2002). *Ten Things You Should Know about Hand and Wrist Pain*. Ann Arbor, MI: F-One Ergonomics.

Hall, J. (2006). Nonverbal behavior, status and gender: How do we understand their relations? *Psychology of Women Quarterly* 30:384–91.

Han, J. Y. (2005). Low-cost multi-touch sensing through frustrated total internal reflection. *Proceedings of Eighteenth Annual Association for Computing Machinery Symposium on User Interface Software and Technology*, Seattle, 115–18.

Hedge, A., Muss, T. M., & Barrero, M. (1999). *Comparative Study of Two Computer Mouse Designs*. Ithaca, NY: Cornell Human Factors Laboratory.

Hummels, C., & Stapers, P. J. (1998). Meaningful gestures for human computer interaction: Beyond hand postures. *Proceedings of Third International Conference on Automatic Face & Gesture Recognition*, Nara, Japan, 591–96.

Jaimes, A., & Sebe, N. (2005). Multimodal human computer interaction: A survey. *Proceedings of IEEE International Workshop on Human–Computer Interaction in Conjunction with International Conference on Computer Vision*, Beijing.

Keates, S., & Robinson, P. (1998). The use of gestures in multimodal input. *Proceedings of Association for Computing Machinery Special Interest Group on Accessible Computing*, 35–42.

Keir, P. J., Bach, J. M., & Rempel, D. M (1998). Effects of finger posture on carpal tunnel pressure during wrist motion. San Francisco: Division of Occupational Medicine, University of California.

Klein, J., Moon, Y., & Picard, R. W. (2002). This computer responds to user frustration: Theory, design, and results. *Interacting with Computers* 14(2):119–40.

Langdon, P., Keates, S., Clarkson, P. J., & Robinson, P. (2000). Using haptic feedback to enhance computer interaction for motion-impaired users. *Proceedings of Third International Conference on Disability, Virtual Reality and Associated Technologies*, Alghero, Italy.

Lee, J., & Kunjii, T. (1995). Model-based analysis of hand posture. *IEEE Computer Graphics and Applications* 15:77–86.

Lin, J., Wu, Y., & Huang, T. S. (2001). Modeling the natural hand motion constraints. *Proceedings of Fifth Annual ARL Federated Laboratory Symposium*, 105–10.

Long, A. C., Landay, J. A., Rowe, L. A., & Michiels, J. (2000). Visual similarity of pen gestures. *CHI Letters: Human Factors in Computing Systems (SIGCHI)* 2(1):360–67.

Loomis, J. M., Klatzky, R. L., Philbeck, J. W., & Golledge, R. G. (1998). Assessing auditory distance perception using perceptually directed action. *Perception & Psychophysics* 60:966–80.

Min, B. W., Yoon, H. S., Soh, J., Ohashi, T., & Ejima, T. (1999). Gesture-based editing system for graphic primitives and alphanumeric characters. *Engineering Applications of Artificial Intelligence* 12(4):429–41.

Moeslund, T. B., & Granum, E. (2003). Sequential Monte Carlo tracking of body parameters in a sub-space. *IEEE International Workshop on Analysis and Modeling of Faces and Gestures*, 84–91.

Moeslund, T. B., Störring, M., Broll, W., Aish, F., Liu, Y., & E. Granum. (2004). The ARTHUR system: An augmented round table. *Proceedings of Eighth World Multi-Conference on Systemics, Cybernetics and Informatics*, Orlando.

Moeslund, T. B., Störring, M., & Granum, E. (2004). Pointing and command gestures for augmented reality. *Proceedings of FGnet Workshop on Visual Observation of Deictic Gestures, Pointing, in Association with Seventeenth International Conference on Pattern Recognition*, Cambridge.

Moustakas, K., Strintzis, M. G., Tzovaras, D., Carbini, S., Bernier, O., Viallet, J. E., Raidt, S., Mancas, M., Dimiccoli, M., Yagdi, E., Balci, S., & Leon, E. I. (2006). Masterpiece: Physical interaction and 3D content–based search in VR applications. *IEEE Multimedia* 13(6): 92–100.

Müller-Tomfelde, C., & Steiner, S. (2001). Audio-enhanced collaboration at an interactive electronic whiteboard. *Proceedings of International Conference on Auditory Display*, Espoo, Finland.

Nielsen, J. (1992). The usability engineering life cycle. *Computer* 25:12–22.

Paggio, P., & Music, B. (2000). Linguistic interaction in staging—A language engineering view. In Qvortrup, L., ed., *Virtual Interaction: Interaction in/with Virtual Inhabited 3D Worlds*. New York: Springer-Verlag, 235–49.

Pantic, M., & Patras, I. (2006). Dynamics of facial expression: Recognition of facial actions and their temporal segments from face profile image sequences. *IEEE Transactions on Systems, Man and Cybernetics, Part B* 36(2):433–49.

Pantic, M., & Rothkrantz, L. J. M. (2003). Toward an affect-sensitive multimodal human–computer interaction. *Proceedings of IEEE* 91(9):1370–90.

Rasmussen, N. T., Störring, M., Moeslund, T. B., & Granum, E. (2006). Real-time tracking for virtual environments using SCAAT Kalman filtering and unsynchronized cameras. *Proceedings of International Conference on Computer Vision Theory and Applications*, Setúbal, Portugal.

Riviera, J-B., & Guitton, P. (2004). Hand posture recognition in large-display VR environments. In *Gesture-Based Communication in Human-Computer Interaction*. Berlin: Springer, 259–68.

Sage, K., Howell, A. J., & Buxton, H. (2003). Developing context sensitive HMM gesture recognition. *Proceedings of Fifth International Workshop on Gesture and Sign Language-Based Human–Computer Interaction*, Genova, Italy.

Shaw, G., & Hedge, A. (2002). The effect of keyboard and mouse placement on shoulder muscle activity and wrist posture. *Ergo Info*, available at *http://www.humanscale.com/ergo_info/articles.cfm?type = studies*.

Steininger, S., Lindemann, B., & Paetzold, T. (2002). Labeling of gestures in SmartKom—The coding system. *Proceedings of Gesture and Sign Language in Human–Computer Interaction, International Gesture Workshop*, London, 215–27.

Störring, M., Andersen, H. J., & Granum, E. (2004). A multispectral approach to robust human skin detection. *Proceedings of Second European Conference on Color in Graphics, Images and Vision.*

Störring, M., Granum, E., & Moeslund, T. B. (2001). A natural interface to a virtual environment through computer vision-estimated pointing gestures. *Proceedings of Workshop on Gesture and Sign Language-Based HCI*, London.

Streitz, N. A., et al. (2001). *Roomware: Towards the Next Generation of Human–Computer Interaction Based on an Integrated Design of Real and Virtual Worlds*. German National Research Center for Information Technology, Integrated Publication and Information Systems Institute.

Tsagarakis, N. G., Gray, J. O., Caldwell, D. G., Zannoni, C., Petrone, M., Testi, D., & Viceconti, M. (2006). A haptic-enabled multimodal interface for the planning of hip arthroplasty. *IEEE Multimedia* 13(6):40–48.

Tse, E., Shen, C., Greenberg, S., & Forlines, C. (2006). Enabling interaction with single user applications through speech and gestures on a multi-user tabletop. *Proceedings of Working Conference on Advanced Visual Interfaces Table of Contents*, Venezia, Italy, 336–43.

Vroomen, J., & de Gelder, B. (2004). Temporal ventriloquism: Sound modulates the flash-lag effect. *Journal of Experimental Psychology: Human Perception and Performance* 30(3): 513–18.

Wachs, J., Stern, H., Edan, Y., Gillam, M., Feied, C., Smith, M., & Handler, J. (2006). A real-time hand gesture interface for a medical image-guided system. *Proceedings of Ninth Israeli Symposium on Computer-Aided Surgery, Medical Robotics, and Medical Imaging.*

Wang, S. B., Quattoni, A., Morency, L-P., Demirdjian, D., & Darrell, T. (2006). Hidden conditional random fields for gesture recognition. *IEEE Computer Society Conference on Computer Vision and Pattern Recognition* 2:1521–27.

4

Locomotion Interfaces

Mary C. Whitton, Sharif Razzaque

This chapter is about locomotion interfaces—interfaces that both enable users to move around in real or virtual spaces and make users feel as if they are moving. Locomotion is a special type of movement: Locomotion, as used by life scientists, refers to the act of an organism moving itself *from one place to another*. This includes actions such as flying, swimming, and slithering. For humans, locomotion is walking, running, crawling, jumping, swimming, and so on. In Figure 4.1, the user moves from one place to another by leaning to control speed and direction of the Segway Personal Transporter (PT); in Figure 4.2, the user is (really) walking on the treadmill (shown in detail on the right) to move through a virtual landscape.

The focus of this chapter is computer-based locomotion interfaces for moving about in computer-generated scenes. One way to think of these interfaces is that they are virtual-locomotion interfaces for virtual scenes. We may occasionally use an example of locomotion in the real world, but when we say locomotion interface, we mean *virtual*-locomotion interface.

Many common virtual-locomotion interfaces take input from only the user's hands; for instance, three-dimensional (3D) computer video games are operated by joystick, keyboard, mouse, and/or game controller. Other interfaces use more of the body. For example, in the real world, Segway PTs, hang gliders, and skateboards require users to lean and shift their weight to control motion. In a virtual-reality system that has a real-walking locomotion interface, people use their bodies in a completely natural fashion to walk about the virtual scene (Figure 4.3). This chapter concentrates on virtual-locomotion interfaces that involve all of, or at least much of, the user's body.

Segway i2 self-balancing Personal Transporter.
The user controls the speed by leaning forward or backward and turns by leaning
to the right or left with the LeanSteer frame. (Courtesy of Segway Inc.)

Even though both arms and both legs may be involved in controlling a vehicle and the whole body may be involved in receiving feedback about the state of the vehicle, simulating movement resulting from operating a vehicle is conceptually different from simulating the movement of a human walking about human-scale spaces such as buildings and ships. In a vehicle, users' intentions of what direction to move and how fast to go are specified unambiguously by their interaction with the controls of the vehicle. The challenge for virtual-locomotion interfaces is to capture the user's intent using data that can be derived from sensing the *pose* (position and orientation) and *movement* of the user's body.

The scope of this chapter is, then, interfaces between human users and computer-generated scenes that enable users to control their locomotion through the (virtual) scene by moving their bodies. The body pose and movement specify how they want to move—both the *direction* they want to move in and *how fast* they want to move.

FIGURE

4.2
University of Utah's Treadport virtual-locomotion interface.
Left: The user walks on a treadmill while viewing the moving virtual landscape on the large projector screens. *Right:* To simulate hills, the entire treadmill can tilt up. To simulate the user's virtual inertia, the Treadport physically pushes or pulls the user via a large rod that connects to a user-worn harness. (Courtesy of University of Utah, School of Computing; © John M. Hollerbach.)

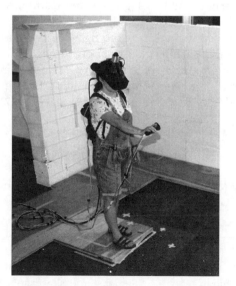

FIGURE

4.3
Real-walking virtual locomotion interface.
To move about the virtual scene (*right*), the user actually walks about the real-world laboratory (*left*). (Images courtesy of the Department of Computer Science, UNC at Chapel Hill.)

There are three dominant metaphors for virtual locomotion. In *real-walking–style* systems, input movements and resulting movement through the space are as natural and as much like really walking as possible. Examples are two-dimensional (2D) treadmills, walking-in-place, and real-walking interfaces. In *vehicle-style* interfaces, input movements and response are similar to driving a vehicle. For example, Fleming and colleagues (2002) developed a joystick interface in which pushing the joystick up (away from you) moves you forward, and pushing the joystick left rotates you left but does not move you forward. To move left, you must first turn (rotate) left using the joystick and then move forward. This is similar to how a driver operates a tank.

Both real-walking and vehicle-style interfaces fall into a category of locomotion techniques that Slater and Usoh (1994) call *mundane*, that is, they attempt to mimic, as closely as possible, how we walk or control vehicles in the real world. In contrast, *magical-style* interfaces are those that permit movements that have no natural corollary in the real world but that serve a useful purpose when you are moving about in a virtual scene. For instance, the ability to teleport between two distant locations is a magical component of a locomotion interface. Similarly, the ability to adapt the length of your virtual stride so that you move miles with each step is a magical property.

Regardless of the choice of metaphor, the interface has to convert the user's physical movement (in the real world) into virtual movement—both direction and speed—in the virtual scene. Table 4.1 gives some examples of body movements and how they might be used to control locomotion direction and speed.

Note that locomotion is distinct from *wayfinding*, which refers to the cognitive task of determining a route for reaching a destination. We avoid using *wayfinding* because it is used in the literature to refer to both wayfinding and locomotion. Other authors have referred to movements such as going from a virtual desk to a virtual bookshelf as *travel* (e.g., Bowman, Koller, & Hodges, 1997), but we prefer

TABLE 4.1   Examples of How Body Movement Might Be Interpreted to Control Direction and Speed of Locomotion

Body Part	Direction Control	Speed Control
Hands	Move in direction user is pointing	Distance between hands controls step size
Feet	Move in direction feet are pointing	Step faster or slower to speed or slow pace
Legs	Swing leg sideways from hip to move sideways	Step faster or slower to speed or slow pace
Arms	Make "banking" gesture to turn like bird	Flap arms faster or slower to speed or slow pace

locomotion[1] because travel implies a large distance—one *travels* from New York to Chicago, but does not travel across a room. Hollerbach's (2002) definition of a locomotion interface requires that the user make repetitive motions such as tapping a key on a keyboard or walking-in-place. We do not use Hollerbach's definition in this chapter.

4.1   NATURE OF THE INTERFACE

The essence of a locomotion interface (for both real and virtual worlds) can be generalized. Users communicate their intentions to the system (computer or vehicle) by making real-world motions—they press keys, move a joystick, move their hands, lean forward, push with a foot, and so on. These user input motions are then converted through an interface system to movement through the real or virtual world. The system may also provide feedback to the users, indicating how they are moving or have moved. Users use the feedback to plan their next motion and, for visual feedback, to help maintain balance and to update their mental model of their surroundings.

4.1.1   The Physiological Nature of the Interface: Perceiving Self-Motion

There are several sensory channels that provide us with information on how and where we are moving. Each of these sensory channels—auditory, visual, vestibular (from the inner ear), and proprioceptive (the sense of body position)—contributes information to awareness of self-motion. Humans rely on sensory cues to determine whether they themselves are moving (self-motion) or if the objects around them are moving (external motion) and for balance and orientation (Dichgans & Brandt, 1977).

Auditory: Humans have the ability to deduce qualities of their environment from the way the environment sounds; for instance, large rooms sound different from small rooms. We also have the ability to localize sound sources, and as we move the perceived location of a stationary sound moves in relation to our head. A moving sound source can, by itself, cause a stationary person to feel as if she is moving (Lackner, 1977).

Tactile, proprioceptive, and podokinetic: Humans can sense movement in their joints, muscles, and viscera, and can sense pressure and slippage on the skin. These cues are important for walking, as they tell a person where her limbs are

---

[1] Researchers may have adopted the term *travel* because the verb form of locomotion, to *locomote*, sounds awkward.

and when her feet are touching the ground. These cues also communicate the relative motion of the person's body (i.e., how the limbs move relative to the torso).

Vestibular: The vestibular system is able to sense motion of the head with respect to the world. Physically, the system consists of labyrinths in the temporal bones of the skull, just behind and between the ears. The vestibular organs are divided into the semicircular canals (SCCs) and the saccule and utricle (Figure 4.4). As a first-order approximation, the vestibular system senses motion by acting as a three-axis rate gyroscope (measuring angular velocity) and a three-axis linear accelerometer (Howard, 1986). The SCCs sense rotation of the head and are more sensitive to high-frequency (quickly changing) components of motion (above roughly 0.1 Hz) than to low-frequency (slowly changing) components. Because of these response characteristics, it is often not possible to determine absolute orientation from vestibular cues alone. Humans use visual information to complement and disambiguate vestibular cues.

Visual: Visual cues alone can induce a sense of self-motion, which is known as *vection.* The kinds of visual processing can be separated into *landmark recognition* (or *piloting*), where the person cognitively identifies objects (e.g., chairs, windows) in her visual field and so determines her location and *optical flow*.

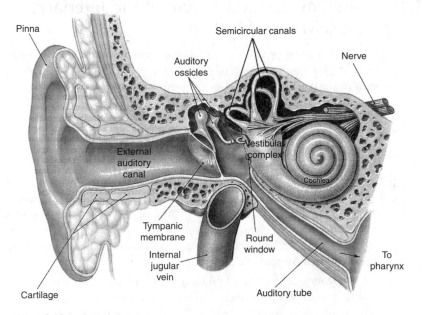

FIGURE

4.4

Vestibular system.

A cut-away illustration of the outer, middle, and inner ear. *Source:* Adapted from Martini (1999). Copyright © 1998 by Frederic H. Martixi, Inc. Reprinted by permission of Pearson Education, Inc.

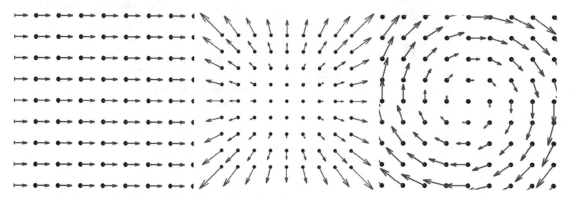

FIGURE

4.5

Three optical flow patterns.
Left: Laminar translation result from turning the head left. *Center:* Radial expansion resulting from moving forward. *Right:* Circular flow resulting from rolling about the forward axis.

Optical flow is a low-level perceptual phenomenon wherein the movement of light patterns across the retina is sensed. In most situations, the optical-flow field corresponds to the motion field. For example, if the eye is rotating in place, to the left, the optical-flow pattern is a laminar translation to the right. When a person is moving forward, the optical-flow pattern radiates from a center of expansion (Figure 4.5). Both optical-flow and landmark recognition contribute to a person's sense of self-motion (Riecke, van Veen, & Bulthoff, 2002; Warren Jr. et al., 2001).

Visual and Vestibular Senses Are Complementary

As mentioned before, the vestibular system is most sensitive to high-frequency components of motions. On the other hand, the visual system is most sensitive to *low*-frequency components of motion. The vestibular and visual systems are complementary (Figure 4.6).[2] The crossover frequency of the two senses (Figure 4.7) has been reported to be about 0.07 Hz (Duh et al., 2004).

Combining Information from Different Senses into a Coherent Self-Motion Model

Each sensory modality provides information about a particular quality of a person's motion. These pieces of information are fused to create an overall sense of

---

[2] This has a striking similarity to the hybrid motion trackers mentioned in Section 4.2.1, which use accelerometers and gyros to sense high-frequency components of motion while correcting for low-frequency drift by using low-frequency optical or acoustic sensors.

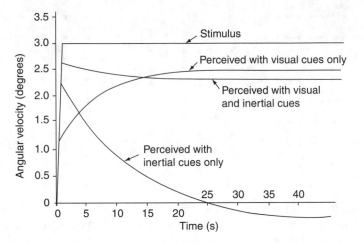

FIGURE

4.6

Contribution of the visual and vestibular (or inertial) systems to perceiving a step function in angular velocity.

The vestibular system detects the initial, high-frequency step, whereas the visual system perceives the sustained, low-frequency rotation. *Source:* Adapted from Rolfe and Staples (1988). Reprinted with the permission of Cambridge University Press.

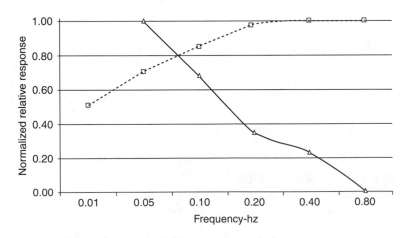

FIGURE

4.7

Crossover frequency of visual and vestibular senses.

Visual (*solid line*) and vestibular (*dashed line*) responses as a function of motion frequency. *Source:* Adapted from Duh et al. (2004); © Dr. Donald Parker and Dr. Henny Been-Lim Duh.

self-motion. There are two challenges in this process. First, the information must be fused quickly so that it is up to date and relevant (e.g., the person must know that and how she has tripped in time to regain balance and footing before hitting the ground). Second, the total information across all the sensory channels is often incomplete. One theory states that, at any given time, a person has a *model* or hypothesis of how she and surrounding objects are moving through the world. This model is based on assumptions (some of which are conscious and cognitive, while others are innate or hardwired) and previous sensory information. New incoming sensory cues are evaluated *in terms of this model*, rather than new models being continuously constructed from scratch. The model is updated when new information arrives.

Consider the familiar example of a person on a stopped train that is beside another stopped train. When the train on the adjacent track starts to move, the person might have a short sensation that *her* train has started moving instead. This brief, visually induced illusion of self-motion (vection) is consistent with all of her sensory information thus far. When she looks out the other side of her train and notices the trees are stationary (relative to her train), she has a moment of disorientation or confusion and then, in light of this new information, she revises her motion model such that her train is now considered stationary. In short, *one tends to perceive what one is expecting to perceive*. This is an explanation for why so many visual illusions work (Gregory, 1970). An *illusion* is simply the brain's way of making sense of the sensory information; the brain builds a model of the world based on assumptions and sensory information that happen to be wrong (Berthoz, 2000).

For a visual locomotion interface to convincingly give users the illusion that they are moving, it should reduce things that make the (false) belief (that they are moving) inconsistent. For example, a small display screen gives a narrow field of view of the virtual scene. The users see the edge of the screen and much of the real world; the real world is stationary and thus inconsistent with the belief that they are moving. A large screen, on the other hand, blocks out more of the real world, and makes the illusion of self-motion more convincing.

Perception is an active process, inseparably linked with action (Berthoz, 2000). Because sensory information is incomplete, one's motion model is constantly tested and revised via interaction with and feedback from the world. The interplay among cues provides additional self-motion information. For example, if a person sees the scenery (e.g., she is standing on a dock, seeing the side of a large ship only a few feet away) shift to the left, it could be because she herself turned to her right or because the ship actually started moving to her left. If she has concurrent proprioceptive cues that her neck and eyes are turning to the right, she is more likely to conclude that the ship is still and that the motion in her visual field is due to her actions. The active process of self-motion perception relies on prediction (of how the incoming sensory information will change because of the person's actions) and feedback. Locomotion interfaces must maintain such a feedback loop with the user.

4.1.2   Locomotion Interfaces as Interaction Loops

In the case of a whole-body virtual locomotion interface, the inputs to the inter-face are the pose of one or more parts of the user's body, and the output of the interface is the point of view (POV) in the virtual scene. The change in eye position between samples is a vector quantity that includes both direction and magnitude. The primary mode of *feedback* in most virtual locomotion systems is visual: What you see changes when you move. In the locomotion system, the change in POV causes the scene to be (re)drawn as it is seen from the new eye position, and then the user sees the scene from the new eye point and senses that she has moved. Figure 4.8 shows the interaction loop that is repeated each time the pose is sampled and the scene redrawn.

Delving a level deeper, a locomotion interface must detect whether it is the user's intent to be moving or not, and, if moving, determine the direction and speed of the intended motion and initiate a change in POV. Figure 4.9 shows this as a two-step process. First, sensors of various types (described in more detail in the next section) are used to capture data such as the position and orientation of parts of the user's body, or to capture information about the speed and/or acceler-ation of the motion of parts of the user's body. These data are processed to gener-ate signals that specify whether the user is moving, the direction of the motion, and the speed of the movement. Those signals are interpreted and changed into a form that specifies how the POV is moved before the scene is next rendered. We show examples of this in the case study (available at *www.beyondthegui.com*).

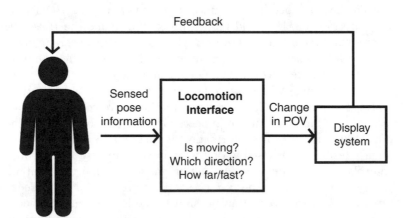

FIGURE

4.8

Data about the user's body pose are inputs to the interface.
The interface computes the direction and distance of the virtual motion from the inputs. The direction and distance, in turn, specify how the POV (from which the virtual scene is drawn) moves. By displaying the scene from a new POV, the interface conveys, to the user, that she has moved. Based on this feedback, she, in turn, updates her model of space and plans her next move.

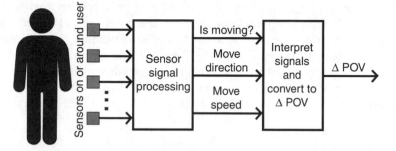

FIGURE

4.9

Details of virtual-locomotion interface processing.
The sensors measure position, orientation, velocity, and/or acceleration of body parts. Signal processing of the sensor data determines whether the user intends to move and determines the direction and speed of her intended movement. Speed and direction are interpreted and converted into changes in the POV.

4.2 TECHNOLOGY OF THE INTERFACE

Whole-body locomotion interfaces require technology to *sense* user body position and movement and to *display* to feed back the results of the locomotion to the user.

4.2.1 Pose and Motion Sensors

Sensors that measure and report body motions are often called *trackers*. Trackers can measure the position and orientation of parts of the body, or can measure body movement (e.g., displacement, rotational velocity, or acceleration). Trackers come in a myriad of form factors and technologies. Most have some pieces that are attached to parts of the user's body (e.g., the head, hands, feet, or elbows) and pieces that are fixed in the room or laboratory. A detailed overview of tracking technology can be found in Foxlin (2002). Common categories of tracking systems are trackers with sensors and beacons (including full-body trackers) and beaconless trackers.

Trackers with Sensors and Beacons
One class of tracker, commonly used in virtual-reality systems, has one or more sensors worn on the user's body, and beacons fixed in the room. Beacons can be active, such as blinking LEDs, or passive, such as reflective markers. Trackers with sensors on the user are called *inside-looking-out* because the sensors look out to the beacons. Examples are the InterSense IS-900, the 3rdTech HiBall-3100, and the Polhemus LIBERTY. Until recently, body-mounted sensors were often connected to the controller and room-mounted beacons with wires carrying power and data; new models are battery powered and wireless.

In inside-looking-out systems, each sensor reports its pose relative to the room. Many virtual-environment (VE) systems use a sensor on the user's head to establish which way the head is pointing (which is approximately the way the user is looking) and a sensor on a handheld device that has buttons or switches that allow the user to interact with the environment. In Figure 4.3, the sensor on the top of the head is visible; the sensor for the hand controller is inside the case. A sensor is required on each body part for which pose data are needed. For example, if a locomotion interface needs to know where the feet are, then users must wear sensors or beacons on their feet. Figure 4.10 shows a walking-in-place

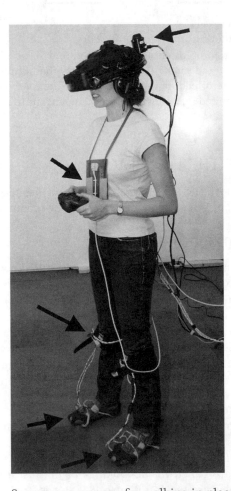

FIGURE

4.10

Sensors on a user of a walking-in-place locomotion interface. This system has a head position and orientation sensor on the back of the headset, a torso orientation sensor hanging from the neck, knee position sensors, and pressure sensors under each shoe. (Photo courtesy of the Department of Computer Science, UNC at Chapel Hill.)

user with sensors on her head, torso, and below the knee of each leg. The tracker on the head determines the user's view direction; the torso tracker determines which way she moves when she pushes the joystick away from her; the trackers at the knees estimate the position and velocity of the heels, and those data are used to establish how fast the user moves.

Sensor beacon trackers can be built using any one of several technologies. In magnetic trackers, the beacons create magnetic fields and the sensors measure the direction and stretch of the magnetic fields to determine location and orientation. In optical trackers, the beacons are lights and the sensors act as cameras. Optical trackers work a bit like celestial navigation—the sensors can measure the angles to three or more stars (beacons) of known position, and then the system triangulates to find the position of the sensor. Acoustic trackers measure position with ultrasonic pulses and microphones.

Another class of trackers is called *outside-looking-in* because the beacons are on the user and the sensors are fixed in the room. Camera-based tracker systems such as the WorldViz Precision Position Tracker, the Vicon MX, and the PhaseSpace IMPULSE system are outside-looking-in systems and use computer vision techniques to track beacons (sometimes called markers in this context) worn by the user. This technology is common in whole-body trackers (also called *motion capture systems*) where the goal is to measure, over time, the path of tens of markers. Figure 4.11 shows a user wearing highly reflective foam balls and three cameras on tripods. Motion capture systems can also be built using magnetic, mechanical (Figure 4.12 on page 121), and other nonoptical technologies. Motion capture systems often require significant infrastructure and individual calibration.

Beaconless Trackers

Some tracking technologies do not rely on beacons. Some, for example, determine the orientation of the body part from the Earth's magnetic field (like a compass), or the Earth's gravitational field (a level-like *tilt sensor*). Others are inertial sensors—they measure positional acceleration and/or rotational velocity using gyroscopes. Some inertial tracking systems integrate the measured acceleration and rotation data twice to compute position or displacement. However, this computation causes any errors in the acceleration measurement (called *drift*) to result in rapidly increasing errors in the position data. Some tracker systems, such as the Intersense IS-900, correct for accelerometer drift by augmenting the inertial measurements with measurements from optical or acoustic trackers. Such systems with two (or more) tracking technologies are called *hybrid trackers*.

Trade-offs in Tracking Technologies

Each type of tracker has advantages and disadvantages. Wires and cables are a usability issue. Many magnetic trackers are susceptible to distortions in magnetic fields caused by metal objects in the room (e.g., beams, metal plates in the floor, other electronic equipment). Some systems require recalibration whenever

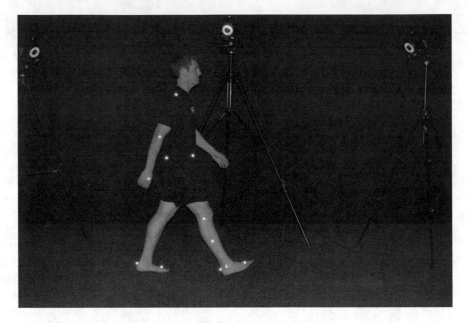

FIGURE

4.11

Optical motion capture system.
In this system, the user wears foam balls that reflect infrared light to the cameras
mounted on the tripods. The bright rings are infrared light sources. Cameras
are embedded in the center of the rings. (Courtesy of Vicon.)

the metal objects in the room are moved. Optical systems, on the other hand,
require that the sensors have a direct line of sight to the beacons. If a user has
an optical sensor on her head (pointing at a beacon in the ceiling) and bends over
to pick something off the floor, the head-mounted sensor might not be able to
"see" the ceiling beacon anymore and the tracker will no longer report the pose
of her head.

Other Motion Sensors

In addition to position-tracking motion sensors, there are sensors that measure other, simpler characteristics of the user's body motion. Examples include
pressure sensors that report if, where, or how hard a user is stepping on the
floor. These sensors can be mounted in the shoes (Yan, Allison, & Rushton,
2004) or in the floor itself. The floor switches can be far apart, as in the
mats used in Konami Corporation's Dance Dance Revolution music video game
(Figure 4.13 on page 122) or closer together to provide higher positional resolution (Bouguila et al., 2004).

FIGURE
4.12

Mechanical motion capture system.
This mechanical full-body system is produced by SONALOG. (Courtesy of
SONALOG.)

4.2.2 Feedback Displays

As users specify how they want to move, the system must provide feedback via a
display to indicate how and where they are moving. Feedback closes the locomotion interface loop (Figure 4.8). *Display* is a general term, and can refer to visual
displays as well as to means of presenting other stimuli to other senses.

Many locomotion interfaces make use of head-worn visual displays, as in
Figure 4.3, to provide visual feedback. These are often called head-mounted displays (HMDs). Using such a display, no matter how the user turns her body or head,
the display is always directly in front of her eyes. On the negative side, many HMDs
are heavy to wear and the cables may interfere with the user's head motion.

Another type of visual display has a large flat projection screen or LCD
or plasma panel in front of the user. These displays are usually of higher image
quality (e.g., greater resolution, higher contrast ratio, and so on) than HMDs.
However, if the user physically turns away from the screen (perhaps to look
around in the virtual scene or to change her walking direction), she no longer
has visual imagery in front of her.

To address this problem, several systems surround the user with projection displays. One example of this is a CAVE, which is essentially a small room where one or
more walls, the floor, and sometimes the ceiling, are display surfaces (Figure 4.14).
Most surround screen displays do not completely enclose the user, but some do.

Konami Dance Dance Revolution sensor pad.
The game player presses switches with her feet (and sometimes hands) to
coordinate with on-screen dance steps. (Image courtesy of the Department of
Computer Science, UNC, Chapel Hill.)

Surround-screen display.
The three-wall-plus-floor projection system is seen (*left*) with the projectors off
and (*right*) displaying a scene with a virtual human on the left and a real human
user on the right. (Courtesy of University College London, Department of
Computer Science.)

In addition to visual displays, there are displays for other senses. As mentioned earlier, people can get the sensation that they are moving from auditory cues alone, without any visuals (Lackner, 1977). While we have seen one virtual-environment system that uses only auditory displays (from AuSIM), most systems use auditory displays as a supplement to the primary visual display. Hearing your own footsteps is one cue that helps you remain oriented during movement, and subtle echoes tell you much about the size of the room you are in and your position in it.

Another display used in whole-body locomotion interfaces is a motion platform—the user is sitting or standing on a platform that is physically moved under computer control. Some platforms tilt along one or more axes, whereas others tilt and translate. Motion platforms are most commonly used in flight simulators. The deck of the ship in Disney's Pirates of the Caribbean ride has a motion platform as well as a surround screen visual display (Schell & Shochet, 2001). The motion platform tilts the deck so that there is a match between the virtual waves the user sees and the real rolling of the ship's deck (Figure 4.15).

FIGURE

4.15

Disney's Pirates of the Caribbean ride.
This ride has surround screens (in the distance) and a motion platform (in the shape of a pirate ship's deck). Players aim the cannons and steer the ship using the physical props. (Courtesy of Disney.)

Finally, a motorized treadmill, where the computer system continually adjusts the speed and direction of the moving belt in response to the user's inputs and position, can also be considered as a kind of motion display. The University of Utah Treadport (Figure 4.2) combines several kinds of displays into a single locomotion interface: It has a surround screen visual display, a motorized treadmill, a motion platform that tilts the entire treadmill up and down (to simulate different slopes), and a force display that pushes and pulls a harness worn by the user (to simulate momentum).

4.3 CURRENT IMPLEMENTATIONS OF THE INTERFACE

This section discusses the current state of the art of locomotion interfaces. For all of the examples described, at least one prototype has been built. Unfortunately, several of them no longer exist (as far as we know).

4.3.1 Flying and Leaning

The most common locomotion technique in virtual-reality systems is flying using a joystick or some other hand controller. When the user pushes a joystick or presses a button, she moves forward in the virtual scene. She can still move about locally by leaning or taking a real step in any direction, assuming her head is tracked by the VE system. The effect is similar to that of walking about on a moving flatbed truck or flying carpet (Robinett & Holloway, 1992). When the user presses the button, the truck moves forward in the virtual scene and the user can simultaneously move about on the truck bed.

There are significant variations in how flying is implemented. Pushing the joystick or the button signals that the user intends to move. The direction of motion can be established in various ways. In some VE systems, the user moves in the direction her head (or nose) is pointing (*head directed*). Other interfaces move the user in the direction in which she is pointing with her hand controller. Still others interpret forward using a vehicle metaphor: Forward is toward the center-front wall of the multiwall projection system.

Similar to flying, leaning techniques move the user in the virtual scene in the direction in which she is leaning (LaViola Jr. et al., 2001; Peterson et al., 1998). Most leaning interfaces also control speed—the farther the user leans, the faster she moves. Leaning has the advantage of not requiring a hand controller, leaving the hands free.

4.3.2 Simulated Walking and Walking

Locomotion interfaces that require users to make walking motions with their legs are attractive because of their similarity to real walking. Two primary methods allow a person to feel he is walking but keep him in a restricted space: treadmills and walking-in-place.

Treadmills and Treadmill-Like Devices

There are a number of techniques that simulate the physical act of walking with treadmills (Brooks 1998; Hollerbach et al., 2000). As users walk forward on the treadmill, they move forward in the virtual scene. Using a passive treadmill (i.e., without a motor) requires a physical effort beyond that of walking in the real world because of the mass and friction of the belt and rollers. Motorized treadmills address this problem and still allow the user to change speed. Motorized treadmills sense the user's position on the belt and vary the belt motor speed to keep the user centered. The disadvantage of motorized treadmills is the noise and potential for injury—many virtual-locomotion treadmills have the users wear a safety tether to prevent them from falling. All of the treadmills discussed thus far have the limitation that the treadmill has a preferred orientation; it is difficult, disorienting, and often impossible to turn on the spot in the virtual scene. The University of North Carolina's treadmill (Brooks, 1988), for example, had handlebars to steer like a bicycle.

To allow turning on the spot, several groups have developed 2D treadmills (Darken, Cockayne, & Carmein, 1997; Iwata, 1999) or omnidirectional treadmills (Virtual Devices) where the user can walk in any direction on the ground plane. As can be seen in Figure 4.16, these devices are large, noisy, and mechanically complex. Darken and colleagues (1997) commented about the omnidirectional treadmill that it was hard to change directions while running because it felt like "slippery ice."

Another approach to 2D treadmill-like locomotion is having the user contained inside a very large sphere (akin to a human-sized "hamster ball") that is held in place while it rotates. There are at least two implementations of this idea—the VirtuSphere and the CyberSphere. The VirtuSphere (Figure 4.17) is passive, whereas the CyberSphere is motorized. As of this writing, several Virtu-Spheres have been installed, but we are unaware of any working CyberSpheres having been built.

Walking Interfaces

When using a walking-in-place (WIP) interface, the user makes walking or step-ping motions, that is, lifting the legs, but stays on the same spot physically. The locomotion interface detects this motion and moves her forward in the virtual scene (Slater, Usoh, & Steed, 1995; Templeman, Denbrook, & Sibert, 1999; Usoh et al., 1999). Like flying, WIP does not require a large lab or tracking space. Like a treadmill, the user makes motions that are similar to real walking. Virtual-reality system users who walk-in-place have a greater sense of presence than those who fly with a joystick or hand controller (Slater et al., 1995; Usoh et al.,

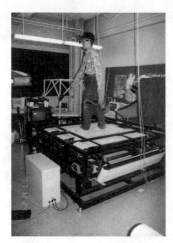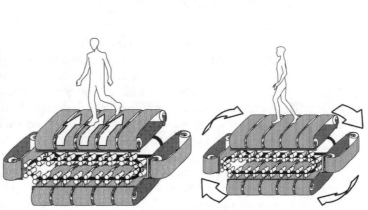

FIGURE   2D treadmills.
              Left: The Torus treadmill, which consists of many small one-dimensional
4.16   treadmills connected together such that they become the links of a giant treadmill.
              Center: When the small treadmills are rotating at the same speed and the large
              treadmill is stopped, the user walks in the direction of the small treadmills.
              Right: When the small treadmills are stopped but the large one is moving, the user
              walks in the direction of the large treadmill. *Source:* From Iwata (1999); © Nivoo
              Iwata.

FIGURE   Transparent version of the VirtuSphere (standard versions are made of a
              mesh).
4.17   The users are wearing wireless headsets that provide visual feedback while
              moving. (Courtesy of VirtuSphere, Inc.)

1999). Walking-in-place is the subject of this chapter's case study and is explored in greater detail there.

If the virtual scene is the same size as or smaller than the tracked space, then real walking is feasible. Here the user's movement in the virtual scene corresponds exactly to her movement in the lab: If she walks 5 m in the lab, she also walks 5 m in the virtual scene. When virtual spaces grow very large, it is sometime necessary to augment a real-walking interface with a magical interface.

4.3.3   Magical and Unnatural Interfaces

When the virtual scene is very large and the task requires the user to visit faraway places, mundane interfaces may be augmented with *magical* interface techniques that allow the user to quickly, or even instantaneously, move over long distances or to a completely different scene. Current techniques include *teleportation* to and from special locations within the scene, going through *special portals* that take the user to a new scene, or a special locomotion mode that switches the user's pace from a normal walking speed to a *hyperdrive* mode.

A concern with magical locomotion techniques is that the user may become disoriented when the scene changes rapidly. Alerting the user that something special is going to happen—for instance, by requiring her to go to a special place or through a doorway—prepares her for an abrupt change of location. Gradually speeding up and slowing down at the start and end of a high-speed move is less disturbing to users than abrupt speed changes. A lovely example of context-preserving movement over great distance is the Google Earth mapping service: The viewpoint zooms away until the whole Earth is visible, the Earth then rotates until the new target location is centered, and then the viewpoint zooms back in.

The Digital ArtForms SmartScene Visualizer product[3] and Miné (1997) have demonstrated techniques where the user can grab the virtual scene and move it toward himself. By repeatedly grabbing points in the virtual scene and pulling them in, the user can locomote from one place to another. Even less like the real world (but no less fun), Stoakley's Worlds-in-Miniature technique has the user manipulate a handheld, dollhouse–sized model of the virtual scene. The user moves in the virtual scene by moving a doll (representing herself) to the desired location in the miniature virtual scene. Then the miniature virtual scene grows to become human scale and the user finds herself in the desired new location in the virtual scene (Stoakley, Conway, & Pausch, 1995). Figure 4.18 illustrates what the user sees with the Worlds-in-Miniature technique.

---

[3] SmartScene was originally developed by MultiGen.

FIGURE

4.18

Worlds-in-Miniature interface as seen by the user.
The small room with the checkerboard floor (in the foreground) is held in the user's hand (invisible here), and is identical to the full-scale room in which the user is standing (background). To move herself in the full-scale room, she moves a tiny doll (not pictured) in the miniature handheld room. *Source:* Adapted from Stoakley et al. (1995); © Randy Pausch.

4.4 HUMAN FACTORS OF THE INTERFACE

Section 4.1.1 showed that humans rely on a complex combination of sensory input to plan and control their movement. The sensory complexity is there whether the motion is controlling a real or a virtual vehicle or controlling real or virtual movement on foot. In this section, we again focus on locomotion on foot, and will discuss factors that make the interface natural (or at least intuitive) to use and elements of system design that minimize the risk of simulator sickness.

4.4.1 Making the Interface Feel Natural

We are a long way from being able to place sensors on users that even approximate the number, variety, and sensitivity of the sensory inputs that control natural human movement. The sensor inputs to the locomotion interface are impoverished when compared to the inputs that control real movement. The job of the interface designer, then, is to make trade-offs to maximize the correspondence of virtual locomotion to real locomotion in terms of the movements the user makes *and* the movements the user perceives through feedback to her various senses.

If you are designing a locomotion interface, you have to decide what motion by which part of the user's body will generate a particular type of locomotion. Factors in this choice are not only how like natural motion it is, but also whether this motion will stimulate other senses—in particular the vestibular system—in the same way that natural motion would.

First, as the interface designer, you have to choose one or more tracker systems, decide what parts of the body to track, and determine how to interpret those tracker readings. Natural forward motion is relatively easy to accomplish in a walking-in-place interface with only a few sensor input signals, such as if trackers on the knees show the feet are going up and down, then move forward. If the design criteria require that the user be able to move sideways and backward while looking forward, it becomes much more difficult to define a natural motion that can be detected from the output of only two sensors. The Gaiter system enabled sideways and backward motion by having the user make distinctive gestures with her legs (Templeman et al., 1999). The user swings her leg sideways from the hip to move sideways and kicks her heels up behind her when stepping to move backward. In fact, users can move in any direction by extending their legs in the desired direction. While these movements are not the way we normally move sideways or backward, the gestures are somewhat consistent with the intended direction of motion.

Walking-in-place provides many of the same sensory cues as really walking, except that the user experiences, and senses, none of the physical forces associated with forward momentum. The part of the interface most critical for making feedback natural are the algorithms executed in the function box in Figure 4.9 labeled "Interpret signals and convert to Δ POV." To date, almost all of these algorithms have been very simple—detect a step, move forward a step. A few interfaces have begun to improve the naturalness of the visual feedback with a more complex model that reflects, frame to frame, the subtle variations in velocity that occur in natural human walking.

4.4.2   Higher-Fidelity Reproduction of Human Gait

When we walk through the real world, the position of our viewpoint changes continuously, but not at a constant speed. Our speed goes through cycles of acceleration and deceleration as we step. The graph in Figure 4.19 (*right*) shows how our speed varies: There is a dominant pattern of speeding up as we start followed by a *rhythmic phase* of near constant speed that lasts until we slow down to stop. The smaller perturbations in speed are the small accelerations and decelerations that occur as we take steps (Inman, 1981). Ideally, if we measured and plotted the speed for users of our virtual-locomotion interface, it would look much like this figure. The ideal is difficult to achieve, but our approximations are getting better.

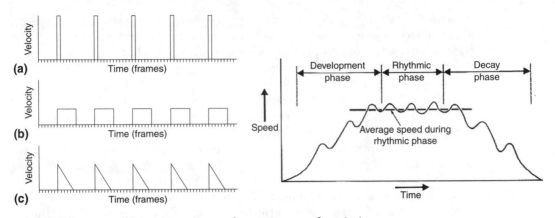

FIGURE
4.19

Walking viewpoint and start-up speed variations.
Left: Three methods of moving the viewpoint over the course of a few virtual steps. (a) All change in eye position is made at a single instant—user appears to snap forward with each step. This makes the motion seem jerky. (b) The change is distributed uniformly over several frames (a few hundred msec). Users found the glide forward/stop/glide forward/stop nature of this technique very disturbing. (c) Change is distributed over several frames, with the amount decreasing over time. Velocity is highest immediately after the foot strike, and then it decays.
Right: Speed variations as a person starts walking, walks at a steady pace, and stops. Note the small variations in speed that are associated with each step, even during the rhythmic phase. *Source:* Adapted from Inman et al. (1981).

We have experimented with three methods of allocating movement to the change in eye position per frame: Insert the entire step length in one frame, divide the step length into uniform segments and insert the parts over several frames, and divide the step length into nonuniform segments and insert the segments over several frames (Figure 4.19, *left*). In informal testing, users preferred the last method. Note that even the last waveform only crudely approximates the pattern of the rhythmic phase of real human walking.

4.4.3   Avoiding Simulator Sickness

Locomotion interfaces, just like flight simulators, have the risk of making the user sick. After exposure to virtual environments, some percentage of users report some sickness symptoms: dry mouth, nausea, dizziness, visual aftereffects (flashbacks), pallor, sweating, ataxia (loss of balance), and even vomiting. VE practitioners commonly refer to this phenomenon as simulator sickness, or *cybersickness*. The occurrence of these symptoms varies widely from person to person and among VE systems. Twenty to forty percent of military pilots suffer from simulator sickness, depending on the simulator (Kolasinski, 1995).

Consequences of Simulator Sickness

Sometimes simulator sickness symptoms linger for days after using a locomotion interface, affect motor control and coordination (Draper, 1998), and can impair a person's performance in the real world (e.g., while driving a car or flying a plane). Kennedy and Lilienthal (1995) and others (Cobb & Nichols, 1998) have proposed quantifying the severity of simulator sickness as a mapping (using measures of a pilot's ability to keep balance) to blood alcohol level—the legal metric of alcohol intoxication used to determine if a person is fit to drive.

Factors That Aggravate Simulator Sickness

It is useful to consider how characteristics of the user and of the system affect the level of sickness suffered. The factors can be divided into those of the individual user and those of the system or simulator. Qualities that correlate with decreased susceptibility to simulator sickness include being in good health (hangovers, medications, stress, illness, and fatigue all increase sickness) and previous experience with simulators or locomotion.

Table 4.2 lists the characteristics of systems divided into two categories. The characteristics in the upper list are technical shortcomings of the equipment. As technologies improve, one expects these shortcomings to be reduced and

TABLE 4.2 Qualities of VE Systems and Flight Simulators That Increase Simulator Sickness

Equipment Shortcomings

✦ Tracker inaccuracies (temporal and spatial)

✦ Low update rate

✦ High latency/lag

✦ Mismatches between display parameters and image generator's parameters (e.g., incorrect FOV setting)

✦ Display flicker

✦ Headset weight

Desirable Functions of System

✦ Stereo display

✦ Long exposure duration

✦ Wide FOV display

✦ Free head movement

✦ Viewpoint motion controlled by someone other than the viewer

✦ High rates of simulated linear or rotational acceleration

Source: Adapted from Kolasinski (1995).

thus decrease the resulting simulator sickness. The characteristics in the lower list are qualities that are often desired in systems. Longer exposures might be required to simulate long missions or to allow users to carry out meaningful tasks. Stereoscopic displays can improve task performance (Pang, Lim, & Quek, 2000) but increase simulator sickness, and higher field-of-view (FOV) displays result in better training and higher levels of user immersion (Arthur, 2000).

There are many theories as to what causes simulator sickness; see LaViola Jr. (2000) for a brief introduction to several theories, and Razzaque (2005) for more in-depth discussion. From a practical point of view, our advice to the designer is to be on the lookout for symptoms of simulator sickness (particularly nausea and dizziness) in your test users, and then to investigate more deeply if you suspect a problem.

4.5 TECHNIQUES FOR TESTING THE INTERFACE

Why are there so many different locomotion interfaces? Each must have its own advantages and shortcomings. However, it is very difficult to make quantitative comparisons among them. Even when the researchers do conduct user studies to compare techniques, the results are often not widely comparable because they study different user populations and different tasks, and evaluate using different metrics. Furthermore, nonlocomotion parameters of the systems, such as frame rate, field of view, latency, and image quality, are different and, in many cases, not measured. These factors can confound any study results, particularly task performance measures.

One particularly impressive comparison of locomotion techniques is a study by Bowman and coworkers (1997). They evaluated several variations of flying by reimplementing each technique to run on the same hardware and having each subject perform the same task using all the flying techniques. These authors evaluated each technique using several criteria: ease of learning, spatial awareness, speed, accuracy, and cognitive load, and pointed out that different applications have different needs. A phobia treatment application may be more concerned with evaluating naturalness and presence, whereas a game may be more concerned with speed and agility so that the user can get to the target quickly. As the application designer, you should first decide which attributes are important, choose a locomotion technique that optimizes the particular attributes important for that application, and then evaluate those attributes.

4.5.1 Guidelines for Testing

One naïve way to test an interface might be to assign a task to a sample of users, and then measure how quickly they perform this task. But faster task performance does not always imply a better interface. For example, imagine a simulator intended to help a patient get over a fear of heights, wherein the user's task is to cross a river using a narrow rope and plank bridge. One interface design might require the user to hold handrails, and look down and slowly step from plank to plank, whereas another interface design might have the user point her hand at the opposite end of the bridge in order to instantly teleport there. Even though the latter has her at the other side of the bridge sooner, it is contrary to the objective of the interface—to expose the patient to the stimulus that evokes the fear of heights response.

Standard practice in usability testing dictates that the designer use test tasks that are appropriate to the goal of the interface, and that the testing must include real users who carry out the entire task for entire sessions. Tasks that seem easy in short sessions can become fatiguing or even undoable in longer sessions.

Many of the specific tests described in the next section result in quantitative data, but are expensive to conduct. While quantitative tests with a large number of users results in more convincing study results, doing more informal tests and/or qualitative tests can result in more insight, and can save time and resources, allowing for more iterations of designing and testing. In this regard, the design and testing of locomotion interfaces is no different than for any other kind of user interface.

4.5.2   Qualities to Consider for Testing and Specific Tests

In this section, we enumerate qualities of a whole-body locomotion interface that you might want to test, and provide specific methods for testing each.

Distraction

The first testing category is simply effectiveness: Can the user move around so that she can perform her tasks, or does the interface interfere with or distract her from the tasks? Distractions may be cognitive—the user has to pay attention to using the interface rather than to doing her task, or they can be physical—the interface device or its cables encumber the user in ways that interfere with task performance.

One example of a cognitive distraction is an interface with poor motion control. Evidence of poor control (in addition to users simply complaining about it) includes the user overshooting her intended target or unintentionally colliding with objects, walls, or the edges of doorways. Such a locomotion interface is not transparent to the user, but constantly requires the user's attention, reducing the cognitive resources available for her primary task. The number of overshoots (where the user goes past the target and must reverse or turn around), the distance of each overshoot, and the number of unintended collisions can be recorded and counted during testing. When the user performs that same task in different versions of the interface, these counts can be used to compare the different versions.

As an example of physical distraction, consider the case of a soldier or game player who is supposed to wield a gun while moving through a virtual building, but must use both hands to control the locomotion interface. In this situation, the user may well not be able aim or shoot her gun effectively while moving. One way to test for this kind of physical distraction is to measure the user's performance on the nonlocomotion part of her task while using different versions of the locomotion interface, and then to attribute the differences in performance to the distraction caused by the locomotion interface itself.

Training Transfer

Some systems that include locomotion interfaces are intended to train users for some task, such as evacuating a building during a fire. One way of testing the effectiveness of such a system is by measuring *training transfer*—how well does it train the user? This is done by using the system (or a prototype) to train a group of users, and then measuring their performance on the real task afterward. This task performance is compared against the performance of other users who were trained using some other system (or a previous version).

As an example, Insko (2001) tested a system (designed to train users to learn a virtual maze) to see if letting the users feel the virtual walls (via haptic feedback) as they moved through the virtual maze would result in their having better learned the maze. It turned out that those users who used the interface with haptics made fewer errors in the real maze. The specific methods to test training transfer, of course, depend on the specific tasks that the interface is intended to train.

Presence

In the context of a virtual scene, *presence* is the feeling, in the user, of being in the virtual scene, as opposed to being in the room where she is physically standing. For applications such as phobia desensitization, presence is the most important quality of a system, and therefore merits testing. One way of testing presence is to use a questionnaire, such as the Slater-Usoh-Steed questionnaire (Slater & Usoh, 1993; Slater et al., 1995; Usoh et al., 1999). Another involves counting break-in-presence (BIP) events (Slater & Steed, 2000). BIPs are defined as events where the illusion of *being somewhere else* is broken and the user's attention is brought back to her physical surroundings.

One example of a BIP event is when the user's virtual body penetrates or intersects a virtual object (Figure 4.20). The user can push a button or call out to the experimenter each time she experience a BIP event, and then the counts can be compared against the counts from users in other interfaces. Using the

FIGURE

4.20

Example of an event that causes a break in presence.
Unnatural events, such as walking through a wall, break the user's illusion that she is "in" the virtual scene, and remind her that she is really in a laboratory. (Courtesy of the Department of Computer Science, UNC at Chapel Hill.)

BIP measures, Slater et al. (1995) determined that the walking-in-place interface results in greater presence than locomotion via flying with hand controller.

We have also explored physiological indicators of presence, such as heart rate and skin conductance. The general idea is to place the user in a simulated stressful situation and record physiological data. The more similar her physical responses in the virtual scene are to responses to a corresponding stressful situation in the real world, the more presence inducing the simulator is. Physiological measures have been used to test the effects of interface design parameters such as the display's frame rate and lag (Meehan et al., 2003, 2005).

One problem with all three ways of testing presence mentioned in this section is that there is large variation between individuals and from session to session. While you might have a *qualitative* feel for which locomotion interface is better after sampling half a dozen users, several dozen to hundreds of users might need to be tested to get a statistically significant *quantitative* indication from physiological data.

4.5.3 Locomotion Realism and Preservation of Spatial Understanding

Another testable quality of a whole-body locomotion interface is how realistic the motion displayed to the user is. Remember that *display* is a generic term and may refer to a visual display, an audio display, or a motion display (motion platform or treadmill). When a user is walking on a treadmill, does visual display match how fast she feels she is walking? With a walk-in-place interface, when the user takes a virtual step, does the step length feel consistent with her expectation? Some researchers have found that when a user is physically pushed in a cart at some acceleration, and the virtual viewpoint moves at the same acceleration in the virtual scene, the user often feels that the virtual acceleration (seen) and the real acceleration (felt) do not match up (Harris, Jenkin, & Zikovitz, 1999). One theory is that the mismatch could be due to the narrow FOV in systems that have the user wear an HMD. Whatever the reasons, you should query your test users as to whether the distance, velocity, and/or acceleration of the displayed movement feels too fast or too slow, and adjust it accordingly.

You can also test how well the locomotion interface preserves the user's spatial awareness. Peterson and colleagues (1998) had users move through a virtual maze and then had them point toward the location in the virtual maze from which they started. He compared the pointing accuracy of two different locomotion interfaces. Iwata and Yoshida (1999) compared the shape of the path that users took when walking in the real world and when using the Torus treadmill.

4.5.4   Simulator Sickness

When designing a whole-body locomotion interface, you should pay careful attention to the extent the interface induces simulator sickness in users, and decide whether a certain amount of sickness is acceptable for a particular application. The most common way of measuring simulator sickness is Kennedy's simulator sickness questionnaire (SSQ) (Kennedy et al., 2003).

The SSQ consists of 16 multiple-choice items, and results in an overall sickness score as well as three subscores. It can be administered to a test user very quickly, but because the scores vary greatly from person to person, large numbers (many dozens) of test users must be used. Also, the experimenter must not administer the SSQ to test users before they use the locomotion interface, as there is evidence that using the SSQ in a pre-exposure/postexposure fashion causes users to report more sickness in the postexposure test than users who have the same virtual locomotion experience, but take the test only after exposure (Young, Adelstein, & Ellis, 2006). The SSQ was developed from an analysis of over a thousand flight simulator sessions by U.S. Navy and Marine pilots. In our experience with whole-body locomotion interfaces, rather than comparing the SSQ scores of test users to the sickness levels reported by military pilots, it is more useful to compare SSQ scores between different locomotion interfaces (or different versions of the same interface).

Researchers have proposed other ways of measuring simulator sickness, such as looking at the change in a user's postural stability—her ability to balance—when measured before and after using the locomotion interface (Stoffregen et al., 2000).

Finally, when you design an interface, you should also consider testing the qualities of the locomotion system that might lead to increases in simulator sickness. For example, the display frame rate and end-to-end system latency (Meehan et al., 2003) can be measured without having to run user studies. We find that a frame rate of 60 frames per second or greater and an end-to-end latency of less than 100 milliseconds are generally acceptable. Values higher than these should cause you to look more carefully for symptoms of sickness in users.

4.6 DESIGN GUIDELINES

General interface design rules apply to locomotion interfaces as well:

1. Always consider the user's goals for the interface as the highest priority.
2. Perform many iterations of the design → test → revise cycle.
3. Always include tests with actual users and the actual tasks those users will perform.

We also suggest guidelines that are specific to whole-body locomotion interfaces.

4.6.1 Match the Locomotion Metaphor to the Goals for the Interface

You should always consider whether the locomotion metaphor suits the goal of the whole-body locomotion interface. For example, if the interface's goal is to simulate real walking, then the interface should require the user to really turn her body to turn (walking metaphor), rather than turning by manipulating a steering wheel or hand controller (vehicle metaphor).

4.6.2 Match the Interface Capabilities to the Task Requirements

Because the user employs the interface to specify locomotion, you must consider how fast the user will need to move, how finely or accurately she must be able to control her locomotion (e.g., will she need to walk through tight spaces?), and if she will be performing other tasks while she is moving. Not being able to move fast enough will result in the user finding the interface tedious, whereas not providing fine enough control will result in unintended collisions and inability to stop exactly where desired.

4.6.3   Consider Supplementing Visual Motion Cues Using Other Senses

The visual display alone may not always communicate some kinds of movement. For example, if the user of a game is hit with bullets, the quick and transient *shock* movement will be imperceptible in a visual display because the visual system is not sensitive to those kinds of motions. An auditory display and motion platform (e.g., a shaker) would work better. On the other hand, a motion platform would not be appropriate for conveying very slow and smooth motion such as canoeing on a still pond.

4.6.4   Consider User Safety

Since the user will be physically moving, consider her safety. One thing you must always consider as a designer is the set of cables that might connect to equipment worn by the user. Will the cables unintentionally restrict her movement or entangle her or cause her to fall? What prevents her from falling? And if she does fall, does the equipment she is wearing cause her additional injury? How will the cables be managed?

4.6.5   Consider How Long and How Often the Interface Will Be Used

As a designer you must also consider the length of time the person will be using the interface and how physically fatiguing and stressful the motions she must make are. For example, if the interface requires her to hold her arms out in front of her, she will quickly tire. If the interface requires her to slam her feet into the ground, repeated use might accelerate knee and joint injury. As discussed above, you should also be on the lookout for symptoms of simulator sickness, and if you suspect that your locomotion interface is causing some users to become sick, investigate ways of changing the interface (e.g., reducing the exposure time, reducing the display's FOV) to address it. For example, the IMAX 3D version of the movie, *Harry Potter and the Order of the Phoenix*, limited the 3D visuals to only the final 20 minutes of the movie.

The applications of whole-body locomotion interfaces are so varied that there are very few things that a design should never do, no matter what the circumstance or application.

4.6.6   Do Not Assume that More Sophisticated Technology (Higher Tech) Is Better

One assumption that we have seen novice interface designers make that often results in a bad interface design is that an interface is better simply because it is "higher tech." You must consider the goals of the user and make design choices based on that.

4.6.7   Do Not Deploy Head-Directed Interfaces

Although easy to implement, a locomotion interface that forces the user to move in the direction in which her head is pointing (or, inversely, forces her to point her head in the direction she wishes to move) is almost always suboptimal. On foot, humans look in the direction they are moving roughly 75 percent of the time (Hollands, Patla, & Vickers, 2002); about 25 percent of the time we are moving in one direction and looking in another. For example, people look around *while* crossing a street or merging into a lane of traffic. If you can only move in the direction you are facing, you cannot look around and move at the same time. Templeman et al. (2007) provide a detailed analysis of why motion direction and head direction must be decoupled if locomotion systems for training soldiers on foot are to be effective. Many designers (including the authors) who first implemented a head-directed locomotion interface then discovered this problem and had to redevelop the interface. On this point, we recommend that you learn from the experience of others and avoid head-directed locomotion interfaces in production systems.

4.6.8   Summary of Practical Considerations

There are myriad practical considerations when designing, implementing, and deploying a locomotion interface. Table 4.3 summarizes some of the factors that should be considered to ensure that an interface will be successful—that users will accept it and that the virtual locomotion interface is effective.

The factors are categorized as functionality, systems and sensor characteristics, and encumbrances and restraints. The Comments column variously explains why a locomotion interface designer should care about the factor, expands on what the factor means, or gives an example of the kinds of problems that can arise if the factor is ignored.

4.7   CASE STUDY

A case study for a walking-in-place locomotion interface can be found at *www. beyondthegui.com.*

TABLE 4.3   Items Considered in Design, Implementation, and Deployment
               of Virtual Locomotion Systems

Practical Considerations	Comments
Functionality	
How hard is it to learn the interface?	If a "natural" interface is hard to use, users will prefer a less natural but easy-to-learn interface.
Is viewing direction independent of motion direction?	Allows you to look around at the same time you are moving.
Are hands used for locomotion?	It is good to have hands free for use during tasks.
Able to move in any direction and change direction easily?	It is difficult to make a sharp corner on some treadmills and with some joysticks.
Able to move in any direction while body faces forward?	Backward? Sideways?
Can base walking speed or step length be set for each individual?	Familiarity with one's own step length is a factor contributing to spatial understanding.
Able to move at different paces, and move quickly (run) or slowly (creep)?	Extends range of applications that can be supported.
Able to adjust step length, that is, take larger or smaller steps?	Without the ability to take small steps, users cannot fine-tune where they stop.
Magical interfaces for fast movement between distant places?	Most often needed for very large virtual scenes; likely used with a mundane-style interface for human-scale movements.
System and Sensor Characteristics	
Are there more than one or two frames of latency between signaling intent to move and visual feedback of movement?	Trade-off between start-up lag and detecting unintended steps.
Does the user move one or more steps further than intended when stopping (stopping lag)?	Trade-off between stopping lag and missing intended steps.
Is tracker technology compatible with the physical environment?	For example, steel raised flooring compromises accuracy of magnetic trackers.
Are tracker sensors immune to other signals in the room?	Are there sources in the room (other than the tracker beacons) that the sensors will detect, thus compromising tracking accuracy?
Are wireless trackers used?	Is there sufficient bandwidth? Is room electrically "noisy" from other devices and wireless hub?
Encumbrances and Restraints	
What parts of the body will have sensors or markers on them?	Will user be able to make appropriate motions with sensors on?
Can cables be managed so they do not interfere with user's motion?	Will user get wound up in the cables when performing the task?
Do you need a fixture to restrict user's real movement in the lab?	This may be a safety/institutional review board issue in addition to keeping the user within range of the tracking system.

4.8 FUTURE TRENDS

Future locomotion interface designers will still face the same problems we do today: how to sense body position and movement and map into locomotion direction and speed while meeting application constraints that may vary from maximizing naturalness to minimizing the range of movement required.

4.8.1 New Tracker Technologies

Our bodies have only a set range of motions; so, to increase the number of parameters that can be controlled with body movement and/or to increase the precision with which those parameters can be controlled, we must have more precise data about the position and motion of the user. The trend to smaller and wireless trackers means that these devices can be put in many more places on the body. Since having more markers can result in slower tracker updates, data-mining techniques are being applied to logs of data from motion capture systems to identify the set of marker locations that are most critical for determining body pose (Liu et al., 2005).

The vision for the future is fully unencumbered tracking of the whole body for several users sharing the same physical space at the same time: no wires, no body suits, and no markers or other devices placed on the user's body. Computer vision algorithms will use as input synchronized frames of video taken from a number of cameras surrounding the space. The output per time step will be an estimation of the body position (pose) of each user that is computed by tracking features between frames and between camera views. There are successful vision-based techniques today that use markers; the challenge for the future is a markerless system.

4.8.2 Reducing the Physical Space Needed for Real-Walking Interfaces

The advantages of the naturalness of a real-walking virtual-locomotion interface are offset by the fact that the virtual scene is limited to the size of the physical space covered by the tracking system. For example, to allow the user to explore a virtual museum while really walking, the tracking space must be as large as the museum. Several efforts are under way to overcome this limitation. Some of the new techniques strive to be imperceptible, such as redirected walking (Razzaque, 2001); some strive to maintain the actual distance walked, such as motion compression (Nitzsche, Hanebeck, & Schmidt, 2004), and others explicitly interrupt users to "reset" them in the physical space while not changing their location in the virtual scene (Mohler et al., 2004).

142

4.8.3   Changing Economics

Historical experience has led practitioners in computer-related technologies to assume continuous improvements in performance and decreases in cost. There have not yet been such trends for technologies supporting whole-body interfaces. The prices of motion trackers, immersive displays, and motion platform output devices have not dropped much in the last 10 years. Since 3D graphics hardware was adopted by mass-market video games, it has become cheap, widely available, and very high performance. As of this writing high-end tracker systems still have five- to six-digit prices (in U.S. dollars), but Nintendo's newest console, Wii, is a mass-market success.

The Wii game console includes handheld controllers that sense hand acceleration and, when pointed at the TV screen, also sense hand orientation. In the game Wii-Sports Tennis, the sensed acceleration is used to control the motion of the racquet (Figure 4.21). The orientation feature enables users to point at objects

FIGURE

4.21

Nintendo Wii game system.
Wii includes rudimentary motion tracking in the handheld controllers. A player swings a virtual tennis racket by swinging her hand and arm. (Courtesy of the Department of Computer Science, UNC at Chapel Hill.)

displayed on the TV. The Wii-Fit game includes a platform that reports the weight and center-of-balance position of the user who is standing on it.

While we have not yet seen any whole-body locomotion interfaces in mass-market video games, we suspect they will soon appear. We ourselves are currently developing interfaces that use Wii controllers. We are hopeful that Wii will be the breakthrough product that will, in a reasonably short time, result in widely available, cheap, and high-performance whole-body trackers for all applications and in particular for virtual locomotion.

REFERENCES

Arthur, K. (2000). Effects of Field of View on Performance with Head-Mounted Displays. CS Technical Report TR00-019. Chapel Hill. Department of Computer Science, University of North Carolina.

Berthoz, A. (2000). *The Brain's Sense of Movement*. Weiss, G., trans. Cambridge, MA: Harvard University Press.

Bouguila, L., Evequoz, F., Courant, M., & Hirsbrunner, B. (2004). Walking-pad: A step-in-place locomotion interface for virtual environments. *Proceedings of Sixth International Conference on Multimodal Interfaces*. State College, PA, ACM, 77–81.

Bowman, D. A., Koller, D., & Hodges, L. F. (1997). Travel in immersive virtual environments: An evaluation of viewpoint motion control techniques. *Proceedings of IEEE Virtual Reality Annual International Symposium*, 45–52.

Brooks, F. P. (1988). Grasping reality through illusion—interactive graphics serving science. *Proceedings of SIGCHI Conference on Human Factors in Computing Systems*, Washington, DC, ACM, 1-11.

Cobb, S. V. G., & Nichols, S. C. (1998). Static posture tests for the assessment of postural instability after virtual environment use. *Brain Research Bulletin* 47(5):459–64.

Darken, R. P., Cockayne, W. R., & Carmein, D. (1997). The omni-directional treadmill: A locomotion device for virtual worlds. *Proceedings of Association for Computing Machinery User Interface Software and Technology*. Banff, Canada, ACM, 213–21.

Dichgans, J., & Brandt, T. (1977). Visual–vestibular interaction: Effects on self-motion perception and postural control. *Perception* 8:755–804.

Draper, M. (1998). The Adaptive Effects of Virtual Interfaces: Vestibulo-Ocular Reflex and Simulator Sickness. PhD diss., University of Washington, Seattle.

Duh, H. B.-L., Parker, D. E., Phillips, J., & Furness, T. A. (2004). "Conflicting" motion cues at the frequency of crossover between the visual and vestibular self-motion systems evoke simulator sickness. *Human Factors* 46:142–53.

Fleming, S. A., Krum, D. M., Hodges, L. F., & Ribarsky, W. (2002). Simulator sickness and presence in a high field-of-view virtual environment [abstract]. *Proceedings of Computer–Human Interaction, Human Factors in Computing Systems*. Minneapolis, ACM, 784–85.

Foxlin, E. (2002). Motion tracking requirements and technologies. In Stanney, K. M., ed. *Handbook of Virtual Environments*. Mahwah, NJ: Erlbaum, 163–210.

Gregory, R. L. (1970). *The Intelligent Eye*. New York: McGraw-Hill.

Harris, L., Jenkin, M., & Zikovitz, D. C. (1999). Vestibular cues and virtual environments: Choosing the magnitude of the vestibular cue. *Proceedings of IEEE Virtual Reality Conference*, Houston, 229–36.

Hollands, M. A., Patla, A. E., & Vickers, J. N. (2002). "Look where you're going!": Gaze behaviour associated with maintaining and changing the direction of locomotion. *Experimental Brain Research* 143(2):221–30.

Hollerbach, J. M. (2002). Locomotion interfaces. In Stanney, K. M., ed., *Handbook of Virtual Environments*. Mahwah, NJ: Erlbaum, 239–54.

Hollerbach, J. M., Christensen, R. R., Xu, Y., & Jacobsen, S. C. (2000). Design specifications for the second generation Sarcos Treadport locomotion interface. *Proceedings of ASME Dynamic Systems and Control Division, Haptics Symposium*, Orlando, 1293–98.

Howard, I. P. (1986). The vestibular system. In Boff, K. R., Kaufman, L., & Thomas, J. P., eds., *Handbook of Perception and Human Performance*, 1:11–26. New York: Wiley-Interscience.

Inman, V. T., Ralston, H. J., & Frank, T. (1981). *Human Walking*. Baltimore/London: Williams and Wilkins.

Insko, B. (2001). Passive Haptics Significantly Enhances Virtual Environments. CS Technical Report TR01-017. Chapel Hill: University of North Carolina, Department of Computer Science.

Iwata, H. (1999). Walking about virtual environments on an infinite floor. *Proceedings of IEEE Virtual Reality Conference*, Houston, 286–95.

Iwata, H., & Yoshida, Y. (1999). Path reproduction tests using a Torus treadmill. *Presence: Teleoperators and Virtual Environments* 8(6):587–97.

Kennedy, R. S., Drexler, J. M., Compton, D. E., Stanney, K. M., Lanham, D. S., & Harm, D. L. (2003). Configural scoring of simulator sickness, cybersickness and space adaptation syndrome: Similarities and differences. In Hettinger, L. J., & Haas, M. W., eds., *Virtual and Adaptive Environments: Applications, Implications, and Human Performance*. Mahwah, NJ: Erlbaum, 247–78.

Kennedy, R. S., & Lilienthal, M. G. (1995). Implications of balance disturbances following exposure to virtual reality systems. *Proceedings of IEEE Virtual Reality Annual International Symposium*, Research Triangle Park, NC, 35–39.

Kolasinski, E. (1995). Simulator Sickness in Virtual Environments. Technical Report 1027. Alexandria, VA: U.S. Army Research Institute for the Behavioral and Social Sciences.

Lackner, J. R. (1977). Induction of illusory self-rotation and nystagmus by a rotating sound-field. *Aviation, Space, and Environmental Medicine* 48(2):129–31.

LaViola Jr., J. J. (2000). A discussion of cybersickness in virtual envronments. *SIGCHI Bulletin* 32(1):47–55.

LaViola Jr., J. J., Feliz, D. A., Keefe, D. F., & Zeleznik, R. C. (2001). Hands-free multi-scale navigation in virtual environments. *Proceedings of Association for Computing Machinery Symposium on Interactive 3D Graphics*, Research Triangle Park, NC, 9–15.

Liu, G., Zhang, J., Wang, W., & McMillan, L. (2005). A system for analyzing and indexing human-motion databases. *Proceedings of Association for Computing Machinery SIGMOD International Conference on Management of Data*, Baltimore, 924–26.

Martini, F. (1999). *Fundamentals of Anatomy and Physiology*, 4th ed. Upper Saddle River, NJ: Prentice Hall.

Meehan, M., Razzaque, S., Insko, B., Whitton, M. C., & Brooks, F. P. J. (2005). Review of four studies on the use of physiological reaction as a measure of presence in a stressful virtual environment. *Applied Psychophysiology and Biofeedback* 30(3):239–58.

Meehan, M., Razzaque, S., Whitton, M., & Brooks, F. (2003). Effects of latency on presence in stressful virtual environments. *Proceedings of IEEE Virtual Reality*, Los Angeles, 141–48.

Miné, M. R. (1997). Exploiting Proprioception in Virtual-Environment Interaction. CS Technical Report TR97-014. Chapel Hill: University of North Carolina, Department of Computer Science.

Mohler, B., Thompson, W. B., Creem-Regehr, S., Pick, H. L., Warren, W., Rieser, J. J., & Willemsen, P. (2004). Virtual environments I: Visual motion influences locomotion in a treadmill virtual environment. *Proceedings of First Symposium on Applied Perception in Graphics and Visualization.* Los Angeles, 19–22.

Nitzsche, N., Hanebeck, U. D., & Schmidt, G. (2004). Motion compression for telepresent walking in large target environments. *Presence: Teleoperators and Virtual Environments* 13(1):44–60.

Pang, T. K., Lim, K. Y., & Quek, S. M. (2000). Design and development of a stereoscopic display rig for a comparative assessment of remote freight handling performance. *Proceedings of Joint Conference of APCHI and ASEAN Ergonomics*, Singapore, 62–67.

Peterson, B., Wells, M., Furness, T., & Hunt, E. (1998). The effects of the interface on navigation in virtual environments. *Proceedings of Human Factors and Ergonomics Society Annual Meeting*, 1496–505.

Razzaque, S. (2005). Redirected Walking. CS Technical Report TR05-018. Chapel Hill: University of North Carolina, Department of Computer Science.

Razzaque, S., Kohn, Z., Whitton, M. (2001). Redirected walking. *Proceedings of Eurographics Workshop*, 289–94.

Riecke, B. E., van Veen, H. A. H. C., & Bulthoff, H. H. (2002). Visual homing is possible without landmarks: A path integration study in virtual reality. *Presence: Teleoperators and Virtual Environments* 11(5):443–73.

Robinett, W., & Holloway, R. (1992). Implementation of flying, scaling and grabbing in virtual environments. *Proceedings of Association for Computing Machinery Symposium on Interactive 3D Graphics*, Cambridge, MA, 189–92.

Rolfe, J. M., & Staples, K. J. (Eds.) (1988). *Flight Simulation.* Cambridge: Cambridge University Press.

Schell, J., & Shochet, J. (2001). Designing interactive theme park rides. *IEEE Computer Graphics and Applications* 21(4):11–13.

Slater, M., & Steed, A. (2000). A virtual presence counter. *Presence: Teleoperators and Virtual Environments* 9(5):413–34.

Slater, M., & Usoh, M. (1993). Presence in immersive virtual environments. *Proceedings of IEEE Virtual Reality Annual International Symposium*, 90–96.

Slater, M., & Usoh, M. (1994). Body-centered interaction in immersive virtual environments. In Mageneat-Thalman, N., & Thalmann, D., eds., *Artifcial Life and Virtual Reality.* New York: John Wiley and Sons, 125–48.

Slater, M., Usoh, M., & Steed, A. (1995). Taking steps: The influence of a walking technique on presence in virtual reality. *ACM Transactions on Computer–Human Interaction* 2(3):201–19.

Stoakley, R., Conway, M. J., & Pausch, R. (1995). Virtual reality on a WIM: Interactive worlds in miniature. *Proceedings of Human Factors and Computing Systems*, 265–72.

Stoffregen, T. A., Hettinger, L. J., Haas, M. W., & Smart, L. J. (2000). Postural instability and motion sickness in a fixed-base flight simulator. *Human Factors* 42:458–69.

Templeman, J. N., Denbrook, P. S., & Sibert, L. E. (1999). Virtual locomotion: Walking in place through virtual environments. *Presence: Teleoperators and Virtual Environments* 8(6):598–617.

Templeman, J. N., Sibert, L., Page, R. C., & Denbrook, P. S. (2007). Pointman—A device-based control for realistic tactical movement. *Proceedings of IEEE Symposium on 3D User Interfaces*, Charlotte, NC.

Usoh, M., Arthur, K., Whitton, M. C., Bastos, R., Steed, A., Slater, M., & Brooks, F. P. (1999). Walking > walking-in-place > flying in virtual environments. *Proceedings of SIGGRAPH*. Los Angeles, 359–64.

Warren Jr., W. H., Kay, B. A., Zosh, W. D., Duchon, A. P., & Sahuc, S. (2001). Optic flow is used to control human walking. *Nature Neuroscience* 4(2):213–16.

Yan, L., Allison, R. S., & Rushton, S. K. (2004). New simple virtual walking method—walking on the spot. Paper presented at Ninth Annual Immersive Projection Technology (IPT) Symposium.

Young, L. R., Adelstein, B. D., & Ellis, S. R. (2006). Demand characteristics of questionnaire used to assess motion sickness in a virtual environment. *Proceedings of IEEE Virtual Reality*, Alexandria, VA, 97–102.

Auditory Interfaces

S. Camille Peres, Virginia Best, Derek Brock, Barbara Shinn-Cunningham, Christopher Frauenberger, Thomas Hermann, John G. Neuhoff, Louise Valgerður Nickerson, Tony Stockman

Auditory interfaces are bidirectional, communicative connections between two systems—typically a human user and a technical product. The side toward the machine involves machine listening, speech recognition, and dialog systems. The side toward the human uses auditory displays. These can use speech or primarily nonspeech audio to convey information. This chapter will focus on the nonspeech audio used to display information, although in the last section of the chapter, some intriguing previews into possible nonspeech audio-receptive interfaces will be presented.

Auditory displays are not new and have been used as alarms, for communication, and as feedback tools for many decades. Indeed, in the mid-1800s an auditory display using Morse code and the telegraph ushered in the field of telecommunications. As technology has improved, it has become easier to create auditory displays. Thus, the use of this technology to present information to users has become commonplace, with applications ranging from computers to crosswalk signals (Brewster, 1994; Massof, 2003). This same improvement in technology has coincidentally increased the *need* for auditory displays.

Some of the needs that can be met through auditory displays include (1) presenting information to visually impaired people, (2) providing an additional information channel for people whose eyes are busy attending to a different task, (3) alerting people to error or emergency states of a system, and (4) providing information via devices with small screens such as PDAs or cell phones that have a limited ability to display visual information. Furthermore, the auditory system is well suited to detect and interpret multiple sources and types of information as will be described in the chapter.

Audio displays (or auditory interfaces) as they are experienced now, at the beginning of the 21st century, are more sophisticated and diverse than the bells and clicks of the past. As mentioned previously, this is primarily due to the increasing availability of powerful technical tools for creating these displays. However, these tools have only recently become accessible to most engineers and designers, and thus the field of auditory displays is somewhat in its adolescence and for many is considered relatively new. Nevertheless, there is a substantial and growing body of work regarding all aspects of the auditory interface. Because this field is young and currently experiencing exponential growth, the lexicon and taxonomy for the various types of auditory displays is still in development. A discussion of the debates and nuances regarding the development of this taxonomy is not appropriate for this chapter, but the interested reader will find details on this topic in Kramer's book on auditory displays (1994), distributed by the International Community on Auditory Displays (ICAD) (*www.icad.org*), as well as other sonification sites (e.g., *http://sonification.de*).

The terms used in this chapter are outlined and defined in the following sections. We have organized the types of auditory displays primarily by the method or technique used to create the display. Additionally, all of the sound examples are available at *www.beyondthegui.com*.

Sonification of Complex Data

Sonification is the use of nonspeech sound to render data, either to enhance the visual display or as a purely audio display. Sonification is routinely used in hospitals to keep track of physiological variables such as those measured by electrocardiogram (ECG) machines. Audio output can draw the attention of medical staff to significant changes in patients while they are otherwise occupied. Other sonifications include the rhythms in electroencephalogram (EEG) signals that can assist with the prediction and avoidance of seizures (Baier, Hermann, Sahle, & Ritter, 2006) and sonification of the execution of computer programs (Berman & Gallagher, 2006). There are various types of sonification techniques, and these will be elaborated throughout the chapter and particularly in Section 5.2.4.

Considerable research has been done regarding questions that arise in designing effective mappings for sonification of data. Among the key issues are *voicing*, *property*, *polarity*, and *scaling/context*. *Voicing* deals with the mapping of data to the sound domain; that is, given a set of instruments, which one should be associated with which variable?

Property deals with how changes in a variable should be represented, such as should changes in a data variable be mapped to pitch, amplitude, or tempo? *Polarity* deals with the way a property should be changed (Walker, 2002), that is, should a change in a variable cause the associated sound property to rise or fall? Suppose weight is mapped to tempo: Should an increase in weight lead to an increase in tempo, or a decrease given that increasing weight generally would be associated with a slower response?

Scaling deals with how quickly a sound property should change. For instance, how can we convey an understanding of the absolute values being displayed, which requires establishing maximum and minimum values displayed, where the starting value is in the scale, and when the data cross zero? How much of a change in an original data variable is indicated by a given change in the corresponding auditory display parameter?

Audification

A very basic type of auditory display, called *audification*, is simply presenting raw data using sound. This will be described in more detail in Section 5.2.4, but essentially everything from very-low-frequency seismic data (Hayward, 1994) to very-high-frequency radio telescope data (Terenzi, 1988) can be transformed into perceptible sound. Listeners can often derive meaningful information from the audification.

Symbolic/Semantic Representations of Information

Similar to other types of interfaces, auditory interfaces or displays are often needed for a task other than data analysis or perceptualization. For instance, GUIs use icons to represent different software programs or functions within a program. This type of visual display is more semantic in nature and does not require analysis on the part of the user.

The auditory equivalents of visual icons are auditory displays known as *auditory icons* and *earcons*. These displays are very useful in translating symbolic visual artifacts into auditory artifacts and will be discussed several times through the chapter. An example of an auditory icon is the paper-"crinkling" noise that is displayed when a user empties the Trash folder. This technique employs sounds that have a direct and thus intuitive connection between the auditory icon and the function or item.

Earcons are a technique of representing functions or items with more abstract and symbolic sounds. For example, just as Morse code sound patterns have an arbitrary connection to the meaning of the letter they represent, the meaning of an earcon must be learned. These types of displays can be particularly appropriate for programs that have a hierarchical structure as they allow for the communication of the structure in addition to the representation of the functions.

A new symbolic/semantic technique for auditory displays that shows some promise is the use of *spearcons*. These are nonspeech cues used in the way that icons or earcons would be. They are created by speeding up a spoken phrase until it is not recognized as speech (Walker, Nance, & Lindsay, 2006). This representation of the spoken phrase, such as a person's name in a phone list, can be slowed down to a recognizable level to facilitate learning the association between the spearcon and the name. Once this association is made, the spearcon can be played using the shortest duration to reduce the amount of time necessary to present the auditory display.

It is very important to understand, and should be clear after reading the entire chapter, that these techniques are not normally and sometimes not even ideally used exclusively. Furthermore, they are not necessarily independent of each other. For instance, in daily weather sonifications (Hermann, Drees, & Ritter, 2005), most data variables were displayed via parameterized auditory icons (e.g., water sounds gave an iconic link to rain, and the duration of the sound allowed the user to judge the amount of rain per unit of time). When designing and testing auditory displays, designers can consider three different dimensions or axes: the interpretation level, from analogic to symbolic; interactivity, from noninteractive to tightly closed interaction; and "hybridness," from isolated techniques to complex mixtures of different techniques. The ability to use these three dimensions gives the designer a wide and intriguing range of tools to better meet users' needs.

5.1   NATURE OF THE INTERFACE

It is important for those who will design and test auditory displays to understand some of the basic perceptual properties of the human auditory system. These properties dictate how humans experience and interpret sounds, and this section describes some of the more fundamental elements of the auditory perception process.

5.1.1   Basic Properties of Sound

Sound arises from variations in air pressure caused by the motion or vibration of an object. Sounds are often described as pressure variations as a function of time and are plotted as a waveform. Figure 5.1 shows the waveform for a pure sinusoid, which is periodic (repeats the same pattern over and over in time) and is characterized by its frequency (number of repetitions per second), amplitude (size of the pressure variation around the mean), and phase (how the waveform is aligned in time relative to a reference point). All complex sounds can be described as the sum of a specific set of sinusoids with different frequencies, amplitudes, and phases.

Sinusoids are often a natural way to represent the sounds we hear because the mechanical properties of the cochlea break down input sounds into the components at different frequencies of vibration. Any incoming sound is decomposed into its component frequencies, represented as activity at specific points along the cochlea. Many natural sounds are periodic (such as speech or music) and contain energy at a number of discrete frequencies that are multiples of a common (fundamental) frequency. Others are nonperiodic (such as clicks or white noise) and contain energy that is more evenly distributed across frequency.

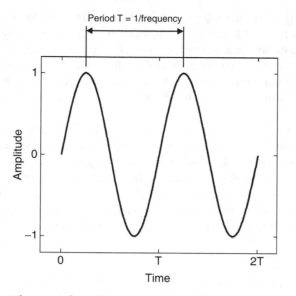

FIGURE

5.1

The waveform for a pure sinusoid.

The elements of a waveform are frequency (number of repetitions per second), amplitude (size of the pressure variation around the mean), and phase (how the waveform is aligned in time relative to a reference point).

5.1.2   Human Sensitivity to Auditory Dimensions

The basic properties of sound give rise to a number of perceptual dimensions that form the basis of auditory displays. The effectiveness of an auditory display depends critically on how sensitive listeners are to changes along the dimensions used to represent the relevant information. The following sections present a brief examination of human sensitivity to the dimensions of frequency and pitch, loudness, timbre, and spatial location. (For an extensive discussion, see Moore, 2003.)

Frequency and Pitch

As mentioned before, sound frequency is a fundamental organizing feature of the auditory system and a natural dimension for carrying information in auditory displays. Effective use of the frequency domain must take into account both the range of frequencies to which human ears respond and how well listeners can distinguish neighboring frequencies within this range.

Human listeners are able to detect sounds with frequencies between about 16 Hz and 20 kHz, with sensitivity falling off at the edges of this range. Frequency resolution (i.e., the ability to distinguish different frequencies) within the audible range is determined by the peripheral auditory system, which acts like a series of overlapping filters. Each filter is responsive to the sound energy in a narrow range of frequencies and has a bandwidth that varies with frequency. For frequencies below 200 Hz, the bandwidth is constant at around 90 Hz, while for higher frequencies it is approximately 20 percent of the center frequency. The width of these filters, or "critical bands," determines the minimum frequency spacing between two sounds required to perceive them separately. Sounds that fall within the same "critical band" will generally be difficult to separate perceptually.

Pitch is the subjective attribute of periodic sound that allows it to be ordered on a musical scale or to contribute to a melody. For pure sinusoids, the pitch is monotonically related to the frequency, with higher frequencies giving rise to higher pitches. For complex periodic sounds, pitch is monotonically related to the fundamental frequency. Most musical instruments produce periodic sounds that have a strong pitch. Periodic sounds without a strong pitch can nonetheless be told apart (discriminated) based on their spectral content.

Loudness

The subjective experience of loudness is related to its physical correlate, the intensity of a sound waveform (or square of the pressure amplitude) (Figure 5.1), and has been described using several different scales (such as the sone scale) (Stevens, 1957). Human listeners are sensitive to a large range of intensities, with sensitivity that is roughly logarithmic so that the smallest detectable change in loudness for wideband sounds is approximately a constant fraction of the reference loudness. Listeners can detect intensity changes of just a few decibels (dB)—defined as 10 times the logarithm of the ratio of the intensities—for many types of stimuli.

Timbre

For a complex sound, energy can be distributed in different ways across frequency and time, giving the sound its quality or timbre. For example, two complex tones of the same pitch and loudness may differ in their timbre due to the detailed structure of the waveform (e.g., differences in the relative magnitudes of the various frequency components). In intuitive terms, these differences are what distinguish instruments (e.g., the bowed violin from the percussive piano) playing the same note or chord from the same location with the same loudness.

Temporal Structure

Most sounds are not stationary, but turn on and off or fluctuate across time. Human listeners are exquisitely sensitive to such temporal changes. Temporal resolution is often quantified as the smallest detectable silent gap in a stimulus, and is on the order of a few milliseconds for human listeners (Plomp, 1964). Temporal resolution can also be described in terms of how well listeners can detect fluctuations in the intensity of a sound over time (amplitude modulation). Modulation detection thresholds are constant for rates up to about 16 Hz, but sensitivity decreases for rates from 16 to 1,000 Hz, where modulation can no longer be detected. For human listeners, relatively slow temporal modulations are particularly important, as they contribute significantly to the intelligibility of naturally spoken speech (Shannon, Zeng, Kamath, Wygonski, & Ekelid, 1995).

Spatial Location

Sounds can arise from different locations relative to the listener. Localization of sound sources is possible due to a number of physical cues available at the two ears (see Carlile, 1996, for a review). Differences in the arrival time and intensity of a sound at the two ears (caused by the head acting as an acoustic obstacle) allow an estimation of location in the horizontal plane. For example, a sound originating from the left side of a listener will arrive at the left ear slightly earlier than the right (by tens to hundreds of milliseconds) and will be more intense in the left ear than in the right (by up to tens of decibels).

In addition, the physical structure of the outer ear alters incoming signals and changes the relative amount of energy reaching the ear at each frequency. This spectral filtering depends on the direction of the source relative to the listener, and thus provides directional information to complement that provided by the interaural cues (e.g., allowing elevation and front–back discrimination). For estimating the distance of sound sources, listeners use a number of cues, including loudness, frequency content, and (when listening in an enclosed space) the ratio of the direct sound to the energy reflected from nearby surfaces such as walls and floors (Bronkhorst & Houtgast, 1999).

The spatial resolution of the human auditory system is poor compared to that of the visual system. For pairs of sound sources presented in succession, human listeners are just able to detect changes in angular location of around 1 degree for sources located in the front, but require changes of 10 degrees or more for discrimination of sources to the side (Mills, 1958). For the case of simultaneous sources, localization is well preserved as long as the sources have different acoustic structures and form clearly distinct objects (Best, Gallun, Carlile, & Shinn-Cunningham, 2007).

5.1.3   Using Auditory Dimensions

When using sound to display information, the available auditory dimensions and human sensitivity to these dimensions are critical factors. However, there are also a number of design questions related to how to map data to these dimensions in an effective way.

As an example, it is not always clear how data polarity should be mapped. Intuitively, increases in the value of a data dimension seem as though they should be represented by increases in an acoustic dimension. Indeed, many sonification examples have taken this approach. For example, in the sonification of historical weather data, daily temperature has been mapped to pitch using this "positive polarity," where high pitches represent high temperatures and low pitches represent low temperatures (Flowers, Whitwer, Grafel, & Kotan, 2001). On the other hand, a "negative polarity" is most natural when sonifying size, whereby decreasing size is best represented by *increasing* pitch (Walker, 2002). To add to the complexity of decisions regarding polarity, in some cases individual listeners vary considerably in their preferred polarities (Walker & Lane, 2001).

As another example, redundant mappings can sometimes increase the effectiveness with which information is conveyed. Recent work (Peres & Lane, 2005) has shown that the use of pitch and loudness in conjunction when sonifying a simple data set can lead to better performance, but other conjunctions may not.

5.1.4   Perceptual Considerations with Complex Displays

Multiple Mappings

With multidimensional data sets, it may be desirable to map different data dimensions to different perceptual dimensions. As an example, the pitch and loudness of a tone can be manipulated to simultaneously represent two different parameters in the information space (see Pollack & Ficks, 1954). However, recent work has shown that perceptual dimensions (such as pitch and loudness) can interact such that changes in one dimension influence the perception of changes in the other (Neuhoff, 2004).

In many cases, the most effective way of presenting multiple data sets may be to map them to auditory objects with distinct identities and distinct spatial locations. These objects can be defined on the basis of their identity (e.g., a high tone and a low tone) or their location (e.g., a source to the left and a source to the right). This approach theoretically allows an unlimited number of sources to be presented, and offers the listener a natural, intuitive way of listening to the data.

Masking

Masking describes a reduction in audibility of one sound caused by the presence of another. A classic example of this is the reduction in intelligibility when speech is presented against a background of noise.

In auditory displays with spatial capabilities, separating sources of interest from sources of noise can reduce masking. For example, speech presented against a background of noise is easier to understand when the speech and the noise are located in different places (e.g., one on the left and one on the right). In such a situation, the auditory system is able to use differences between the signals at the ears to enhance the perception of a selected source. In particular, it is able to make use of the fact that one of the two ears (the one nearest the speech target) is biased acoustically in favor of the target sound due to the shadowing of the noise by the head (Bronkhorst, 2000).

Auditory Scene Analysis and Attention

Another crucial consideration when delivering multiple signals to a listener is how the auditory system organizes information into perceptual "streams" or "objects." A basic principle of auditory scene analysis (Bregman, 1990) is that the auditory system uses simple rules to group acoustic elements into streams, where the elements in a stream are likely to have come from the same object. For example, sounds that have the same frequency content or are related harmonically are likely to be grouped into the same perceptual stream.

Similarly, sounds that have synchronous onsets and offsets and common amplitude and frequency modulations are likely to be grouped together into a single perceptual object. In addition, sounds that are perceived as evolving over time from the same spatial location tend to be perceived as a related stream of events. Grouping rules can be used in auditory display design when it is desirable that different signals be perceived as a coherent stream, but unwanted groupings can lead to disruptions in the processing of individual signals.

Confusion about which pieces of the acoustic mixture belong to which sound source are quite common in auditory scenes containing sounds that are similar along any of these dimensions (Kidd, Mason, & Arbogast, 2002). Related to this issue, sources in a mixture can compete for attention if each source is particularly salient or contains features that may be relevant to the listener's behavioral goals. By making a target source distinct along one of the perceptual dimensions discussed previously (e.g., by giving it a distinct pitch or spatial location), confusion can be reduced because the listener's attention will be selectively directed along that dimension. For example, when a listener must attend to one voice in a mixture of competing voices, the task is much easier (and less confusion occurs) when the target voice differs in gender from its competitors (Darwin & Hukin, 2000).

Auditory Memory

In complex auditory displays, the capabilities and limitations of auditory memory are important considerations. The auditory system contains a brief auditory store ("immediate" or "echoic" memory) where a crude representation of the sensory stimulus is maintained, normally for no longer than two seconds (Neisser, 1967). This store makes it possible to retain a sound temporarily in order to make comparisons with later-arriving sounds, as well as to process simultaneous stimuli in a serial fashion (Broadbent, 1958). When stimuli are processed in more detail (such as the semantic processing of speech or the learning of a sound pattern in the environment), there is the possibility for more permanent, categorical representations and long-term storage.

5.2 TECHNOLOGY OF THE INTERFACE

This section provides a brief overview of the technology required to create auditory displays. A typical auditory display system encompasses the various components sketched in Figure 5.2. These components establish a closed-loop system that integrates the user and can be recreated in almost any auditory interface. It is often possible for the user to interact with the system—this can be a very simple interaction like starting the sonification playback, or it can involve more complex continuous interactions, which are the particular focus of interactive sonification (described later).

5.2.1 Auditory Display Systems

The loudspeakers or "physical" displays are the only visible (and audible) part in this chain and are often referred to as the "front end" of the auditory display system (ADS). Most of the technology for the ADS is hidden behind the curtain of computation, algorithms, and signal processing. The current section focuses mostly on these "back-end" or invisible parts. (For further reading on the sound engineering and hardware of ADSs, the reader should consult specialized literature such as Miranda, 1998.)

Concerning the sound hardware (or front end of the ADS), the choice of whether to use headphones or speakers is determined basically by issues such as privacy, mobility, practicality, isolation, and/or user goals. Loudspeakers, for instance, do not require the listener to wear any electronic equipment and thus

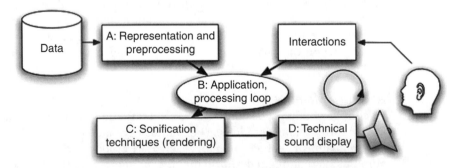

FIGURE 5.2 Sketch of the information flow in a typical auditory display system. A: Data representation. B: Main application (processing loop), which uses the data to determine when sounds should be created. C: Auditory display techniques to render an acoustic signal (digital audio signal) based on the data. D: Technical sound display systems such as sound cards, mixers, amplifiers, headphones, or loudspeakers to convert the digital audio signals to audible vibrations at the user's eardrums.

increase the user's mobility, yet their sounds are audible to everyone in the room. Headphones, in contrast, allow a personal auditory display yet may be impractical since they may interfere with auditory perception of the natural surroundings and thus isolate the user from his or her environment.

Loudspeaker and headphone systems also differ in their ability to communicate the source location of sound. This aspect of auditory interfaces is highly relevant for applications where the goal is to direct the user's focus of attention. Multispeaker systems (Pulkki, 1997) are well suited if the user can be assumed to be stationary and located in a "sweet spot." Typically, headphones lead to the user perceiving the source within his or her head; however, three-dimensional (3D) spatialization with headphones can be achieved by modeling the filter effect of the outer ear (see Spatial Location in Section 5.1.2). This is done mathematically with an adjustment of the source sound using head-related transfer functions (HRTFs) (Carlile, 1996). To be convincing, however, the head position and orientation of the listener have to be tracked so that the perceived sound source position can be kept constant while the user moves his or her head.

The technology for the front end of the ADS (D in Figure 5.2) is well developed and continues to advance due to the work of sound and product engineers. However, the larger back end of the ADS, consisting of the technologies to compute sound signals for auditory displays (A through C in Figure 5.2), is much younger and thus not as well developed. This back end will be the focus of Sections 5.2.2 through 5.2.4. A note for the human factors/usability practitioner: In order to describe the technology of the auditory interface, the information that will be displayed to the user must often be described in terms of programming code or computer science. Thus, for the benefit of readers wanting to create auditory displays, in these technology-oriented sections we will use mathematical descriptions and program code examples. However, readers less familiar with programming may want to skip the mathematics/programming details.

5.2.2   Data Representation and Processing

Let us start the discussion of technologies from the point in the information circle where information is created, measured, or becomes available within a computer system (depicted in Figure 5.2 as A). Information may have a highly different appearance, from a symbolic (textual) level (e.g., an alarm condition that a cooling system failed) to a more analogic level of raw measurements (e.g., temperature values). Data are frequently organized as a table of numbers, each column representing a different feature variable, each row representing measurements for a single record. In a census data set, for instance, columns could be features such as income, weight, and gender, while rows could represent different persons. We call such a representation data set X, and in most cases data or information can be recoded into such a form. We will refer to rows of X as $\vec{x}^{\alpha}$, and to the i-th feature as x_i^{α}.

Many auditory displays—ranging from applications in process monitoring and exploratory data analysis sonifications to table viewers for blind users—operate using this type of data representation. Communication of single events (e.g., to signal that an e-mail arrived) is also common. The message can be characterized by a set of feature values (e.g., sender, e-mail length, urgency, existing e-mail attachments, etc.). These values form a row vector $\vec{x}^\alpha$ following the previous representation. An auditory display technique should be able to represent all possible feature vectors using the systematic transformation of event data to sound.

Statistics and data mining (Fayyad, Piatetsky-Shapiro, Smyth, & Uthurusamy, 1996) are the disciplines concerned with explaining all peculiarities of how data are structured and summarized, and a thorough introduction would exceed the scope of this chapter. Central aspects of data, though, are the range (minimum/maximum values), whether the data are discrete or continuous, and whether a variable is nominal or ordinal. It is important to identify these aspects of data because certain auditory variables are often considered better suited to represent certain data types; for example, timbre is a better match for nominal variables whereas frequency fits well with ordinal variables. Data features with a zero value (e.g., velocity of a car) match with acoustic variables that have a zero value (e.g., loudness, vibrato frequency, pulsing rate, etc.).

5.2.3   Sound Synthesis

Sound synthesis is the technological basis for controlling sound characteristics with data. In rare cases, it might be possible to simply record sounds for every possible condition or event. But whenever full control over all sound characteristics is wished or needed, sound synthesis is essential.

Digital sound signals are vectors of numbers that describe the sound pressure at every moment. Real-time sound computing is thus computationally quite demanding and scales with the number of independent audio channels. Powerful programming systems for sound synthesis are available. The following code examples derive from SuperCollider (McCartney, 1996), a versatile, compact, powerful, and open-source textual programming system. Pure Data is another graphical engine (also cross-platform and open-source) (Puckette, 1997).

Additive/Subtractive Synthesis

Additive synthesis is the creation of complex sounds from simple ingredients. These ingredients are the building blocks in a bottom-up approach. The building blocks are simple signals (such as sine waves $b_\omega(t) = sin(\omega t + \varphi)$), and their superposition represents the result of additive sound synthesis

$$s(t) = w(t) \cdot \sum_{i=1}^{N} a_k\, b_{\omega_k}(t) = w(t) \cdot \sum_{i=1}^{N} a_k\, sin(\omega_k t + \varphi_k)$$

To obtain harmonic timbres, frequencies ω_k are chosen as integer multiples of a fundamental frequency ω_1. The coefficients a_i determine how strongly each component contributes to the sound. An example of achieving additive synthesis in SuperCollider follows: For only two partial tones of 440 Hz, the designer would use

```
{SinOsc.ar(440, mul: 0.4) + SinOsc.ar(880, mul: 0.2)}.play
```

Sound example S1 (at *http://sonification.de/publications/BeyondGUI*) provides one second of the sound. Sound examples S2 to S4 demonstrate some typical additive synthesis sounds.

Subtractive synthesis takes the opposite (top-down) approach and creates complex timbres by removing material from a spectrally rich source such as sawtooth-shaped signals (refer to Section 5.1 for more on signal shapes), pulse trains, or spectrally rich noise signals. A good introduction to subtractive synthesis can be found in Moore (1990). Sound examples S5 to S8 demonstrate some typical sounds of subtractive synthesis, and sound examples S9 to S12 demonstrate different filters.

Wavetable Synthesis

One of the most practically relevant synthesis techniques is wavetable synthesis, in which a recorded version of a real-world sound is used to generate the synthesized output. Mixing and manipulating the synthesis algorithm allows the audio designer to create novel sounds or to play a sample at different musical notes. Many commercial sound synthesizers rely on this technique, and most of the auditory icons (discussed later) are produced using wavetable synthesis.

In SuperCollider, a wavetable is represented by a `Buffer`, and a `PlayBuf` unit generator can be used to play the buffer at arbitrary speed, as demonstrated in the following code example (sound example S13), where the playback rate is slowly modulated by the sine oscillator:

```
b = Buffer.read(s,"soundsample.wav") ;
{PlayBuf.ar(1, b.bufnum, SinOsc.kr(0.2, mul: 0.4, add: 1)*
BufRateScale.kr(b.bufnum), loop: 1)}.play
```

Other Synthesis Techniques

Other synthesis techniques include granular synthesis, physical modeling sound synthesis, FM-synthesis, and nonlinear synthesis. Discussion of these techniques easily fills books, and the interested reader should look to Moore (1990), Roads (2001), and Cook (2002) for more information.

5.2.4   Auditory Display Techniques in a Nutshell

This section focuses on various auditory display techniques. In general, these techniques are algorithms that connect the data to be displayed to sound synthesis techniques (described in Section 5.2.3). As mentioned previously, auditory display techniques can roughly be characterized as symbolic or analogic. We start here with symbolic sonification techniques.

Auditory icons: Auditory icons, as mentioned earlier, represent specific messages via an acoustic event that should enable the quick and effortless identification and interpretation of the signal with respect to the underlying information. These sounds need to be selected from a database of recordings, or synthesized according to the data features, which is practically achieved by adapting appropriate sound synthesis algorithms (see Section 5.2.3).

Earcons: Different from auditory icons, earcons use musical motifs to represent messages and require the user to learn the meaning for each earcon. As a benefit, earcons can inherit structural properties from language as a more abstract and highly symbolic form of communication (Blattner, Papp, & Glinert, 1994). These sounds can be built using concatenation, which allows the designer to compose more complex messages from simple building blocks.

Audification: In audification, the data "speak for themselves" by using every data value as a sound sample in a sound signal $s(t)$. Since only variations above 50 Hz are acoustically perceptible (see Section 5.1), audifications often consume thousands of samples per second. Thus, the technique is suitable only if (1) enough data are available, (2) data can be organized in a canonical fashion (e.g., time-indexed measurements), and (3) data values exhibit variations in the selected feature variable. Mathematically, audification can be formalized as the creation of a smooth interpolation function going through a sample of (time, value) pairs (t^α, x^α) for all data items α. The simplest implementation of audification, however, is just to use the measurements directly as values in the digital sound signal by setting $s[n] = x^n$. Some sound examples for audifications demonstrate the typical acoustic result (S14, S15). S14 is an audification of EEG measurements—one electrode measures the brain activity of a beginning epileptic attack (roughly in the middle of the sound example). S15 plays the same data at lower time compression. Clearly the pitch drops below the well-audible frequency range and the epileptic rhythm is perceived as an audible rhythm of events.

Parameter mapping sonification (PMS): This is the most widely used sonification technique for generating an auditory representation of data. Conceptually, the technique is related to scatter plotting, where features of a data set determine graphical features of symbols (such as x-position, y-position, color, size, etc.) and the overall display is a result of the superposition of these graphical elements. For example, imagine a data set of measurements for 150 irises. For each flower, measurements of the petal length, sepal length, petal width, and sepal width are listed. A parameter mapping sonification (S16) could, for instance, map the petal length to the onset time of sonic events, the sepal length to the pitch of sonic events, the petal width to brilliance, and the sepal width to duration. The resulting sonification would allow the listener to perceive how the data are organized in time or change with time. Each sound event represents a single flower, while the PMS displays the entire data set of measurements of all 150 flowers!

Model-based sonification (MBS): This is a structurally very different approach from PMS (Hermann, 2002). In PMS, data directly control acoustic attributes, whereas in MBS the data are used to create a sound-capable dynamic model. The result of this is that a sonification model will not sound at all unless excited by the user and thus puts interaction into the fore. The set of rules regarding how to create a virtual sound object from data is called a sonification model, and the rules can be designed in a task-oriented way. For example, imagine that every text message in a mobile phone is like a marble in a box. By shaking the phone, the marbles move, interact, and thereby create an acoustic response from which you can infer how many text messages, size, and so on, have arrived (Williamson, Murray-Smith, & Hughes, 2007). The excitation here is "shaking," the dynamics are the physical laws that describe the marble motion and interactions, and so forth. The sonification model simulates the entire physical process, and thus creates an interactive and informative sonification.

Interactive sonification: This is a special focus in auditory interfaces and can be used with many types of sonification techniques. Often a particular benefit results from tightly closed interaction loops between the user and a sonification system (Hunt & Hermann, 2004). All sonification techniques can be modified to be more interactive; for instance, for audification, interactive sonification can enable the user to actively navigate the data while generating the sound and so on. The rationale behind interactive sonification is that people typically get latency-free acoustic responses as by-products of their interaction activity, and they use the acoustic feedback continuously to refine their activity, be it within a search, scan, discovery, or any other task.

5.3   CURRENT INTERFACE IMPLEMENTATIONS

The following section provides descriptions of when and why auditory interfaces are currently utilized, both for sighted and visually impaired users.

5.3.1   A Bit of History

Sound often helps direct our focus and describes what is going on. Listening to a car engine or the spinning of a hard drive can offer vital clues about whether the car is in good mechanical condition or the computer is finished saving a file. In early computing, such incidental sounds were often used—for example, beeps were introduced to indicate errors in the program or the beginning of a new iteration of a program loop. There is a long tradition of audio indicating warnings and alerts. Sound has the potential to be used in many more sophisticated ways, ranging from short sounds to indicate specific events to fully immersive spatial sound environments.

5.3.2   Why Sound Is Used

In addition to those mentioned in the introduction of this chapter, there are numerous reasons why sound may be used in an interface. A very common one is to reinforce a visual message, such as an alert. Other reasons are outlined next.

Reducing visual overload: Visual interfaces tend to be busy, filling as much of the screen as possible with information. Constantly changing visual information can be distracting and can limit the amount of information that reaches the user. Where applicable, the cognitive load can be reduced by channeling information to the ears (Brown, Newsome, & Glinert, 1989).

Reinforcing visual messages: Sending the same information to more than one sense can ensure that the user receives the information, making the interface more effective. For example, when entering a personal identification number (PIN) at an automated teller machine (ATM), the machine beeps for each number entered and the visual interface uses an asterisk to represent each number. This dual feedback can reassure users that their input was received.

When eyes are elsewhere: Since sound can be perceived from all directions, it is ideal for providing information when the eyes are otherwise occupied. This could be where someone's visual attention should be entirely devoted to a specific task such as driving or surgically operating on a patient (Recarte & Nunes, 2003).

When audio is more informative: Humans are very good at hearing patterns in sound. This means that at times it is easier to understand information when it is sonified (Bly, 1982). Two prime examples of this are seismic data (Hayward, 1994) and medical monitoring data (Baier & Hermann, 2004). Users can very quickly notice a change that may not be as easily noticed when looking at numbers or a graph.

Small or no visual display: Unlike visual displays, where the size of the interface is determined by the size of the device, audio is limited only by the sound quality that the device can provide. It is therefore a good candidate for augmenting or replacing a visual interface.

Conveying emotion: The aesthetics of sound can have a great impact on the user's impression of an application. This is particularly obvious in video games, where the sound design is carefully orchestrated to make players enjoy the game and to impact the amount of tension the player experiences.

5.3.3 Drawbacks to Using Sound

Some of the major disadvantages to using sound are annoyance and lack of privacy. There are also dangers to using too much sound. If sound is poorly designed or used at the wrong time, users are very quick to turn it off. Some of the drawbacks are outlined next:

Annoyance: Sound is very good at drawing the user's attention. However, the urgency of the sound should be relative to the importance of the information. The use of obtrusive sounds to represent something of low importance can quickly annoy the user.

Privacy: Sound is omnidirectional and thus cannot be directed at a single user unless they are using a headset or a handset. Therefore, if headsets are not employed, an auditory interface can publicize what a user is doing in an undesirable manner.

Auditory overload: Displaying too much information in sound can result in that information losing meaning and being interpreted as noise.

Interference/masking: As mentioned in Section 5.1.3, sounds can interfere or mask one another. Like auditory overload, this can result in loss of information. Environmental sounds can also cause interference and/or masking.

Low resolution: With visual displays, objects can be located very precisely on the screen; with spatial sound interfaces, there is a greater area that can be used but the resolution is lower (see Section 5.1.2). This difference is evident, for example, in visual and audio games. In a visual game, it is possible to pinpoint a target with great accuracy on a computer screen. In an auditory game, however, although the auditory display may be presented in an area much larger than a computer screen, the lower resolution that hearing affords in locating audio sources greatly reduces the accuracy with which they can be located.

Impermanence: Unlike visual displays, where the information on the screen remains, audio is serial and once played is easily forgotten (see Section 5.1).

Lack of familiarity: Sounds take a while to get to know and can be initially confusing to users until they know what they are listening to. As with visual icons, the sounds in an audio interface need to be easy to learn.

5.3.4 Advanced Audio Interfaces

The availability of increasingly sophisticated audio hardware and software has provided the possibility for more widespread use of audio in interfaces. This sophisticated technology has contributed to using audio to support richer human–computer interactions as well as increased the quality of the audio that can be used, which can be the equivalent of that used in a film or radio production. This more sophisticated use of audio may of course be part of a multimodal interface or the basis of an audio-only display. Although much of this potential remains underexploited, in this subsection we examine several application areas where these improved audio capabilities have been used to good effect.

Audio for Monitoring

Audio is a well-known method for monitoring in specialized environments such as hospitals and environmental monitoring facilities such as for weather and seismic activity. Other applications are stock market and network monitoring.

Accentus (*http://www.accentus.com*) makes Sonify!, a product for traders to (among other things) monitor stock price changes and progress toward a target (Janata & Childs, 2004). Sonify! uses real instruments and short melodies to create pleasant tones that are informative yet remain unobtrusive. The use of sound is particularly appropriate as traders are usually surrounded by numerous screens and work in a very visually intensive environment. They are often required to monitor data while talking on the phone. The nonspeech audio can reduce the visual overload and can be heard in the background of other auditory events such as telephone conversations.

Numerous programs that use sound to monitor network traffic have been developed. The Sheridan Institute in Canada has developed *i*SIC (Farkas, 2006), which uses mathematical equations on network traffic data to create music. The Sound of Traffic (Weir, 2005) uses MIDI sounds to track traffic to specific ports on a computer. These network monitors allow administrators to hear real-time data about their sites and networks without having to purposely go to an interface or log to check activity; they allow monitoring to be a background process. While these programs are not commercially available, they show a growing trend in using audio in diverse applications.

Audio in Games

Audio plays a very important role in video games, using music to set the mood and sound effects to bring realism to the action. Audio in most modern games is a multilayered production used to render complex auditory scenes, creating an immersive environment. Sound usually takes the form of effects such as one would hear in film—known as Foley effects[1]—and human dialog to imbue scenes with realism. This audio is increasingly synthesized dynamically rather than prerecorded. A particular problem faced by game designers is the fact that, while broached early on in the design phase, often sounds are only incorporated into the game after the whole of the interactive graphic design has been implemented. Thus, proper sound testing is done at a very late stage. The end result

---

[1] Jack Foley pioneered the craft of creating sound effects, using all kinds of materials to imitate sounds, in the early days of the "talkies" for Universal Studios.

is that the sound design is mostly aesthetic as opposed to informative and not nearly as powerful as it could be.

Audio-Only Games

Audio-only games are often, but not exclusively, targeted at the visually impaired. Evolving from mostly speech and simple sound effects, today's audio games, such as Shades of Doom (first-person shooter) and Lone Wolf (submarine simulator), use sophisticated sound, transitioning smoothly between multilayered and immersive soundscapes. Many of today's audio-only games also provide a visual interface for collaborative game play between visually impaired and sighted gamers. A typical issue that audio game designers face is supporting orientation and navigation in a 3D world (Andresen, 2002)—for example, which way is the player facing, where are the walls/entrances/exits, and what is the state of other players/objects? The AudioGames website (*http://www.audiogames.net*) is dedicated to blind-accessible computer games and includes issues of the audio gaming magazine *Audyssey* (2006).

5.3.5   Applications of Auditory Interfaces to Accessibility

Nonspeech sound has significant potential for improving accessibility, either on its own or as a complement to other media, particularly for users with visual impairments, because of its ability to convey large amounts of information rapidly and in parallel. Imagine a visually impaired teacher monitoring the progress of students taking an electronic multiple-choice test. The teacher could get a rapid "auditory glance" of approximately how far through the test students were and the proportion of correct answers, individually and as a whole, using the data sonification concepts described at the beginning of the chapter.

Accessibility and the Desktop

Visually impaired users employ software called a *screen reader*, which uses synthetic speech to speak text such as menus, the state of GUI widgets, and text in documents. The most popular screen readers used worldwide are Jaws for Windows (JFW) by Freedom Scientific (2005) and Window-Eyes by GW Micro (2006). The use of nonspeech audio has seen relatively little commercial uptake. The release of JFW version 5 in 2003 represented the first significant use of nonspeech sound in the form of the ability to customize feedback. These customizations, called "behaviors," are defined in schemes. Each scheme can be associated with a specific application. Examples include focus on a particular type of widget, the state of a checkbox, upper/lower case, degree of indentation, and values of HTML attributes.

The inclusion of nonspeech sound is intended to improve efficiency and effectiveness, such as in navigation of the GUI or screen-based proofreading of lengthy documents. However, significant effort is required by users or interface developers to associate specific sounds with events or symbols, and further to develop a coherent set of these associations into an overall sound scheme for an application.

Other uses of nonspeech sound to improve accessibility have been reported in the ICAD literature (*http://www.icad.org*). These include supporting the navigation of structured documents and the Web, auditory progress bars, auditory graphs, auditory widgets, auditory access to the widgets in a word processor, and use of audio to review data in spreadsheets.

Mobile Accessibility

Nonspeech sound has been employed in mobility devices, such as giving longer-range warnings to a blind user of an approaching obstacle than can be obtained using a white cane. A study carried out by Walker and Lindsay (2004), where participants were guided along routes by sound beacons, showed good performance even in the worst case, proving that the nonspeech auditory interface could successfully be used for navigation.

The K-Sonar (Bay Advanced Technologies, 2006), intended to supplement a white cane, is like a flashlight that sends out a beam of ultrasound instead of light. Using the ultrasound waves reflected from the objects in the beam's path, K-Sonar provides complex audible representations of the objects that users can learn to recognize and use to build mental maps of their environment.

Despite the success of these applications, problems can arise with the use of audio in such mobility and orientation devices. Sounds that are fed back to the user can be masked or made less clear by ambient sound such as traffic or construction noise. Although some users might choose to reduce this problem with earphones, many visually impaired users are wary of anything that limits their perception of their environment. This problem can be resolved through the use of a small speaker positioned just a few inches away from one of the user's ears.

The Trekker, developed by Humanware *(www.humanware.com),* a largely speech-based geographic positioning system (GPS) mobility and orientation system for visually impaired users, uses such a system. This approach keeps the audio output audible and relatively private without masking other environmental sounds. Because of the problems of sound interference, a number of other ultrasound or infrared systems employ haptic feedback rather than sound, using the strength and type of vibrations to provide information about the size and proximity of nearby objects. A promising approach to the sound interference problem is offered by the use of bone conduction headphones. Rather than going through the eardrum, bone conduction headphones convert sounds into mechanical vibrations going through the skull straight to the auditory nerve.

5.4 HUMAN FACTORS DESIGN OF AN AUDITORY INTERFACE

Sound is commonly thought of as a way to enhance the display of visual information, but as the preceding sections in this chapter have demonstrated, it also has a range of informational capacities and advantages that make it particularly useful in a variety of interactive settings, and, in some contexts, make it the best or only choice for conveying information. Thus, the design problem is not just one of evaluating criteria for the use of sound, but is also a question of determining how sound will be used to its best effect.

However, it is important that designers do not use auditory displays unless the auditory displays are beneficial and/or necessary. There are many circumstances where auditory displays *could* be utilized, but it is only through the utilization of sound user–centered design principles that the practitioner can determine whether an auditory display *should* be utilized.

As with all good user-centered design, the decision to use sound as a display technique should be based on a comprehensive task and needs analysis. This analysis will inform design specification and the ultimate implementation. This section provides information about some of the unique issues and challenges that practitioners and designers may face when considering auditory interfaces.

5.4.1 Task Analysis—How Does It Differ for an Auditory Interface?

Task analysis is a common technique used as part of the user-centered design process for standard GUIs. However, the unique nature of auditory interfaces requires special considerations in the development and utilization of the task analysis.

Description of the User

In addition to the usual variables such as gender, age, experience with information technology, and so on, that developers should consider, the development of auditory displays needs to take specific account of the experience of audio of the target user population for the application as a whole. This in turn breaks down into a number of distinct variables, including the following:

Musical expertise: What is the cultural background, level, and range of musical expertise of participants? Users with musical training are better in discriminating variations of pitch or temporal features like rhythm.

Familiarity with auditory displays: Irrespective of musical ability, what level of experience will the users have with using auditory displays? (See also Section 5.6.1.)

Hearing abilities: Will the user/group of users have a higher likelihood of hearing impairment than other populations (e.g., workers who operate loud machinery on a regular basis)? Do they have limited spatial resolution because of hearing loss in one ear? Are the users capable of analytical listening?

Work the User Must Accomplish

The goal of articulating the specific tasks the user will be performing is twofold. First, it develops an account of both what the user is expected to do in the targeted activity and how that is to be accomplished. Second, it provides detailed descriptions of the information that must be available to the user in order to perform the tasks. Generally, this process entails creating detailed descriptions of the procedures that users must follow for accomplishing certain tasks. For auditory displays, however, these procedures are not necessarily visual or psychomotor (e.g., look at the information displayed on the speedometer and adjust the speed accordingly with either the brake or the gas pedal). Thus, additional elements of the task must be considered.

What Tasks Will the User Perform Using the Auditory Display? If an auditory display will be part of a larger multimodal interface, which is more often than not the case, the task analysis must account for all facets of the user's task, not just the aural component.

Exploring data with interactive sonification techniques, for example, usually entails more than just interactive listening (Hermann & Hunt, 2005). Many users need iterative access to the data as well as ways to stipulate and revise the manner of auditory interaction, be this variable parameter mappings or a sonification model. Each of these aspects of the task must be identified and described procedurally. In particular, the use and role of any nonaudio interface components, such as a command line or graphical display, common or novel input devices, and so on, should be spelled out in the analysis, as well as the kind of cognitive, perceptual, and motor actions that the user must perform to define and interact with the sonification. The resulting description of the complete task then serves as the primary basis for specifying the compositional details of the user interface, that is, what the user will actually hear, see, and physically manipulate.

In addition, some auditory tasks, particularly those driven by external events or data, may be critical components of larger operations that are subject to variable rates of activity or priority. When this is likely to be an issue, identifying potential performance boundary conditions is important. It may be possible, for instance, for event pacing to exceed the abilities of users to respond to the demands of the task or for the user's attention to be overwhelmed by the priority of other stimuli during periods of high operational activity. Concern with performance limits and task priorities is closely related to other concerns that complex user environments raise, but is easily overlooked. By making note of how and when these factors may arise, the task analysis helps to identify where the auditory interface design is likely to require the greatest effort. Indeed, in some situations, auditory displays are used to reduce cognitive load at points in the task when the cognitive load is the highest.

What Information Should the Sound Convey? A detailed account of the information that sound will be used to convey in the auditory interface must also be developed in conjunction with the task analysis. This information analysis specifies what the user must know to achieve his or her application goals, and is especially important in the design of auditory displays because of the somewhat unique challenges that auditory information design poses. Unlike other representational techniques that are commonly used in the design of information tasks, particularly those used in the design of visual displays, the mapping of data to, and the perception of meaning in, sound can be subject to a range of individual performance differences (see, in particular, Sections 5.1 and 5.3).

The information analysis is best organized to correspond to the organization of the task analysis. Descriptions of the information involved at each step in the task should be detailed enough to address the representational and perceptual considerations that will arise at the auditory design stage, many of which have been described in preceding sections. In particular, the information to be conveyed to the user should be characterized in the following ways:

✦ Application to elements of the task (one, several, many, etc.) and/or its conceptual purpose

✦ Class and organization: qualitative (nominal, categorical, hierarchical); spatial (direction, distance); temporal; quantitative (nominal, binary, ordinal, integral, ratio, etc.); range

✦ Expected degree of user familiarity (Is this information that the user will easily recognize or will training be required?)

✦ Meta-level description of what the information will be used for (e.g., when the information is relevant to other aspects of the task, and when it is redundant or reinforces other displayed information, or has ramifications in some broader context)

An additional facet of the information analysis that is often needed for the auditory information design is an explicit specification of the underlying data (see Section 5.2.2). This is generally the case for qualitative information and for task-specific subsets of numerical information that are nonlinear (e.g., noncontiguous) in one or more ways.

Context and Environment of the Auditory Display

Since certain operational environments have implications for how an auditory display should be organized and presented, the task analysis should also consider the following questions.

First, will the task be performed in a single or shared user space? This question bears directly on how the auditory display is rendered for individual or group use, that is, through headphones or loudspeakers.

Second, will the user be engaged in more than one task, and if so what is the role of concurrency? The most important question here is how auditory information in the task being analyzed is likely to interact with the performance of another task the user is expected to perform and vice versa, especially if sound is also used by the other task. Particular attention should be given to the potential for conflicting or ambiguous uses of sound.

Third, will external sound be present in the environment, and if so what is its informational role? This concern is somewhat like the previous consideration, only at a larger scale. If the auditory task will be embedded in a larger environment in which noise or intentional uses of sound are present, the task analysis should address the potential for masking or other disruptive impacts on the auditory task.

5.4.2   Constraints on Interface Design

After user, task, and environmental considerations, designers of auditory interfaces must consider the practical limits associated with the capacities and costs of current computer and audio technologies. The primary issues that designers must confront are processing power and mode of rendering, particularly in middle- and lower-tier platforms. There are some environments that will be conducive to a computationally intensive auditory display, such as a system that has been engineered to meet the requirements of a virtual immersive environment in a laboratory or a research and development facility. However, many handheld devices and portable computers do not have the capacity for a display that requires much computational power, and this will likely be the case for a number of years to come. For instance, many portable devices currently lack the throughput and/or the processing requirements for real-time 3D auditory rendering of multiple sources. Furthermore, widely available solutions for tailoring binaural processing for individual listeners are still several years away. Thus, if the task requires that the display device be portable, the designer should avoid creating an auditory interface that depends on conveying accurate spatial information to the listener.

Other practical considerations that are unique to auditory interface designs are factors associated with loudspeaker and headphone rendering, personalization, and choice of sound file format. Accurate perception of auditory detail in loudspeaker rendering, for instance, such as location, motion, and high-frequency information, requires that the listener be ideally positioned in relation to the output of the loudspeaker system. Rendering with headphones and/or earbuds may be inappropriate in some operational settings because this can defeat the listener's ability to hear other important sounds in the immediate environment. Designers also must consider how users will access and control parameters such as loudness, equalization, program selection, and other features of an auditory interface that the user may wish to individualize. A related, ergonomic issue that designers must be aware of is the risk of hearing loss associated with repeated exposure to excessively loud sounds.

Finally, designers should understand how MIDI (musical instrument digital interface) files work and what the perceptual advantages and functional trade-offs are between uncompressed auditory file formats, such as the WAV and AIFF specifications, and compression formats, such as WMA and MP3 (e.g., see Lieder, 2004). If the user of an auditory interface must be able to exchange sounds with collaborators, cross-platform compatibility is a crucial consideration. When this is the case, the designer should ensure that the interface supports a range of sound file formats that are appropriate for the perceptual requirements of the task. The MP3 format, for example, is adequate for auditory alerts and icons but inferior to the WAV format for many scientific applications.

5.4.3   When to Use an Auditory Interface

Design motivations for auditory interfaces can be broadly organized into four functional categories: managing user attention, working with sound directly, using sound in conjunction with other displays, and using sound as the primary display modality. Although each of these has been mentioned and described in previous sections, in this subsection we briefly identify specific human factors relevant for each application type. Many of these human factors have been described in some detail in Section 5.1, but the usability practitioner may require more information when actually designing and implementing these applications. Consequently, additional issues and sources of information are provided here.

Managing User Attention

Various auditory materials can be used to manage attention, and the manner in which information is conveyed can be either discrete or continuous, and may or may not involve the manipulation of auditory parameters. For instance, a substantial body of human factors research relating the parameters of auditory signals to perceived urgency exists for the design of auditory warnings, which fall under this design category (see Stanton and Edworthy, 1999, for a review). Similarly, changes in the character of continuous or streaming sounds allow listeners to peripherally or preattentively monitor the state of ongoing background processes while they attend to other functions. Whenever designers create sounds for attentional purposes, the perceptual strengths and weaknesses of the auditory materials must be carefully evaluated in the context of the larger task environment. In addition, a range of related nonauditory human factors may also need to be considered. These include interruptions (McFarlane & Latorella, 2002), situation awareness (Jones & Endsley, 2000), time sharing (Wickens & Hollands, 2000), and stress (Staal, 2004).

Working with Sound Directly

Many activities involve working with sound itself as information, as a medium, or both. In this design category, perception and/or manipulation of sound at a meta-level is the focus of the task. For example, a user may need to be able to monitor or review a live or recorded audio stream to extract information or otherwise annotate auditory content. Similarly, sound materials, particularly in music, film, and scientific research, often must be edited, filtered, or processed in specific ways. Interfaces for sonifying data, covered at length previously, also fall under this heading. Knowledge of human auditory perceptual skills (Bregman, 1990) and processes in auditory cognition (McAdams & Bigand, 1993) are both important prerequisites for the design of any task that involves end-user development or manipulation of auditory materials.

Using Sound in Conjunction with Other Display Modalities

Sound is perhaps most frequently called on to complement the presentation of information in another modality. Although haptic interfaces are beginning to make use of sound, (see, e.g., McGookin & Brewster, 2006a), more often than not sound is used in conjunction with a visual display of some sort. Like its function in the real world, sound not only reinforces and occasionally disambiguates the perception of displayed events but also often augments them with additional information. In addition to previously mentioned human factors issues, aesthetics (Leplatre & McGregor, 2004), multisensory processing (Calvert, Spence, & Stein, 2004), and the psychology of music (Deutsch, 1999) could be relevant for these types of applications.

Using Sound as a Primary Display Modality

In a range of contexts, sound may be the most versatile or the only mode available for representing information. In these auditory applications, a user population with its own, often unique, set of interaction goals and expectations is targeted. Many of the human factors considerations that are relevant for the preceding design categories can also be applicable here, particularly those relevant to working with sound and those relevant to using sound in larger operational contexts.

5.4.4   Specifying Requirements for an Auditory Interface

An important step in the development process is to turn the products of the contextual task analysis and iterative prototype testing into a coherent specification. Depending on the size and scope of the project, formal methods (e.g., Habrias & Frappier, 2006) or a simple software requirements document can be used for this purpose. However, the value of this exercise and its importance should not be underestimated. The specification is both a blueprint for the interface and a road map for the implementation process.

In whatever form it takes, the specification should detail how the interface will be organized, how it will sound and appear, and how it will behave in response to user input well enough to prototype or implement the auditory task to a point that is sufficient for subsequent development and evaluations. The interface should organize and present the actions and goals enumerated in the task analysis in a manner that is intuitive and easy to understand. Any auditory materials, sound-processing specifications, and examples that were developed for the auditory design should be referenced and appended to the specifications document.

In many cases, it will also be effective to sketch the sound equivalent of a visual layout with an auditory prototyping tool. The designer should take care to ensure that the specified behavior of the interface is orderly and predictable, and, because audio can be unintentionally too loud, the interface ideally should cap the amplitude of auditory presentations and should always include a clear method for the user to cancel any action at any point. A number of auditory prototyping tools, ranging from simple sound editors to full-blown application development environments, are available both commercially and as open-source projects that are freely available on the Internet for downloading and personal use.

Section 5.2.3 provided programming examples for sonifying data in the open source SuperCollider environment (McCartney, 1996), and also mentioned Pure Data, another popular and powerful open-source programming environment that can be used for prototyping audio, video, and graphical processing applications (Puckette, 1997). A recent, commercially available tool for developing immersive virtual auditory environments is VibeStudio (VRSonic, 2007).

Last, a good understanding of current auditory display technology is important for achieving good human factors in auditory tasks. The designer should weigh the advantages of commercial versus open-source audio synthesis and signal-processing libraries and should also give attention to the implications of various audio file formats and rendering methods for auditory tasks. Sounds rendered binaurally with nonindividualized HRTFs, for instance, are perceived by most listeners to have functional spatial properties but are accurately localized by only a small percentage of listeners. Technical knowledge at this level is integral to developing effective specification and ensuring that usability concerns remain central during the iterative stages of implementation and formative evaluation that follow.

5.4.5   Design Considerations for Auditory Displays

One of the goals of this chapter's sections on the nature, technology, and current implementations of auditory interfaces is to give the reader a tangible sense of the exciting scope and range of challenges that auditory information designs pose, especially with regard to human factors. Auditory design is still more of an art than a science, and it is still very much the case that those who choose to implement auditory interfaces are likely to find they will have to do a bit of trailblazing.

Sections 5.2 and 5.6 provide detailed development and design guidelines that the designer and practitioner should find useful when developing sounds for auditory interfaces. Because auditory interfaces use sounds in different paradigms than many (indeed most) people are familiar with, the current section is devoted to providing the reader with a way of thinking about sound from an informational perspective and how the design and production of sounds should be influenced by this different way of thinking.

Thinking about Sound as Information

In the process of designing auditory materials to convey task-related information, it is important to keep in mind a number of conceptual notions about sound as information. First, sound can be usefully categorized in a number of ways, non-speech versus speech, natural versus synthetic, nonmusical versus musical, and so on. Listeners generally grasp such distinctions when they are obvious in the context of an auditory task, so this sort of partitioning—in an auditory graphing application, for instance—can be quite useful as a design construct.

Another valuable design perspective on auditory information, introduced in Kramer (1994), is the auditory analogic/symbolic representation continuum mentioned at the beginning of the chapter. Sounds are analogic when they display relationships they represent, and symbolic when they denote what is being represented. Much of the displayed sound information that people regularly experience falls somewhere between, and typically combines, these two ideals. The classic Geiger counter example can be understood in this way—the rate of sounded clicks is analogic of the level of radiation, while the clicks themselves are symbolic of radiation events, which are silent in the real world.

The analogic/symbolic distinction can also be usefully conceptualized in terms of semiotics (i.e., the study of symbols and their use or interpretation). When sound is designed to convey information to a listener, the intended result, if successful, is an index, an icon, and/or a symbol (e.g., see Clark, 1996). Indices work by directing listeners' attention to the information that they are intended to signal, and icons work by aurally resembling or demonstrating the information they are meant to convey. Both of these types of sounds function in an analogic manner. A remarkable example of an auditory signal that is both indexical and iconic is a version of Ulfvengren's (2003) "slurp," which, when rendered spatially in an airplane cockpit, is intended to draw the pilot's attention to and resemble a low-fuel condition. Note how the informational use of this sound in this context also makes it a symbol. Sounds that are symbols work by relying on an associative rule or convention that is known by the listener. Outside of a cockpit, most listeners would not readily associate a slurping sound with their supply of fuel!

A final meta-level aspect of sound as information that should be kept in mind as the auditory design process begins is people's experience and familiarity with the meanings of everyday and naturally occurring sounds (Ballas, 1993). Much of the information conveyed by these classes of sounds is not intentionally signaled but is a perceptual by-product of activity in the real world, and is understood as such. Footsteps, the sound of rain, the whine of a jet, the growl of a dog—all have contextual meanings and dimensions that listeners readily comprehend and make sense of on the basis of lifelong experience and native listening skills. The inherent ease of this perceptual facility suggests an important range of strategies for auditory designers to explore.

Fitch and Kramer (1994) used analogs of natural sounds in a successful patient-monitoring application to render concurrent, self-labeling auditory streams of physiological data. Even more innovative is work by Hermann et al. (2006) in which pathological features in EEG data that are diagnostic of seizures are rendered with rhythms and timbres characteristic of human vocalizations.

In contrast, sounds that are unusual or novel for listeners, including many synthetic and edited sounds as well as music, have an important place in auditory design, primarily for their potential for contextual salience and, in many cases, their lack of identity in other settings. In general, though, unfamiliar uses of sounds require more training for listeners.

Designing the Sound

Turning now to practice, once the information to be conveyed by sound has been analyzed, the designer should begin the process of selecting and/or developing appropriate sounds. A useful perspective on the design problem at this point is to think of the auditory content of an interface as a kind of sound ecology (Walker & Kramer, 2004). Ideally, the interface should be compelling, inventive, and coherent—it should tell a kind of story—and the sounds it employs should have a collective identity that listeners will have little difficulty recognizing, in much the same way that people effortlessly recognize familiar voices, music, and the characteristic sounds of their daily environments. Good auditory design practice involves critical listening (to both the users of the sounds and the sounds themselves!) and strives above all to accommodate the aural skills, expectations, and sensibilities that listeners ordinarily possess. It is easier than many people might think to create an auditory interface that is unintentionally tiresome or internally inconsistent or that requires extensive training or special listening skills.

Once some initial thought has been given to the organization and character of the listening environment, the first component of the auditory design process is to work out how sound will be used to convey the task-related information that is identified in the task analysis. Often, it is also useful to begin developing candidate sounds for the interface at this time, because this can help to crystallize ideas about the design; however, this may not always be possible. The mapping from information to sound should, in many cases, be relatively straightforward, but in other cases—for instance, with complex data relations—it will generally be necessary to experiment with a number of ideas. Several examples follow:

+ Event onsets intuitively map to sound onsets.
+ Level of priority or urgency can be represented systematically with a variety of parameters, including rhythm, tempo, pitch, and harmonic complexity (e.g., Guillaume, Pellieux, Chastres, & Drake, 2003).
+ Drawing attention to, or indexing, a specific location in space—a form of deixis (Ballas, 1994)—can be accomplished with three-dimensional audio-rendering techniques.
+ Emotional context can be conveyed with music or musical idioms.
+ Distinct subclasses of information can be mapped to different timbres; ranges can be mapped to linearly varying parameters.
+ Periodicity can be mapped to rhythm.

Many more examples could be given. Often, there will be more than one dimension to convey about a particular piece of information and in such instances auditory parameters are frequently combined. An auditory alert, for example, can pack onset, event identity, location, level(s) of urgency, duration, and confirmation of response into a single instance of sound (Brock, Ballas, Stroup, & McClimens, 2004).

Producing the Sound

As mappings and candidate sounds for the interface are developed, another factor the auditory designer must address is how the final sounds will be produced, processed, and rendered. Although an introduction to the technology of sound production was given in Section 5.2, the emphasis there was primarily on computational techniques for the synthesis of sound. Other means of sound production include live sources and playback of recorded and/or edited material. In addition, many auditory applications require sounds to be localized for the listener, usually with binaural filtering or some form of loudspeaker panning. And some tasks allow or require the user to control, manipulate, assign, or choose a portion or all of its auditory content.

Consequently, vetting an auditory design to ensure that its display implementation will function as intended can range from assembling a fixed set of audio files for a modest desktop application to specifying a set of audio sources and processing requirements that can have substantial implications for an application's supporting computational architecture. In the latter situation, one would reasonably expect to be part of a collaborative project involving a number of specialists and possibly other designers.

Choosing between one construct and another in rich application domains and knowing what is necessary to or most likely to meet the user's needs is not always a matter of just knowing or going to the literature. All of these advanced considerations, however—the production, filtering, augmentation, timing, and mixing of various types and sources of sound—are properly part of the auditory design and should be identified as early as possible in a complex design project because of their implications for the subsequent implementation and evaluation phases of the auditory interface.

5.4.6   Iterative Evaluation

The final and indispensable component of the auditory design process is formative evaluation via user testing (Hix & Hartson, 1993). Targeted listening studies with candidate sounds, contextual mock-ups, or prototypes of the auditory interface, designed to demonstrate or refute the efficacy of the design or its constituent parts, should be carried out to inform and refine iterative design activities. For more on evaluation, see Section 5.5.

5.5   TECHNIQUES FOR TESTING THE INTERFACE

As with every interactive system, the evaluation of auditory displays should ensure a high degree of usability. For auditory displays, finding methods to evaluate usability is not a trivial task. The following sections highlight some of the specific approaches and issues relevant to the evaluation of auditory displays.

5.5.1   Issues Specific to Evaluation of Auditory Displays

From prototyping to data capture, a number of issues unique to displays that use sound need to be considered when evaluating an auditory display.

Early Prototyping
There are few generally available tools for the early prototyping of concepts in auditory displays, but the desirability of obtaining early feedback on auditory interface

designs is, if anything, even more important than on prototyping visual displays because many users are relatively unfamiliar with the use of audio. Wizard of Oz techniques (Dix, Finlay, Abowd, & Beale, 2004) and the use of libraries of sounds available on the Internet (FindSounds, 2006) can provide the basis of ways around this dilemma. The quality and amplitude of the sounds employed in prototypes must be close to those anticipated in the final system in order to draw conclusions about how well they are likely to work in context.

For instance, research by Ballas (1993) shows that the way an individual sound is interpreted is affected by the sounds heard before and after it, so accurate simulation of the pace of the interaction is also important. One early-stage technique is to vocalize what the interface is expected to sound like; for example, Hermann et al. (2006) used such a technique to develop sonifications of EEG data. By making it possible for people to reproduce the sonifications, users were able to more easily discuss what they heard and these discussions facilitated the iterative testing process.

Context of Use

Evaluation of the display in the environment where the system will be used and in the context of users performing normal tasks is particularly important for auditory displays. The primary factors associated with context are privacy and ambient noise. For instance, although the task analysis (described in Section 5.4.1) may have determined that privacy is necessary and thus that headphones are the best delivery method for the sounds, an evaluation of the display in the environment may determine that wearing headphones interferes with the users' task. Conversely, if privacy is not an issue and speakers are being used for the auditory display, an evaluation of the display in the context where it will be used could determine the appropriate placement and power of speakers.

Cognitive Load

If an auditory display is being used to reduce cognitive load, the evaluation process should confirm that load reduction occurs. One way to measure cognitive load is Hart and Staveland's (1988) NASA Task Load Index (TLX). If this measure is not sensitive enough for the tasks associated with the auditory display, it may be better to use accuracy and/or time performance measures as indirect measures of whether the display has reduced the cognitive load.

Choice of Participants

When conducting evaluations of auditory displays, as with all evaluations of all types of displays, the characteristics of the participants should match those of the intended users as closely as possible. Some of the obvious variables that should be considered are gender, age, and experience with information technology. Furthermore, evaluators also need to match participants on the specific variables associated with audition (listed in Section 5.4.1).

When conducting evaluations, there are also dangers of making false assumptions concerning the applicability of evaluation data across different user types. For example, in the development of systems for visually impaired users, it is not unusual for sighted users who have had their view of the display obscured in some way to be involved in the evaluation. However, in a study involving judgments concerning the realism of sounds and sound mappings, Petrie and Morley (1998) concluded that the findings from sighted participants imagining themselves to be blind could not be used as a substitute for data from participants who were actually blind.

Finally, evaluators need to be particularly diligent about determining whether participants have any hearing loss. Obviously, hearing losses are likely to impact participants' interactions with the system and thus should be taken into account.

Data Capture

As might be expected, problems can arise with the use of think-aloud protocols for capturing the results of formative evaluations of auditory displays. Participants are likely to experience problems when asked to articulate their thoughts while at the same time trying to listen to the next audio response from the interface. This is not to rule out the use of think-aloud protocols altogether. For example, Walker and Stamper (2005) describe, in the evaluation of Mobile Audio Designs (MAD) Monkey, an audio-augmented reality designer's tool, that there may be situations where the audio output is intermittent and allows sufficient time for participants to articulate their thoughts in between audio output from the system. Another alternative would be to use what is commonly known as a retrospective think-aloud protocol, in which participants describe the thoughts they had when using the system to the evaluator while reviewing recordings of the evaluation session.

Learning Effects

Improvement in performance over time is likely to be important in most systems, but there is currently little known about learning effects observed in users of auditory displays, other than the fact that they are present and need to be accounted for. In experiments conducted by several of the authors, significant learning effects have frequently been seen early in the use of auditory displays as users transition from never having used a computer-based auditory display to gaining some familiarity in reacting to the display. For instance, a study by Walker and Lindsay (2004) of a wearable system for audio-based navigation concluded that "practice has a major effect on performance, which is not surprising, given that none of the participants had experienced an auditory way-finding system before. Thus it is critical to examine performance longitudinally when evaluating auditory display designs."

Heuristic Evaluations

It is one of the great advantages of sound that the auditory cues employed can be designed to be background noises; hence, auditory displays are often used as ambient displays. Mankoff and colleagues (2003) developed heuristics for revealing usability issues in such ambient displays. Heuristic evaluation of user interfaces is a popular method because it comes at very low cost. For example, Nielsen and Molich (1990) found that a panel of three to five novice evaluators could identify 40 to 60 percent of known issues when applying heuristic evaluation. However, doubts have been expressed about the results of some studies investigating its effectiveness and some usability professionals argue that heuristic evaluation is a poor predictor of actual user experience (e.g., see *http://www.usabilitynews.com/news/article2477.asp*).

5.5.2   Example of a Cross-Modal Collaborative Display: Towers of Hanoi

One common concept in accessibility is that, given a collaborative situation between sighted and visually impaired users, the difference in interaction devices can cause problems with the interaction. Winberg and Bowers (2004) developed a Towers of Hanoi game with both a graphical and an audio interface to investigate collaboration between sighted and nonsighted workers. To eliminate any problems associated with having different devices for sighted and blind users, both interactions were mouse based, employing a focus feature to enable the mouse to track the cursor. The sighted player worked with a screen, and the blind one had headphones. In order to encourage collaboration, there was only one cursor for the two mice.

Testing Setup

To keep things equal, neither player had access to the other's interface. The sighted player had a screen and could see the blind participant but not his or her mouse movement; both could hear each other, as this was necessary for collaboration. Each player was trained independently and had no knowledge of the other's interface.

Evaluation

The entire interaction between the two players was videotaped and the game window was recorded. The video enabled Winberg and Bowers (2004) to study the entire interaction, and the screen capture allowed them to see the state of the game at all times. The players played three games with three, four, and five disks, respectively.

Analysis

In this experiment on cross-modal collaboration, Winberg and Bowers (2004) studied the following aspects of the interaction: turn taking, listening while moving, monitoring the other's move, turn-taking problems and repair, reorientation and re-establishing sense, engagement, memory and talk, and disengagement. The major method used to evaluate the interaction was conversation analysis (tenHave, 1999). The transcription and subsequent study of the conversation paired with the players' actions provided an in-depth qualitative analysis of problems in the interaction. Any problems with the interfaces became apparent from stumbling and confusion in the conversation. Actions were also timed, and this helped to pinpoint problems with the direct manipulation in the auditory interface.

Conclusions

The system developed and evaluated by Winberg and Bowers (2004) enabled the examination of some basic issues concerning the cooperation between people of different physical abilities supported by interfaces in different modalities. They concluded that sonic interfaces could be designed to enable blind participants to collaborate on the shared game: In their evaluation, all pairs completed all games. The auditory interface enabled blind players to smoothly interleave their talk and interactions with the interface. The principle of continuous presentation of interface elements employed in the game allowed blind players to monitor the state of the game in response to moves as they were made. This enabled the blind player to participate fully in the working division of labor. Both blind and sighted collaborators therefore had resources to monitor each other's conduct and help each other out if required. However, problems were seen when the blind players stopped manipulating the display and listening to the consequent changes. In these situations the state of the game became unclear to the blind players and difficulties were experienced in reestablishing their understanding of the current state of the game.

   Important findings resulting from the study by Winberg and Bowers (2004) can be summarized as follows:

1. The manipulability of an assistive interface is critical, not only for the purpose of completing tasks but also to enable a cross-modal understanding of the system state to be established. This understanding can become compromised if the linkage among gesture, sound, and system state becomes unreliable.

2. When deciding whether to implement functionality in sound, the availability of other channels of communication and the appropriateness of audio for representing the function should be kept in mind. For example, there could be situations where the sonification of an interface artifact may simply take too long and may be better replaced by talk between collaborators.

3. It is not enough to design an assistive auditory interface so that it facilitates the development of the same mental model as the interface used by sighted individuals. Additionally, it is essential to examine how the assistive interface will be used and how this usage is integrated with the various things that participants do, such as "designing gestures, monitoring each other, establishing the state of things and one's orientation in it, reasoning and

describing" (Winberg and Bowers, 2004). In order to do this effectively, it becomes essential to study how people use assistive interfaces in collaborative situations.

5.6   DESIGN GUIDELINES

Although the potential of audio as an interaction modality in HCI is high and many applications have shown this (e.g., Brewster, 2002), the efficient design of audio remains something of a mysterious process and guidance is often scarce. Hence, in this section we attempt to describe existing guidelines, principles, and design theories.

5.6.1   Analysis and Requirement Specification

In auditory design, some aspects of the requirement specifications demand special attention. As mentioned in Section 5.4.1, it is important to have a clear understanding of the users' listening background and abilities. In addition to the issues listed in Section 5.4.1, designers should consider users' openness to an alternative display modality. Audio as part of human–technology interaction is a comparatively new field, and thus users have little experience with using it. This also means that designers may encounter skepticism and prejudice against using audio, not only from users but from all stakeholders in the design process. Although there might be strong arguments for using audio in a specific application, a client might still request a visual solution because he or she cannot imagine an auditory solution.

A valuable concept for auditory information design in this early phase was proposed by Barrass (1997) and is called "TaDa" analysis. It is a method for describing the task and the data to be represented in a formal way including a story about the usage and properties that are indicators for auditory cues like attention levels or data types. Barrass (1997) used these TaDa descriptions as a starting point for the selection of sounds, and then employed these descriptions to match them with sounds stored in a database (EarBender).

The EarBender database contains a large number of sounds tagged with semantic and other properties that can be matched with the requirements from a TaDa analysis (Barrass, 1997). The author also proposed the creation of auditory design principles based on principles for generic information design, such as directness or the level of organization. Barrass (1997) linked these with the properties of auditory perception to create auditory design principles.

The TaDa technique has been used by a number of auditory designers. Notably, at the Science by Ear workshop that took place at the Institute of Electronic Music (IEM) in Graz in 2006, the technique was used to formalize requirements for a number of case studies for which multidisciplinary teams were formed to design data sonifications. The case studies included data drawn from particle physics, electrical power systems, EEG data, global social data, and rainfall data. The TaDa technique proved helpful in providing a standard format for representing the information requirements of each sonification to be designed for use by the multidisciplinary teams. Examples of the case studies employed at the workshop can be found in the papers by de Campo (2007) and others presented at the International Conference on Auditory Displays (ICAD, 2007).

Of course, other methods and guidelines can and should be applied at the requirement specification stage of the design. Task analysis, user scenarios, personae, and other concepts have been successfully applied to visual design and do not involve the need to specify any interaction modality. Examples where these concepts have been applied in auditory designs are the use of rich user scenarios in the design of an auditory web browser (Pirhonen, Murphy, McAllister, & Yu, 2006) and the first stage in the design methodology proposed by Mitsopoulos (2000). Both approaches are elaborated in the next section.

5.6.2   Concept Design

Concept design is when high-level design decisions are made while leaving most details still unspecified. This phase links the design problem with concepts of auditory displays. The first and foremost task in this phase is to decide which parts of the user interface audio will be used and which auditory concepts match the requirements and constraints defined in the requirements phase.

Brewster (1994) addressed this issue in a bottom-up approach: Find errors in individual parts of an existing interface and try to fix them by the addition of sound. He adopted the event and status analysis and extended it to be applicable to different interaction modalities (i.e., to accommodate audio). This was an engineering approach to reveal information hidden in an interface that could cause errors. Brewster (1994) suggested using sound to make this information accessible and linked the output of the analysis to his guidelines for the creation of earcons.

As mentioned in the previous section, Pirhonen et al. (2006) proposed a design method that linked user tasks and auditory cues through rich use case scenarios. The use case was developed with a virtual persona that represented the target group and told the story of how this persona carried out a specific task. The story was enriched with as much detail about the environment and the background of the user as possible to create a compelling scenario; the authors proposed that "the use scenario should have qualities that enable the interpreter (to) identify him/herself with the character" (Pirhonen et al., 2006:136). Then a panel of four to five designers went through this scenario and tried to produce descriptions of sounds to support the task. After creating the sounds as suggested by the designers, another panel was organized and went through the use case scenario that was enriched by the initial sound designs. This procedure was iterated until a working design was identified. The method stressed the importance of linking user tasks with the design, but also relied heavily on the availability of expert designers for a panel and on their experience and ideas.

Another tool for concept design is design patterns (Frauenberger, Holdrich & de Campo, 2004). However, there are not yet enough successful implementations or methodological frameworks for auditory displays that incorporate design patterns in the auditory design process. Nevertheless, this tool most likely will prove beneficial in the future.

Another, more theoretical approach has been proposed by Mitsopoulos (2000), who founded his methodology on a framework for dialog design. Mitsopoulos's methodology consists of three levels: (1) the concept level in which the "content" of the interface is specified in terms of semantic entities, (2) the structural level in which sounds are structured over time, and (3) the implementation level in which the physical features of the sound are determined. Mitsopoulos (2000) proposed guidelines for each of these levels that are derived from theories of auditory perception and attention (e.g., Arons, 1992; Bregman, 1990). By applying these theories, Mitsopoulos intended to narrow the design space by eliminating designs that would violate psychological principles. Notably he argued for two fundamental modes of presentation of information by audio: fast presentation, that is, "at a glance," and the interactive presentation for more detailed user interaction. Each representation is defined in all three levels. Although his methodology and guidelines are properly founded in theory, it is important to note that the approach requires a steep learning curve.

In general, decisions in the concept design phase are crucial for successful auditory design, but there is little guidance available that may help novice designers. It is important to note that most flaws in auditory designs derive from decisions made during the concept phase as novices tend to be overly influenced by visual thinking. Good auditory design emphasizes the characteristics and strengths of audio and adopts visual concepts only if there is evidence that they work in the auditory domain.

5.6.3   Detailed Design

Many specifications resulting from the previous design stage describe the sounds vaguely or describe only certain properties. In the detailed design stage, these specifications are mapped onto physical properties of sound.

Auditory Display: Sonification, Audification and Auditory Interfaces, edited by Kramer (1994), is often seen as a landmark publication in auditory display design. The volume contains the proceedings of the first meeting of the ICAD in 1992, and includes a companion CD of audio examples illustrating many of the psychological phenomena, techniques, and applications discussed. Several chapters in the book present principles for use in representing information in audio. The book presents a number of methods for associating perceptual issues in auditory display with techniques for their practical implementation. This work introduces fundamental sonification techniques such as the direct representation of data in sound (or audification), as well as a number of approaches to mapping data variables into a range of sound parameters such as pitch, loudness, timbre, and tempo. The book also provides an overview of many other relevant issues in auditory display design such as the following:

✦   As would be expected, concurrency is an issue in auditory display design. Clearly there is a limit to how much auditory information human beings can perceive and process concurrently. Nevertheless, concurrency is potentially a powerful tool in auditory display design, as evidenced by the ease with which even untrained musicians can detect a single instrument playing out of tune in an entire orchestra of players.

✦   Metaphor, as in the rest of user interface design, can be an effective mechanism for developing and supporting the user's mental model of the system. See the use of the group conversation metaphor to support multitasking described in the clique case study *(www.beyondthegui.com)* as a particularly effective example of this.

Much of the book focuses on applications involving the design of sonifications of complex data, that is, applications representing either raw data or information to be presented in sound. Additionally, there is a good deal of valuable guidance in the book for those involved in the design of more symbolic auditory interface elements.

A contributor to *Auditory Display*, Gaver (1994) provides a clear user interface focus. He presents techniques to create auditory icons for user interfaces in computing systems. As mentioned previously, auditory icons are based on our everyday hearing experience; thus, familiarity and inherent meaning make them highly efficient auditory cues. Hence, when creating auditory icons, they are not described in the usual dimensions of sound like pitch or timbre, but rather according to properties of the real-world object that causes the sound. With regard to detail design, auditory icons can be parameterized by dimensions such as material, size, or force, and when synthesizing auditory icons, designers seek to use algorithms that allow them to influence these parameters instead of the physical properties of the sound directly (e.g., pitch, loudness, etc.). Gaver (1994) provides a wide range of such algorithms for impact sounds, breaking, bouncing, and spilling effects, scraping, and other machine sounds.

Blattner et al. developed guidelines for constructing earcons based on visual icons (Blattner, Sumikawa, & Greenberg, 1989). In these authors' terminology, representational earcons are similar to auditory icons and are built on metaphors and inherent meaning. For abstract earcons, they use musical motifs (a brief succession of ideally not more than four tones) as a starting point, and define rhythm and pitch as the fixed parameters. Timbre, register, and dynamics are the variable parameters of the motif. By systematically altering the fixed parameters, designers could create distinctive earcons, while altering the variable parameters would produce earcons with perceivable similarity and may be used to create related families of earcons. In their guidelines, Blattner et al. (1989) suggest choosing the tones according to the cultural background of the target group (e.g., Western tonal music), and they elaborate on exploiting hierarchical structures in earcon families for better learnability. Such compound earcons can be created through combination, transformation, and inheritance of single-element earcons.

In his work on guidelines for creating earcons, Brewster (1994) refines the guidelines mentioned previously and provides more specific guidance regarding rhythm, timbre, pitch, among other properties. Key guidelines given by Brewster follow:

✦   Use musical instrument timbres to differentiate between earcons or groups of earcons, as people can recognize and differentiate between timbres relatively easily.
✦   Do not use pitch or register alone to differentiate between earcons when users need to make absolute judgments concerning what the earcon is representing.
✦   If register must be used on its own, there should be a difference of two or three octaves between earcons.
✦   If pitch is used, it should not be lower than 125 Hz and not higher than 5 kHz to avoid the masking of the earcon by other sounds, and must be easily within the hearing range of most users.
✦   If using rhythm to distinguish between earcons, make the rhythms as different from each other as possible by putting different numbers of notes in each earcon.
✦   Intensity (loudness) should not be used to distinguish between earcons, as many users find this annoying.
✦   Keep earcons short in order not to slow down the user's interaction with the system.
✦   Two earcons may be played at the same time to speed up the interaction.

Brewster (1994) also investigated the concurrent use of earcons, and McGookin and Brewster (2006b) summarized some of the issues with using concurrent audio presentations in auditory displays. Lumsden and colleagues provided guidelines for a more specific scenario, that is, the enhancement of GUI widgets, such as buttons, by earcons (Lumsden, Brewster, Crease, & Gray, 2002). Although a detailed description of the design guidelines of these additional considerations for earcons is beyond the scope of this chapter, the studies by Brewster with McGookin and Lumsden et al. are good resources for the earcon designer.

For more on auditory information and interface design, a number of excellent resources can easily be found on the Web, including some mentioned in this section. De Campo (2007) presents a useful design space map for data sonification, and references numerous examples that are available at the SonEnvir project website (*http://sonenvir.at*). A new online edition of *The Handbook for Acoustic Ecology* (Truax, 1999) provides an invaluable glossary for acoustic concepts and terminology, as well as hyperlinks to relevant sound examples. Additionally, a wealth of research papers and other resources for auditory design as well as an active online design community can be found at the ICAD website.

5.7 CASE STUDIES

Two case studies of auditory interfaces can be found at *www.beyondthegui.com*.

5.8 FUTURE TRENDS

The material presented in this chapter illustrates that auditory interfaces are a versatile and efficient means for communicating information between computer systems and users. Listening is perhaps the communication channel with the highest bandwidth after vision, and certainly a channel whose characteristics are so different from visual perception that the combination of visual and auditory displays in particular covers a very wide range of interface possibilities. Whereas vision is a focused sense (we only see what we look at), sound surrounds the user; while vision stops at surfaces, sound allows us to discover "inner" worlds, beyond visible limits; whereas vision is persistent, sound is intrinsically aligned to changes over time and dynamic interaction, and our auditory skills are particularly good at discerning these properties in sound.

Looking at auditory interfaces from a more removed perspective, we see that two directions are possible: sound as a display or as an input modality. This chapter focused on the display mode. However, the input mode can be equally compelling. For example, an interface could employ nonspeech sounds and vocalizations as auditory inputs. This has potential for the development of auditory interactions involving sonification. A user could analogically specify or select the tempo of a sonification by tapping or uttering a rhythm into a microphone. Similarly, musical idioms could be queried and/or selected by humming. Another avenue for nonsymbolic, vocal auditory input in sonification systems is already being explored. Hermann et al. (2006) describes the vocal sonification of EEGs, inspired by the exceptional human capacity for discerning and memorizing structure in vocal sounds. Because humans are easily able to mimic vocal patterns, they can use vocalizations to actively reference certain patterns in the sonification. This process may simplify the communication of data patterns.

Auditory interfaces involving analogic and symbolic input and output are likely to gain relevance in future interaction design for several reasons. First, the technological development is just starting to enable these interactions at a sufficient level of sophistication. Second, there are many situations where the visual sense is not available or otherwise used. Finally, we are simply wasting an excellent and highly developed communication channel if sound is neglected in the user interface.

Additional future trends that may develop in auditory displays include (1) interactive sonification—a better closure of interaction loops by interactive sonification, and (2) an amalgamation of auditory display with other modalities such as visual display and tactile/haptic interfaces that would result in truly multimodal interfaces.

Interactive sonification (Hermann & Hunt, 2005) bears the potential to create intuitive control of systems at a level beyond a rational (logic) analysis of steps, more as an intuitive, creative, and synergetic approach to solving problems. For instance, in data analysis via interactive sonification, discovering patterns would turn from a step-by-step analysis into a continuous movement through the data space or sonification space. The user would integrate any locally gained insight regarding the structure of the data into his or her exploratory activity in a continuous way, without disrupting the activity and experience. Such a continuous, interruption-free mode may better create the experience of *flow*, the dissolving of a user in the activity, which in turn may give a better access to the user's often covered creative potential.

Multimodal interfaces, on the other hand, will allow the designer to combine the strengths of various modalities, so that the communication of data is simplified and achieves a better match with the user's perception capabilities. For instance, if the user's visual focus is already highly used, multimodal display engines can automatically select auditory components, or even just emphasize them against the visual counterpart, so that the communication between the user and a computer system is optimized.

Another trend in auditory interfaces that is gaining momentum is the concept of intelligent or adaptive auditory environments and displays. Advances in rendering, signal processing, user modeling, machine listening (Wang & Brown, 2006), and noninvasive user and contextual monitoring technologies mean that in the relatively near future many interactive devices and environments will transparently adapt the audio component of their information displays to match the needs of users, much like people rely on each other's listening skills in social settings to coordinate the aural dimension of conversation and other shared forms of sound information. Three areas in which adaptive sound technology is already being explored are mobile telephony, pervasive computing, and social robotics.

Mobile phones have arguably become the most common auditory interface people encounter in their day-to-day lives. To compensate for noise in dynamic environments, new wireless headsets are already being marketed that adaptively alter a mobile phone's outgoing and incoming audio signals to improve speech communications (e.g., see Mossberg, 2006).

As mobile phones move beyond telephony into areas as diverse as Internet access, personal entertainment, content creation, and interaction with so-called smart and pervasive computing environments, exciting new opportunities for intelligent auditory presentation behaviors arise. In recent pervasive-computing research, for instance, users intuitively navigated their way to undisclosed outdoor locations using a context-dependent, directionally adaptive auditory display (Etter & Specht, 2005). The underlying system uses global positioning data and a geographical information system to infer the mobile user's geographical context. Navigation cues are then rendered by adaptively panning and filtering music selected by the user to correspond with his or her direction of travel.

Intelligent, adaptive auditory displays are also expected to be an important technology for social and service robots. Recent human–robot interaction work by Martinson and Brock (2007) explores several strategies for adaptively improving a robot's presentation of auditory information for users, including user tracking, ambient noise level monitoring, and mapping of auditory environments.

In summary, auditory interfaces are a rapidly evolving mode for human–computer interaction, with a huge potential for better use of people's communicative skills and perceptual resources.

Acknowledgments

We want to acknowledge the primary sources of contribution for each section. We also wish to thank the reviewers for their insights and feedback on the chapter.

Introduction: S. Camille Peres

Section 5.1: Virginia Best, Barbara Shinn-Cunningham, and John Neuhoff

Section 5.2: Thomas Hermann

Sections 5.3, 5.5, 5.6, and 5.7: Tony Stockman, Louise Valge∂ur Nickerson, and Christopher Frauenberger

Section 5.4: Derek Brock and S. Camille Peres

Section 5.8: Thomas Hermann and Derek Brock

REFERENCES

Andresen, G. (2002). Playing by ear: Creating blind-accessible games. Gamasutra—available at *http://www.gamasutra.com/resource_guide/20020520/andresen_01.htm*.

Arons, B. (1992). A review of the cocktail party effect. *Journal of the American Voice I/O Society* 12:35–50.

Audyssey Magazine. (1996–2006). See *http://www.audiogames.net/page.php?pagefile=audyssey*.

Baier, G., & Hermann, T. (2004). The sonification of rhythms in human electroencephalo-gram. *Proceedings of International Conference on Auditory Display*, Sydney.

Baier, G., Hermann, T., Sahle, S., & Ritter, H. (2006). Sonified epileptic rhythms. *Proceedings of International Conference on Auditory Display*, London.

Ballas, J. A. (1993). Common factors in the identification of an assortment of brief everyday sounds. *Journal of Experimental Psychology: Human Perception and Performance* 19:250–67.

Ballas, J. A. (1994). Delivery of information through sound. In Kramer, G., ed., *Auditory Display: Sonification, Audification and Auditory Interfaces*. Reading, MA: Addison-Wesley, 79–94.

Barras, S. (1997). Auditory information design. PhD diss., Australian National University.

Bay Advanced Technologies. (2006). BAT K-Sonar—available at *http://www.batforblind.co.nz/*.

Berman, L., & Gallagher, K. (2006). Listening to program slices. *Proceedings of International Conference on Auditory Display*, London.

Best, V., Gallun, F. J., Carlile, S., & Shinn-Cunningham, B. G. (2007). Binaural interference and auditory grouping. *Journal of the Acoustical Society of America* 121:1070–76.

Blattner, M. M., Sumikawa, D. A., & Greenberg, R. M. (1989). Earcons and icons: Their structure and common design principles. *Human–Computer Interaction* 4(1):11–44.

Blattner, M., Papp, A., & Glinert, E. P. (1994). Sonic enhancement of two-dimensional graphics displays. In Kramer, G., ed., *Auditory Display: Sonification, Audification, and Auditory Interfaces*. Reading, MA: Addison-Wesley, 447–70.

Bly, S. (1982). Presenting information in sound. *Proceedings of SIGCHI Conference on Human Factors in Computing Systems*.

Bregman, A. S. (1990). *Auditory Scene Analysis: The Perceptual Organization of Sound*. Cambridge, MA: MIT Press.

Brewster, S. A. (1994). Providing a structured method for integrating non-speech audio into human–computer interfaces. PhD diss., University of York. Available at *http://www.dcs.gla.ac.uk/~stephen/papers/Brewster_thesis.pdf*.

Brewster, S. A. (2002). Nonspeech auditory output. In *The Human–Computer Interaction Handbook: Fundamentals, Evolving Technologies and Emerging Applications Archive*. Mahwah, NJ: Erlbaum, 220–39.

Broadbent, D. E. (1958). *Perception and Communication*. New York: Oxford University Press.

Brock, D., Ballas, J. A., Stroup, J. L., & McClimens, B. (2004). The design of mixed-use, virtual auditory displays: Recent findings with a dual-task paradigm. *Proceedings of International Conference on Auditory Display*. Sydney.

Bronkhorst, A. W., & Houtgast, T. (1999). Auditory distance perception in rooms. *Nature* 397:517–20.

Bronkhorst, A.W. (2000). The cocktail party phenomenon: A review of research on speech intelligibility in multiple-talker conditions. *Acustica* 86:117–28.

Brown, M. L., Newsome, S. L., & Glinert, E. P. (1989). An experiment into the use of auditory cues to reduce visual workload. *Proceedings of SIGCHI Conference on Human Factors in Computing Systems*.

Calvert, G. A., Spence, C., & Stein, B. E., eds. (2004). *The Handbook of Multisensory Processes*. Cambridge, MA: MIT Press.

Carlile, S. (1996). The physical and psychophysical basis of sound localization. In Carlile, S., ed., *Virtual Auditory Space: Generation and Applications*. Austin, TX: RG Landes; New York: Chapman & Hall (distributor).

Clark, H. H. (1996). *Using Language*. Cambridge: Cambridge University Press.

Cook, P. R. (2002). *Real Sound Synthesis for Interactive Applications*. Wellesley, MA: A. K. Peters Ltd.

Darwin, C. J., & Hukin, R. W. (2000). Effectiveness of spatial cues, prosody and talker characteristics in selective attention. *Journal of the Acoustical Society of America* 107:970–77.

de Campo, A. (2007). Toward a sonification design space map. *Proceedings of International Conference on Auditory Display*, Montreal.

Deutsch, D., ed. (1999). *The Psychology of Music*, 2nd ed. San Diego: Academic Press.

Dix, A., Finlay, J., Abowd, G. D., & Beale, R. (2004). *Human–Computer Interaction*, 3rd ed. London: Prentice Hall Europe.

Etter, R., & Specht, M. (2005). Melodious walkabout: Implicit navigation with contextualized personal audio contents. *Adjunct Proceedings of Third International Conference on Pervasive Computing*, Munich.

Farkas, W. (2006). iSIC—Data Modeling through Music—available at *http://www.vagueterrain. net/content/archives/journal03/farkas01.html*.

Fayyad, U. M., Piatetsky-Shapiro, G., Smyth, P., & Uthurusamy, R. (1996). *Advances in Knowledge Discovery and Data Mining*. Cambridge, MA: AAAI/MIT Press.

FindSounds. (2006). Search the Web for Sounds. Home page—available at *http://www. findsounds.com/*.

Fitch, W. T., & Kramer, G. (1994). Sonifying the body electric: Superiority of an auditory display over a visual display in a complex, multivariate system. In Kramer, G., ed., *Auditory Display: Sonification, Audification and Auditory Interfaces*. Reading, MA: Addison-Wesley, 307–26.

Flowers, J. H., Whitwer, L. E., Grafel, D. C., & Kotan, C. A. (2001). Sonification of daily weather records: Issues of perception, attention and memory in design choices. *Proceedings of International Conference on Auditory Display*, Espoo, Finland.

Frauenberger, C., Höldrich, R., & de Campo, A. (2004). A generic, semantically based design approach for spatial auditory computer displays. *Proceedings of International Conference on Auditory Display*, Sydney.

Freedom Scientific. (2005). Jaws R for Windows—available at *http://www.freedomscientific. com/fs_products/JAWS_HQ.asp*.

Gaver, W. W. (1994). Using and creating auditory icons. In Kramer, G., ed., *Auditory Display: Sonification, Audification and Auditory Interfaces*. Reading, MA: Addison-Wesley, 417–46.

Goffin, V., Allauzen, C., Bocchieri, E., Hakkani-Tür, D., Ljolje, A., Parthasarathy, S., Rahim, M., Riccardi, G., & Saraclar, M. (2005). The AT&T WATSON speech recognizer. *Proceedings of IEEE International Conference on Acoustics, Speech and Signal Processing*, Philadelphia.

Guillaume, A., Pellieux, L., Chastres, V., & Drake, C. (2003). Judging the urgency of nonvocal auditory warning signals: Perceptual and cognitive processes. *Journal of Experimental Psychology* 9(3):196–212.

GW Micro. (2006). Window-Eyes—available at *http://www.gwmicro.com/Window-Eyes/*.

Habrias, H., and Frappier, M. (2006). *Software Specification Methods: An Overview Using a Case Study*. London: ISTE Publishing.

Hart, S. G., and Staveland, L. E. (1988). Development of NASA TLX (Task Load Index): Results of empirical and theoretical research. In Hancock, P. A., & Meshkati, N., eds., *Human Mental Workload*. Amsterdam: North-Holland, 139–83.

Hayward, C. (1994). Listening to the earth sing. In Kramer, G., ed., *Auditory Display: Sonification, Audification and Auditory Interfaces*. Reading, MA: Addison-Wesley, 369–404.

Hermann, T. (2002). Sonification for exploratory data analysis. PhD diss., Bielefeld University.

Hermann, T., & Hunt, A. (2005). An introduction to interactive sonification. *IEEE Multimedia* 12(2):20–24.

Hermann, T., Baier, G., Stephani, U., & Ritter, H. (2006). Vocal sonification of pathologic EEG features. *Proceedings of International Conference on Auditory Display*, London.

Hermann, T., Drees, J. M., & Ritter, H. (2005). Broadcasting auditory weather reports—a pilot project. *Proceedings of International Conference on Auditory Display*, Boston.

Hix, D., & Hartson, H. R. (1993). Formative evaluation: Ensuring usability in user interfaces. In Bass, L., & Dewan, P., eds., *User Interface Software*. Chichester, England; New York: Wiley.

Hunt, A., & Hermann, T. (2004). The importance of interaction in sonification. *Proceedings of International Conference on Auditory Display*, Sydney.

International Community for Auditory Display (ICAD). (2007). Home page—*http://www.icad.org/*.

Janata, P., & Childs, E. (2004). Marketbuzz: Sonification of real-time financial data. *Proceedings of International Conference on Auditory Display*, Sydney.

Jones, D. G., and Endsley, M. R. (2000). Overcoming representational errors in complex environments. *Human Factors* 42(3):367–78.

Kidd Jr., G., Mason, C. R., & Arbogast, T. L. (2002). Similarity, uncertainty, and masking in the identification of nonspeech auditory patterns. *Journal of the Acoustical Society of America* 111(3):1367–76.

Kramer, G., ed. (1994). *Auditory Display: Sonification, Audification and Auditory Interfaces*. Reading, MA: Addison-Wesley.

Leplatre, G., & McGregor, I. (2004). How to tackle auditory interface aesthetics? Discussion and case study. *Proceedings of International Conference on Auditory Display*, Sydney.

Lieder, C. N. (2004). *Digital Audio Workstation*. New York: McGraw-Hill.

Lumsden, J., Brewster, S. A., Crease, M., & Gray, P. D. (2002). Guidelines for audio-enhancement of graphical user interface widgets. *Proceedings of People and Computers XVI—Memorable Yet Invisible: Human Computer Interaction*, London.

Mankoff, J., Dey, A. K., Hsieh, G., Kientz, J., Ames, M., & Lederer, S. (2003). Heuristic evaluation of ambient displays. *Proceedings of SIGCHI Conference on Human Factors in Computing Systems*, Fort Lauderdale.

Martinson, E., & Brock, D. (2007). Improving human–robot interaction through adaptation to the auditory scene. *Proceedings of International Conference on Human–Robot Interaction*, Washington, DC.

Massof, R. W. (2003). Auditory assistive devices for the blind. *Proceedings of International Conference on Auditory Display*, Boston.

McAdams, S., & Bigand, E., eds. (1993). *Thinking in Sound: The Cognitive Psychology of Human Audition*. Oxford; New York: Oxford University Press.

McCartney, J. (1996). SuperCollider Hub—available at *http://supercollider.sourceforge. net*.

McFarlane, D. C., & Latorella, K. A. (2002). The scope and importance of human interruption in HCI design. *Human–Computer Interaction* 17(1):1–61.

McGookin, D. K., & Brewster, S. A. (2006a). Haptic and audio interaction design. First International Workshop, HAID 2006, Glasgow. *Proceedings (Lecture Notes in Computer Science)*. New York: Springer-Verlag.

McGookin, D. K., & Brewster, S. A. (2006b). Advantages and issues with concurrent audio presentation as part of an auditory display. *Proceedings of International Conference on Auditory Displays*, London.

Mills, A. W. (1958). On the minimum audible angle. *Journal of the Acoustical Society of America* 30(4):237–46.

Miranda, E. R. (1998). *Computer Sound Synthesis for the Electronic Musician*. Oxford: Focal Press.

Mitsopoulos, E. N. (2000). a principled approach to the design of auditory interaction in the non-visual user interface. PhD diss., University of York.

Moore, B. C. J. (2003). *An Introduction to the Psychology of Hearing*, 5th ed. London: Academic Press.

Moore, F. R. (1990). *Elements of Computer Music*. Upper Saddle River, NJ: Prentice Hall.

Mossberg, W. (2006). New earphone devices let you go cordless on iPods, cellphones. Personal Technology. *The Wall Street Journal Online*, December 21; available at *http:// ptech.wsj.com/archive/ptech–20061221.html*.

Mynatt, E. (1995). Transforming graphical interfaces into auditory interfaces. PhD diss., Georgia Institute of Technology, Atlanta.

Neisser, U. (1967). *Cognitive Psychology*. New York: Appleton-Century-Crofts.

Neuhoff, J. (2004). Interacting auditory dimensions. In J. Neuhoff, ed., *Ecological Psychoacoustics*. Boston: Elsevier/Academic Press.

Nielsen, J. (1994). Ten usability heuristics—available at *http://www.useit.com/papers/ heuristic/heuristic_list.htm*.

Nielsen, J., & Molich, R. (1990). Heuristic evaluation of user interfaces. *Proceedings of SIGCHI Conference on Human Factors in Computing Systems*.

Peres, S. C., & Lane, D. M. (2005). Auditory graphs: The effects of redundant dimensions and divided attention. *Proceedings of International Conference on Auditory Displays*, Limerick, Ireland.

Petrie, H., & Morley, S. (1998). The use of non-speech sounds in non-visual interfaces to the MS-Windows GUI for blind computer users. *Proceedings of International Conference on Auditory Displays*, Glasgow.

Pirhonen, A., Murphy, E., McAllister, G., & Yu, W. (2006). Non-speech sounds as elements of a use scenario: A semiotic perspective. *Proceedings of International Conference on Auditory Displays*, London.

Plomp, R. (1964). Rate of decay of auditory sensation. *Journal of the Acoustical Society of America* 36:277–82.

Pollack, I., & Ficks, L. (1954). Information of elementary multidimensional auditory displays. *Journal of the Acoustical Society of America* 26:155–58.

Puckette, M. S. (1997). Pure Data. *Proceedings of International Computer Music Conference*, 224–27.

Pulkki, V. (1997). Virtual sound source positioning using vector base amplitude panning. *Journal of the Audio Engineering Society* 45(6):456–66.

Recarte, M. A., & Nunes, L. M. (2003). Mental workload while driving: Effects on visual search, discrimination, and decision making. *Journal of Experimental Psychology, Applied* 9(2):119–37.

Roads, C. (2001). *Microsound*. Cambridge, MA: MIT Press.

Sawhney, N., & Schmandt, C. (2003). Nomadic Radio: Wearable Audio Computing. Cambridge, MA: Speech Interface Group, MIT Media Laboratory; available at *http://web.media.mit.edu/~nitin/NomadicRadio/*.

Schmandt, C. (1993). Phoneshell: The telephone as computer terminal. *Proceedings of Association for Computing Machinery International Conference on Multimedia*.

Shannon, R. V., Zeng, F., Kamath, V., Wygonski, J., & Ekelid, M. (1995). Speech recognition with primarily temporal cues. *Science* 270(5234):303–304.

Staal, M. A. (2004). Stress, cognition, and human performance: A literature review and conceptual framework. National Aeronautics and Space Administration (NASA) Technical Memorandum, TM-2004-212824. Moffett Field, CA: Ames Research Center.

Stanton, N. A., & Edworthy, J., eds. (1999). *Human Factors in Auditory Warnings*. Aldershot, Hampshire, England: Ashgate Publishers.

Stevens, S. S. (1957). On the psychophysical law. *Psychological Review* 64(3):153–81.

tenHave, P. (1999). *Doing Conversation Analysis: A Practical Guide*. London: Sage.

Terenzi, F. (1988). Design and realization of an integrated system for the composition of musical scores and for the numerical synthesis of sound (special application for translation of radiation from galaxies into sound using computer music procedures). Physics Department, University of Milan, Milan, Italy.

Truax, B., ed. (1999). *Handbook for Acoustic Ecology*; see *http://www2.sfu.ca/sonic-studio/handbook/*.

Ulfvengren, P. (2003). Design of natural warning sounds in human machine systems. PhD diss., Royal Institute of Technology, Stockholm.

VRSonic. (2007). VibeStudio—available at *http://www.vrsonic.com*.

Walker, B. N. (2002). Magnitude estimation of conceptual data dimensions for use in sonification. *Journal of Experimental Psychology: Applied* 8(4):211–21.

Walker, B. N., & Kramer, G. (2004). Ecological psychoacoustics and auditory displays: Hearing, grouping, and meaning making. In Neuhoff, J., ed., *Ecological Psychoacoustics*. New York: Academic Press, 150–75.

Walker, B. N., & Lane, D. M. (2001). Psychophysical scaling of sonification mappings: A comparison of visually impaired and sighted listeners. *Proceedings of International Conference on Auditory Display*, Espoo, Finland.

Walker, B. N., & Lindsay, J. (2004). Auditory navigation performance is affected by waypoint capture radius. *Proceedings of International Conference on Auditory Display*, Sydney.

Walker, B. N., & Stamper, K. (2005). Building audio designs monkey: An audio augmented reality designer's tool. *Proceedings of International Conference on Auditory Display*, Limerick, Ireland.

Walker, B. N., Nancy, A., & Lindsay, J. (2006). Spearcons: Speech-based earcons improve navigation performance in auditory menus. *Proceedings of International Conference on Auditory Display*, London.

Wang, D., & Brown, G. J. (2006). *Computational Auditory Scene Analysis: Principles, Algorithms, and Applications*. Hoboken, NJ: John Wiley & Sons.

Weir, J. (2005). The sound of traffic—available at *http://www.smokinggun.com/projects/soundoftraffic/*.

Wickens, C. D., & Hollands, J. G. (2000). *Engineering Psychology and Human Performance*, 3rd ed. Upper Saddle River, NJ: Prentice Hall.

Williamson, J., Murray-Smith, R., & Hughes, S. (2007). Shoogle: Multimodal excitatory interfaces on mobile devices. *Proceedings of SIGCHI Conference on Human Factors in Computing Systems*.

Winberg, F., & Bowers, J. (2004). Assembling the senses: Towards the design of cooperative interfaces for visually impaired users. *Proceedings of Association for Computing Machinery Conference on Computer Supported Cooperative Work*.

Voice User Interfaces

Susan L. Hura

A voice user interface (VUI) is the script to a conversation between an automated system and a user. This script contains all the utterances that the automated system will speak to the user and the logic to decide which utterances to speak in response to user input. Underlying the voice user interface is speech recognition technology that has the ability to capture and decode the user's spoken input to allow the system to "understand" what the user has said. The majority of these automated conversations between a user and a VUI take place over the phone, as most speech technology is currently deployed in speech-enabled interactive voice response interfaces (see Chapter 7). Increasingly, VUIs are being added to the user experience of mobile and handheld devices, in-vehicle navigation systems, and desktop computer applications.

Users come into automated conversations with a set of expectations about how spoken conversation should work and the appropriate way to behave as a cooperative speaker and listener. The overwhelming majority of users' experience comes from unscripted human-to-human speech, a feat that is far outside the capabilities of today's speech technology. This limitation is largely unknown to most users, a great many of whom approach VUIs with the *Star Trek* model in mind: To such users, speech-enabled systems are an always-listening intelligence awaiting one's every command. What Captain Picard requests is always understood, and his spoken interactions with "Computer" appear effortless. Therefore, it is no surprise that users' expectations are somewhat out of sync with the way today's VUIs actually behave (Hura, 2005).

The expectation mismatch is particularly problematic because it leads users to produce unpredictable speech input that is often handled poorly by speech systems. In this chapter, I offer my experiences as an evaluator and designer of VUIs.

The unifying theme for the chapter is the expectations mismatch and how to minimize it. I also provide an overview of speech technology from the perspective of a VUI designer with the goal of achieving effortless spoken interaction with automated systems.

6.1 AUTOMATED CONVERSATION: HUMAN VERSUS MACHINE

Commercial applications of speech recognition are largely bottom-up systems that achieve understanding via statistically based matching of the user's spoken input with stored acoustic models of speech sounds, words, and pronunciations (see Section 6.2 for more detail). Human beings, by contrast, rely heavily on top-down knowledge about meaning and context when recognizing speech (Massaro, 2006). That is, when human beings are listening to spoken language, we are not trying to match the incoming speech signal against any possible word in the language. Instead, we automatically and unconsciously expect to hear words that "fit" into the overall conversational context. Context includes microlevel information such as which phoneme combinations are permissible, rules about sentence structure that determine what sort of word is likely to come next, and nonlinguistic world knowledge that rules out nonsensical interpretations of unclear words. Because we know how the world works, we know that "Banjo" is a highly unlikely response to the question "What do you want for lunch?" no matter how much the acoustic signal points us toward that interpretation. Automated speech recognition systems have no knowledge of the world and must rely solely on the acoustic signal and the bits of intelligence built in by VUI designers.

Even without the benefit of human-like top-down knowledge, many speech recognition systems perform quite well, and through clever VUIs appear to be carrying on their half of the conversation admirably. When interactions with automated speech systems go well, users assume that their interlocutor (the system) is conversing in the same manner they are—performing the tasks of conversation in a way similar to how they do it and playing by the same rules of conversation. Grice (1975) defined a set of maxims of conversation that describe the unconscious rules by which we participate in conversations. According to the maxims of conversation, we expect utterances to be of high quality (truthful and honest), to offer an appropriate quantity of information, to be relevant to the conversation, and to be spoken in a clear, unambiguous manner. Automated speech systems often do not follow these cooperative principles—accidentally or by design—yet users tend to give the systems the benefit of the doubt and assume that they are following the rules. This misapprehension on the part of users is the fundamental challenge in designing effective VUIs.

Why do users expect so much of automated speech systems? In part it is because VUIs sound very real. Most commercial speech applications rely on a set of recorded prompts, spoken by a voice actor who has been carefully coached to produce each utterance as if it were part of an ongoing conversation (see additional information on prompts in Section 6.1.3). The bigger reason behind users' high expectations is the vast experience they have with spoken language. Language is a hugely complex system governed by myriad rules, but people are largely unaware of their own linguistic prowess. Human beings are constantly immersed in spoken language. Exposure begins before birth and there is evidence that human infants come into the world with the ability and experience to begin immediately making sense of their linguistic environment. Language is acquired automatically by human infants through simple exposure and requires no adult intervention, other than simply conversing with the child. Language is often characterized as an innate ability of human beings (Pinker, 1994; Reeves & Nass, 1996); thus, it is difficult for many people to understand why simple conversation can be so daunting for automated systems. Talking and listening are overlearned tasks for humans—we converse effortlessly, moving our vocal musculature to produce speech sounds arranged into fluent utterances without conscious effort. It's *so easy* for us—why is it so hard for computers?

Another reason that users expect a lot from speech systems is that they view the ability to engage in conversation as evidence of the inherent intelligence of the speech system. VUIs often seem smart, friendly, and even human—they understand what you say and respond in ways that a person might in that circumstance. When users discuss their experiences with VUIs, they are often somewhat disconcerted when they notice themselves attributing personal characteristics (smart, friendly) to what they know to be a computer system. These embarrassed users do not believe they are actually talking to a person, but they cannot help talking about the VUI as "she" rather than "it."

When users think this way, they are succumbing to attribution theory (Heider, 1958). Attribution theory describes the phenomenon of viewing an individual's actions as evidence of his or her inherent characteristics. That is, we see the actions of others as evidence of stable factors that are part of their personalities. Human beings have a strong propensity to overattribute individual behaviors to the inherent character of an individual. The pull of attribution theory is so intense and so unconscious that we attribute personalities even to nonhumans with whom we interact, such as computers and VUIs. Research by Nass and colleagues (Reeves & Nass, 1996; Nass & Brave, 2005) demonstrates that we tend to interact with nonhuman actors according to the same social principals we apply in human-to-human interactions. The fundamental attribution error is relevant to VUIs in that users tend to see every action of a VUI as evidence of the personality and abilities of the speech system. When a VUI offers relevant information and gives appropriate responses to users, they tend to attribute this to the intelligence and sensibility of the system. When a VUI says please and thank you, we believe that the system is polite and friendly.

6.1.1   VUI Persona

Voice user interface designers often capitalize on users' tendency to attribute characteristics based on behavior in order to attach a particular *persona* to the VUI. Here, *persona* is not used to describe user characteristics (Pruitt & Adlin, 2006) in the more familiar sense of the term in the general human factors community. Instead, in the world of speech technology, *persona* refers to the personality of a speech interface inferred by users based on the behavior of the VUI. Factors that contribute to the perception of persona are the style and tone of prompts, the flow of dialog, and the responsiveness of the system. Some VUI designers conceive of persona as an imaginary customer service representative being emulated by the VUI. The concept of persona has been a continuing source of controversy in the VUI community. VUI designers disagree vigorously on the importance of persona in overall VUI design (Klie, 2007; Rolandi, 2007). Some designers deny that persona exists at all, and others suggest that persona is the most significant characteristic of an application.

The reality of persona almost certainly lies somewhere between these two extremes. Attribution theory suggests that it may be impossible for users *not* to infer a persona when participating in a spoken conversation, whether with a person or with an automated system. Part of the human language faculty involves the process of sizing up your interlocutor based on the way she speaks. This is not something that humans can turn on and off; thus, it is impossible for there to be "no persona" in a VUI. VUIs supposedly designed with no persona tend to sound either schizophrenic (as if a variety of individuals are speaking rather than one coherent voice), robotic (so devoid of normal human speech patterns that they remind you of bad 1970s science fiction), or both. The relationship of persona to VUI is akin to the relationship of color to the overall GUI design of a web page (Hura, Polkosky, & Gilbert, 2006). Color is not the single defining quality of users' perceptions of the web page, but their impressions are affected in significant ways by color.

There is also evidence that persona and the usability of a VUI are tightly coupled. Users attribute persona through evidence they observe in the voice of the system (e.g., pitch, pitch range, loudness, loudness range, articulatory precision), the linguistic style of prompts (e.g., word choice, syntax, content, and structure of error recovery prompts), and the overall organization of a VUI. Research by Polkosky (2005) suggests that these factors predict an impression of dialog efficiency, interactive ease, and affective response in the user. There is strong similarity here to the characteristics of general usability presented in the standard ISO 9241:11.

6.1.2   Jekyll and Hyde Personas

Because the tendency to attribute a personality to a VUI is so strong, one might think that changing a user's views on this personality would be difficult. Instead, users' perceptions of a VUI are notoriously fragile. One bad turn of the conversation and users can lose all confidence in the system as an intelligent actor in the conversation. Speech recognition failures can cause this erosion of confidence, but this is not always the case. Similarly, prompts that elicit unexpected responses can, but do not always, cause user perceptions to change significantly. The examples that follow demonstrate these two classes.

The first case is when the user speaks an appropriate response clearly and audibly, but the system does not recognize it. Users have a good sense of how intelligible their utterances are, and when a speech system fails to recognize well-formed responses users can become frustrated. The frustration can be exacerbated by long and generic error prompts that do not allow for quick repair of the conversation. As an example, consider the following:

SYSTEM: Please select one of the following: account balances, transaction history, transfers...

USER: Account balances.

SYSTEM: I'm sorry. I did not understand your response. Please choose one of the following: You can say: account balances, transaction history...

This error prompt belabors the recognition failure and draws the user's attention to it.

In fact, in human-to-human conversation this sort of misunderstanding is common, and conversations are repaired and continued with only minimal interruption. The difference is that humans use brief conversational repair techniques rather than taking a whole step back in the dialog. If a friend asks, "Where are you going on vacation?" and fails to understand your response, the friend will likely say, "Where?" as a follow-up rather than repeating the original question in its entirety. Well-designed VUIs mimic conversational repair by using quick reprompts that allow the conversation to get back on track quickly.

The second case is when a prompt elicits unexpected responses. The system asks an open-ended question, and based on previous good experience with the VUI, the user responds appropriately. The system, however, is incapable of dealing with the response and goes into error prompting. This is essentially a case of fooling the user into thinking the VUI is smarter than it really is (Rolandi, 2002). The typical situation here is that a user has been successfully interacting with a simple menu-based VUI for several turns of the conversation (the systems offers a number of discrete choices, and the user responds by speaking one of these choices). The user is then confronted by a more open-ended prompt, like the one in the following example. Because the previous interaction has been so successful, the user attributes intelligence and cooperation to the system, and cooperatively answers the question being asked.

SYSTEM: What else can I help you with? [*pause*]

USER: I need to get the balance on another account.

SYSTEM: I'm sorry. Please choose one of the following . . .

In the first prompt, the open-ended question is intended simply as a friendly and natural-sounding introduction to the list of choices that would have followed the pause. But the user assumes that he should just answer the question being asked and barges in before hearing the choices. This is not a case of the user incorrectly having high expectations for a system. On the contrary, the user's expectations have rightly been built up through a series of positive interactions. The problem here is that asking the open-ended question leads the user to believe that the system will be able to understand his response.[1]

The type of interaction shown here can be particularly damaging because it erodes the trust and confidence the user has built up with the system. Asking a question and then not being able to deal with the answer makes users feel as if they have been deliberately tricked into doing something foolish.

Once an error like this occurs, it can be extremely difficult to get the interaction flowing smoothly again. Once users become aware of the previously hidden limitations of the system, they become attentive for other instances in which the system shows its lack of conversational ability. As the user builds up positive interactions with a system, the conversation seems more and more realistic and the system seems more human. But before the user can reach the uncanny valley (in which almost-human computer systems go from arousing positive emotions to negative ones [Mori, 1970]), the illusion often shatters due to the types of issues described in this section. The end result is too often what has been called the "conversational death spiral" (Attwater et al., 2005) in which the system and the user both try unsuccessfully to repair the conversation via disparate strategies.

6.1.3 The Presentation Layer: Prompts

Users experience the VUI primarily through its prompts. Prompts are what the VUI says and are the entirety of the evidence users receive about their interlocutor. Through prompts, users experience the voice of the system, as well as linguistic and prosodic features that contribute to the attribution of a persona. The best way to appreciate the impact of prompts is to experience them in an automated conversation. Imagine the following situation. A user, Nick, is having trouble with his satellite television service—his television screen goes blank intermittently on

---

[1] There are speech recognition systems designed to handle responses to open-ended questions. The system described here is typical of grammar-based recognition systems that recognize only items preloaded into the grammar.

several channels. Nick goes online and reads the troubleshooting information available there, but does not find anything useful. He then decides to call his satellite provider at the number helpfully provided on the web page. Table 6.1 contains an account of his phone call.

Nick's call is typical of the interactions that many users have with VUIs. Users begin the conversation cooperatively, and when something goes wrong they turn to conversation repair techniques that work in other conversations (e.g., speaking more slowly and clearly as in Step 12 in the table). Ironically these repair

TABLE 6.1   Hypothetical Call to a Satellite Television Provider

Step	Speaker	Utterance	User's Thoughts and Actions/ Comments
1	System	Hello and thank you for calling Best Choice Satellite *(branding sound)*, the best choice for all your satellite television needs.	*So far, so good.*
2	System	You can also contact us via our website, w-w-w dot best choice dot com. Our website offers complete information about setting up satellite service at your home and offers valuable installation coupons worth over $40.	*This is really annoying! I just spent 10 minutes on the website without solving my problem.*
3	System	I can help you with the following five options. You can interrupt me and speak your choice at any time. Please select one of the following: Sign up for new service, add features to my service, move my existing service, problems with my satellite service, or ask a billing question.	*I'm glad the guy behind the counter at the local satellite TV office doesn't talk this much.... Oh wait, which one do I want?*
4	Nick	Service problem.	*Finally!*
5	System	I'm sorry. I did not understand your response. I can help you with the following five options. You can interrupt me and speak your choice at any time. Please select one of the following: Sign up for new service, add features to my service, move my existing service, problems with my satellite service, or ask a billing question.	*Huh? Better listen again to figure out exactly what to say.*
6	Nick	Problems with my satellite service.	

(Continued)

TABLE 6.1   *Cont'd*

Step	Speaker	Utterance	User's Thoughts and Actions/ Comments
7	System	Please say or enter your account number.	*I guess she heard me. But I don't know my account number! Where am I going to find it?*
8	Nick	*(mutters)* Just a second ...	*(Nick frantically shuffles papers trying to find his last satellite TV bill and mutters "Just a second" under his breath.)*
9	System	I'm sorry. The number you entered is not a valid account number. Please say or enter your account number.	*Why did she think that was an account number? Anyone can tell that "Just a second" is not an account number! Whatever, I'll just try again.*
10	Nick	0-0-0-0-dash-4-3-5-7-2-2-7-6	*(Nick reads what he believes to be his account number from his printed billing statement.)*
11	System	I'm sorry. The number you entered is not a valid account number. Please say or enter your account number.	*(System fails to recognize because recognized account numbers are eight digits and do not include leading zeros and the hyphen. System goes to a reprompt.)* *Give me a break. I just told you the account number.*
12	Nick	0... 0... 0... 4... 3... 5... 7... 2...	*This machine is idiotic. I'm going to go extra slow so she can't mess up again.*
13	System	I'm sorry, that is not a valid account number. One moment while I transfer your call.	

techniques often exacerbate the original problem or create new ones. In the following, the conversation is analyzed turn by turn to highlight what went wrong and how to avoid the problem. The design guidelines in Section 6.6 detail many of the principles introduced here.

Step 1 of the conversation actually begins with a useful practice. The recorded greeting incorporates a nonspeech branding sound (e.g., the Intel noise). This serves the useful function of demonstrating that the user is interacting with an automated system, not a real person. Without the nonspeech audio, the user may be initially unsure whether he is talking to a live agent faithfully reciting the official company greeting or listening to a recording. Some systems choose to announce their nonhuman status (e.g., "You have reached ABC Bank's 24-hour automated account access system"). Although a greeting like this can serve the same informative purpose, non-speech audio in the greeting avoids immediately arousing the negative emotions some users experience when they hear they will be talking to a system.

Step 2 demonstrates the common practice of using the moment just following the greeting to insert advertisements or give instructions on how to use the system. The idea here is that users should be given this information early, before they start using the system, just in case they need it later. However, information presented to users out of context will not seem relevant and is unlikely to be retained. When users begin an interaction with an automated system, they tend to be extremely task focused—they are thinking about what they need to accomplish and nothing else. My experience from usability testing suggests that users ignore information presented before they locate the functionality they need to accomplish their task. In the case of advertising messages, it is clear that the placement of the message is ineffective because the audience is primed to ignore it until they reach their goal. This reasoning tends to be persuasive when negotiating with marketing departments—simply ask if they intend their ad to actively annoy the user.

The fictional menu shown in Step 3 is based on the menus one often hears in VUIs:

> I can help you with the following five options. You can interrupt me and speak your choice at any time. Please select one of the following: sign up for new service, add features to my service, move my existing service, problems with my satellite service, or ask a billing question.

There are only five menu options, but the menu still seems quite long. (See Section 6.6.1 for a discussion of menus and memory.) One reason is that there are three sentences of preamble before the options are actually presented.

As with advertisements discussed in the previous section, instructions presented before the main menu are typically ignored by users. These introductory prompts are often thought to be necessary to prepare users who may be using a speech system for the first time and who are more accustomed to touch-tone systems. VUI best practices suggest that most users will be able to give satisfactory spoken responses as long as the prompts include wording like "You can say..." before menu options are presented. Another problem with the menu presented in Step 3 is the menu option names themselves. Each option is long and syntactically complex, and the option names are partially overlapping (four use the term "service"). This makes it difficult for users to locate and remember the menu option they want.

The user's response at Step 4—"Service problem"—demonstrates the sort of responses generated by long, complex menu option names. The user accurately identified the option he wanted, and produced an abbreviated version of it as his response, which was unfortunately not recognized by the system. Because the user accurately selected the appropriate option, the menu option in itself is not fully at fault. Instead, this represents a failure in the grammar for this menu. A grammar is the set of all items that can be recognized at a specific recognition state, as defined by the VUI designer. It is easy for a person to understand that the user's response "Service problem" was intended as a synonym for the menu option "Problems with my satellite service." However, the speech system is unable to make the same judgment unless that particular synonym is built into the grammar. The longer and more complex the menu option names, the more possible synonyms there are, and thus longer and more complex the grammars are required to support the menu options.

Step 5 is an example of ineffective error handling:

> I'm sorry. I did not understand your response. I can help you with the following five options. You can interrupt me and speak your choice at any time. Please select one of the following: sign up for new service, add features to my service, move my existing service, problems with my satellite service, or ask a billing question.

This reprompt blames the user for the error by suggesting that he failed to correctly choose an item from the menu, when in fact the user is giving a predictable variant of a menu option name that should have been covered in the grammar. The reprompt then repeats the entire initial prompt, including the introductory sentences, but does not offer any information to the user about specifically how to give a well-formed response. Simple repetition is an effective reprompting strategy only for those cases in which the user fails to respond because he did not hear the choices initially. Here, the user clearly tried to make a response, so the reprompt is simply annoying.

Why is the user feeling uncomfortable in Step 7?

> Please say or enter your account number.

In part it is because he is being asked for information that he does not have at hand. Even before this, though, the user is uncomfortable with the flow of the conversation because he is not sure whether his response in the previous step was accepted. What's missing is a seemingly unimportant connector word, acknowledging the user's response. Simply adding a word or phrase such as "All right" before asking for the account number would have let the user know that he was heard and understood, and that the conversation would be moving on to a new topic.

These connector words, known as *discourse markers,* serve many functions in spoken conversation, including acknowledgment (okay, all right), changing topics (otherwise, instead), emphasizing or reinforcing (also, above all), and serving as sequence markers (first, next, finally) (Schiffrin, 1987). Discourse markers convey vital information about the structure and flow of the overall dialog. Without discourse markers, a conversation feels incomplete, and will leave users uncomfortable without knowing exactly why.

Step 8, in which the user mutters "Just a second," is both an indication of the ineffectiveness of the account number prompt in Step 7 and an illustration of the blatant lack of real-world knowledge of speech recognition systems. To any person, it is obvious that the user was not intending "Just a second" as a response to the account number prompt. He was simply mumbling as he tried to locate his account number. The system cannot at face value distinguish system-directed speech from side speech (defined as speech audible over the telephone connection that is not intended as part of the telephone conversation). The fact that the user is muttering here is a sign that there is a problem with the account number prompt. The user hesitates because he is unprepared to give the information he is being asked to provide. The best way to prevent this sort of problem is to avoid asking for information that the user will need to look up, in favor of something the user will have memorized such as a telephone number. When this is impossible, give the user practical advice for where they can find and recognize the requested information.

Step 9 is another poorly designed reprompt:

I'm sorry. The number you entered is not a valid account number. Please say or enter your account number.

First, the user did not in fact enter an account number, so the prompt should not accuse him of entering an invalid one. More importantly, the reprompt fails to offer the user assistance in finding the account number. Especially when requesting specific pieces of information, it is vital to write reprompts that give users additional instruction that will help them provide a well-formed response. The user would have been more successful and more comfortable if the prompt had said, for example, "Please say or enter your eight-digit account number. You can find your account number in the blue box under your name on your statement."

Step 10 shows the results of giving the user insufficient instruction about what constitutes an account number. Nick located the number and said, "0-0-0-0-dash-4-3-5-7-2-2-7-6" because he was not instructed to speak only the last eight digits. The grammar contributed to the problem at this step because it did not accommodate the full 12-digit version of the account number.

Step 11 is the identical reprompt played in Step 9. What was annoying and unhelpful on first presentation can become infuriating with repetition. Repeating the identical reprompt, with the identical cheerful tone of voice, is completely out of sync with the conversation. The user has just tried multiple times to give a piece of information and failed in each case. When the system calmly repeats the identical prompt, the user can feel as if he is being patronized. When writing reprompts, be mindful of the user's potential frustration level. Two tries at any given recognition state is about the maximum that users can tolerate without significant annoyance, in my experience. Earlier I recommended brief reprompting for a reprompt at a menu (see Section 6.1.2). Here, and for other states when the user is inputting data (account or ID numbers, etc.), reprompts should actually be more expansive than the initial prompt, adding information to help the user provide a well-formed response.

In Step 12 when Nick slowly says, "0... 0... 0... 4... 3... 5... 7... 2...," we see the expectations mismatch in action. In frustration, the user falls back on a strategy for giving information that works in human-to-human conversations under these circumstances—speaking slowly and clearly. In the automated conversation, however, this strategy makes the recognition problem worse. Acoustic models used in recognition are built on normal conversational speech, not on the deliberate, slow, overenunciated speech spoken here; thus, recognition may suffer. In addition, the user is speaking so slowly that he has exceeded the system's time limit for giving a response. Consequently, the system interrupts the user before he is done speaking, adding a turn-taking error (Sachs, Schleghoff, & Jefferson, 1974) to the recognition problems. The problem can be reduced by adjusting the parameter that specifies the amount of time the user has to respond before the system times out at a dialog state (sometimes labeled "max speech timeout"). It is far preferable to avoid causing the user frustration in the first place by limiting the number of retries and using more effective reprompts.

In Step 13 the user is released from speech technology jail and sent to wait in a queue for a live representative after several minutes of torture and no closer to solving his problem. The sort of frustration and poor customer service experienced by the imaginary user here is the motivation for real users to advocate for IVR reform (for both speech and touch-tone) through the "gethuman" initiative (*http://www.gethuman.com*).

6.2   TECHNOLOGY OF THE INTERFACE

The technology behind VUIs is automatic speech recognition (ASR). The system is often a speech-enabled IVR, but increasingly may also be a personal computer, in-car navigation system, or other speech-enabled handheld device. All these systems depend on the same basic technology to recognize speech input. In each case, the speech signal is captured, digitized, segmented, and then compared against a set of stored acoustic models for speech sounds. The sounds are then built up into potential words that are compared with a grammar of words that are to be recognized at that point in the dialog. The recognition process is acoustically driven—there is no top-down semantic or logical analysis to help the system decode user input. Thus, the way that computer systems recognize speech is fundamentally different from the way that humans do. Most ASR systems in commercial use today rely on Hidden Markov Models (Rabiner, 1989) to recognize words.

In the 1980s and 1990s, a great deal of research was focused on building more human-like speech recognitions systems (e.g., McClelland & Elman, 1986). These systems aimed at mimicking the multiple sources of nonacoustic information used by humans when decoding the acoustic speech signal. Such systems contained huge amounts of linguistic information about the structure of words, sentence structure, and semantics (meaning at the level of words and sentences). Even this was insufficient to model everything that humans bring to the task of speech recognition, so models of dialog structure and world knowledge were also incorporated. The complexity and computational requirements of such systems have generally kept them in the laboratory rather than allowing them to migrate into commercial applications.

6.2.1 The Recognition Process

Figure 6.1 depicts the steps in the recognition process. User input is captured via a microphone (either built into the phone or connected to the computer or other device being used). The first major technical hurdle for speech systems is encountered at this very early stage of the process: What counts as user input? The process of making this decision is called end-pointing. A microphone will capture any sound in the environment, not just those deliberately uttered by the user to the system. In noisy environments, the quality of the audio signal is poorer than the signal captured in a quiet environment. We have all tried using mobile phones in noisy environments (in a car in traffic, at a crowded airport) and experienced problems related to background noise, even when speaking to another human. For speech recognition systems, this problem is amplified because systems cannot automatically distinguish speech from nonspeech. This issue can be improved by simply making the threshold for recognition of any sound higher, so that low-level background noises are not detected. There is a trade-off, however, in this method because users who speak softly will simply not be heard by a system with too high a threshold.

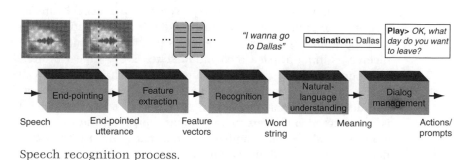

FIGURE
6.1

Speech recognition process.
Sequence of steps in automatic speech recognition. *Source:* From Cohen, Giangola, and Balogh (2004).

In addition to background noise, other classes of sound complicate the capture process. Nonspeech "mouth sounds" such as coughing, clearing the throat, and laughing are particularly hard to distinguish because they share some of the acoustic qualities of speech in that they are produced by the human vocal tract. The most disruptive to conversations with speech systems is side speech—speech that is within range of the microphone but is not directed at the system. Side speech may be spoken by the user or by another person in the room with the user. Side speech is disruptive not only because is begins a round of error handling, but also because it emphasizes the limitations of the system. When the system cannot distinguish between speech that is logically part of the conversation and speech that is meant to be ignored (e.g., Nick's "Just a second" in Table 6.1), the user is suddenly acutely aware of how "stupid" the system really is. Something users perceive as completely effortless—knowing what is and is not part of the current conversation—is beyond the capabilities of the automated system, so users begin to doubt the system's abilities overall.

Once the signal is captured, it is digitized and divided into a set of short segments for acoustic analysis. Each segment is transformed into a feature vector—a numerical representation of the speech signal that contains information relevant to recognition. The sequence of feature vectors is the basis for recognition at the level of words.

The elements that contribute to the recognition model are acoustic models, dictionaries, grammars, and a search algorithm. Acoustic models specify the pronunciation of phonemes—the speech sounds that comprise a language—in terms of sequences of feature vectors. Acoustic models are generally built by training them against large sets of transcribed and labeled acoustic data. Thus, acoustic models only support pronunciations that are well represented in the training data. This is the reason that unusual accents, or particular voices, are sometimes poorly recognized by speech systems.

A dictionary in a speech recognition system is a list of words and associated pronunciations. Pronunciations are represented using a phonetic alphabet, and there may be several alternate pronunciations for a single word. Pronunciation of words may vary depending on factors like dialect, rate of speech, and speaking style. For a specific pronunciation to be recognized, it must be added to the dictionary.

A grammar is the list of utterances that the recognition system is expected to receive as input at each recognition state. Every time we intend the recognizer to accept incoming speech, we must define what it is that callers are likely to say. A grammar should list all the likely variants of responses that users might produce at a given state (recall Step 4 in Nick's conversation, Table 6.1). There is a cost to adding items to a grammar, however; the more items there are in a grammar and the more similar the grammar items are acoustically, the more difficult recognition will be. Grammars that are written by a VUI designer or developer are called rule-based grammars, and are used in the majority of recognition applications in commercial use.

A second sort of grammar is defined statistically rather than by rule. Statistical language models (SLMs) are created automatically from transcribed speech data. The SLM looks at all the possible sequences of words and computes the probability of words appearing in particular contexts. SLMs allow users to give relatively free-form responses rather than simply repeating menu options offered by the system. This makes SLMs attractive to many organizations deploying speech technology. (Remember that *Star Trek* model? Speech recognition software sales folks can convince you that SLMs almost make it possible.) To be effective, however, SLMs require vast amounts of training data (tens or hundreds of thousands of utterances), which must all be transcribed and labeled by a person. SLMs are therefore several orders of magnitude more time consuming and costly than rule-based grammars. Only large organizations tend to see a positive return on investment from SLM-based speech recognition.

A final type of grammar to note is slotted grammars, in which the user can speak several individual pieces of information in a single utterance. Slotted grammars are often used for simple transactions such as transferring funds between accounts or requesting a travel reservation between two cities. User utterances for slotted grammars can seem quite free-form, similar to the possible responses for statistical language models, such as "I want to transfer $3,000 from Market Growth Fund to International Equity Fund." However, slotted grammars are much more limited because they are designed to collect multiple discrete chunks of information that can be connected by filler words. In the simple transfer example above, a slotted grammar would be built to search the incoming audio signal to fill an amount slot, a from-fund slot, and then a to-fund slot. If a user response deviates from this amount $\rightarrow$ from-fund $\rightarrow$ to-fund model, a slotted grammar will fail to recognize it, showing the very limited range of input flexibility possible for slotted grammars. This limitation makes slotted grammars a good fit for tasks where user responses are formulaic and follow a set pattern (such as simple transfers), but less suitable for situations in which users might phrase their response in many different ways.

Using the acoustic models, dictionaries, and grammars, the speech recognition system then looks for matches between the sequence of feature vectors (that represent incoming speech) and possible words and pronunciations in the grammar. The best matching path through the sequence of feature vectors is the word that is recognized. Most speech recognizers have the ability to return not just the best matching path, but n-best paths (the n-best matching paths rather than just the top of the list). Associated with each item in the n-best list will be a confidence score, a measure of how confident the recognizer is that the match is correct. In practice, the n-best list is a rank-ordered list of words and the probability for each. The method for searching for best paths used most often in current speech recognition systems is called Hidden Markov Models (HMMs). HMMs are a statistical technique for computing the probability of a match at a state and between states.

6.2.2   Some Historical Notes on Speech Recognition

Historically, speech recognition systems were constrained by the lack of processing speed and capacity. Speech recognition is computationally intensive, and as recently as the 1990s, the way speech systems functioned was limited by lack of processing power. (See Baber and Noyes [1993] and Lea [1980] for illuminating discussions of the future of speech technology written in the recent past.) One way of dealing with processing limitations was to restrict recognition to the utterances of a single individual. These *speaker-dependent* systems are built to recognize the speech of an individual, and thus require training data (in the form of spoken utterances) from this individual in order to function.

In contrast, *speaker-independent* systems do not require the user to provide training data before being recognized. Instead, speaker-independent recognition proceeds from a set of training data from a large number of individuals who may or may not be users of the system. The advantage of speaker-dependent systems was that they could support larger vocabularies than speaker-independent ones, given a fixed amount of processing power. If processing power is limited, there is a direct trade-off between the number of words in a system's vocabulary and the number of different voices that can be recognized. Today, this distinction has all but disappeared, and the vast majority of commercially deployed speech systems are speaker independent and recognize a limited number of items (although this number may be in the thousands) for any recognition state.

One important counterexample of speaker dependence in use currently is speaker verification systems (Campbell, 1997; Furui, 1996). These systems are trained to identify users by the unique characteristics of their voices. The user's voice therefore becomes a method of authentication that can be used alongside less reliable authenticators like passwords or user IDs. For speaker verification, users are enrolled (typically over the phone) by speaking a particular utterance multiple times. Representations of these utterances (sometimes called *voice prints*) are stored by the organization deploying the verification system and are used as one method of authenticating the user when she next calls in to the system. Speaker verification has been shown to be robust and difficult to fool, and is advantageous because users no longer need to remember passwords. Instead, by simply speaking a piece of identifying information such as an account number, the system is able to both recognize the data being spoken and recognize the voice itself. There are many possible combinations of data and voice prints possible to provide different levels of security for speech applications.

Another historical distinction is between systems that recognize isolated words and those that recognize connected (or continuous) speech. Previously, recognition was much more accurate for isolated words because isolated words eliminate the need to segment the incoming speech signal into words. It can be notoriously difficult to accurately locate word boundaries in connected speech, and eliminating this distinction significantly improved recognition accuracy. The majority of speech systems in use today still rely on isolated word recognition (although the systems can now recognize longer phrases, not just words). Statistical language models, which are trained on continuous speech, offer the nearest approximation to continuous speech recognition in commercial use today. For an excellent history of speech technologies, see appendix A in Weinshenk and Barker (2000).

6.3   CURRENT IMPLEMENTATIONS OF THE INTERFACE: ON THE PHONE

Voice user interfaces are overwhelmingly encountered over the phone. Speech recognition technology has been adopted in the call center as the successor to touch-tone IVR systems. Speech has a number of inherent advantages over touch-tone. Allowing users to speak their responses rather than key them in on the telephone keyboard makes the interaction easier in eyes-busy, hands-busy situations. The prevalence of cordless and mobile phones also gives an advantage to speech input over touch-tone. Speech does not require that users pull the handset away from their ear, press a key, and then return the phone to the ear repeatedly.

Speech also enables functionality that is awkward or impossible using touch-tone. Entering names, cities, fund names, or any string that contains both numbers and digits is very difficult in touch-tone, which requires users to either spell words on the telephone keypad or choose from very long lists of items. Speech recognition simplifies tasks such as collecting name and address, transferring funds between accounts, and making travel reservations. Speech also allows for shallower menu hierarchies, as the interaction is no longer constrained by the limits of the 12-digit telephone keypad. VUIs have grown and matured in the telephony world, and much of VUI design is totally phone-centric. Salespeople also pitch speech systems as more "natural" than touch-tone, since we all know how to talk. (I sincerely hope that this chapter will convince the reader of the fallacy of this sales pitch.)

Voice user interfaces can be found outside the call center, although in much smaller numbers. One application that is rapidly becoming more widespread is speech-enabled in-vehicle navigation systems. These systems allow the driver to control the navigation system and various features of the vehicle (sound system, air conditioning) through voice commands. These vehicle-based systems are multimodal in that there is a screen available throughout the interaction. This makes the speech part of the user interface somewhat less rich than what is seen in telephony applications.

Speech recognition on the personal computer desktop actually preceded widespread use of speech in the call center, but desktop speech recognition use has yet to become the norm for the vast majority of users. Dictation programs have been available for decades and some users are committed to them. These programs are built to recognize free-form input from a single user rather than a small grammar of words spoken by any user. Because of this, dictation programs must be trained to the voice of the user, which entails an initial time investment before the program is usable. Dictation programs can work quite well, but suffer from a poor reputation of being time-consuming and inaccurate. Less commonly used (though now widely available through Windows Vista) are desktop command and control programs. With these programs, users can speak commands to perform the same tasks they usually accomplish with keyboard and mouse. Again, the voice component of the interface here is rather minimal—the user has a set of commands that can be spoken. There is no "conversation" as there is in telephone systems.

Speech is also found is a number of tiny niche markets. There are dedicated speech recognition systems in fields such as medical dictation in which there is a complex but limited specialized vocabulary. Speech systems are also used by workers in warehouses to assist with inventory and order fulfillment tasks. Although these niche applications are on the fringe of VUI today, they are likely to proliferate in coming years. Niche markets are conducive to speech technology because a niche is a limited domain that constrains the vocabulary (thus enabling smaller grammars and better recognition performance), and because niche users tend to be more willing to live with limitations that the general population might not tolerate. Niche users may be willing to wear headsets or other devices to enable speech recognition, and may be amenable to having to learn how to use the system (as opposed to the general population, who tend to demand immediate ease of use for speech technologies.)

6.4   HUMAN FACTORS DESIGN OF THE INTERFACE

Voice user interfaces often suffer from a lack of human factors involvement during design and development. The reasons for this are logistical and technical: Speech systems are most often deployed in a telephony environment, typically in the call center, and VUIs rely on a different technology than GUIs and thus require different developer skill sets. Speech system developers are often not accustomed to working with human factors professionals and may have difficulty adjusting to new procedures and timelines imposed by user-centered design practices. In large organizations, there may be an existing staff of human factors professionals working on website and software issues. However, the call center is not the typical realm for human factors specialists, and IVR managers may be unaware of the human factors skills available to them. Consequently, VUI design often proceeds without any human factors involvement at all. Sadly, even when human factors folks are involved in speech projects, most lack expertise in speech technologies and human conversational behavior and thus do not contribute as much as they do in GUI projects.

In the vast majority of cases, the sole representative of human factors on speech system projects is the VUI designer (VUID). Much like the early human factors pioneers, VUIDs are a heterogeneous group made up of ex-developers, wayward linguists, telephony managers, assorted psychologists, and many others with little or no relevant formal training in speech technology or human factors. There are currently no formal degree programs specifically in VUI design; however, Auburn University has an engineering program with a strong VUI emphasis under the direction of Juan Gilbert. What VUIDs may lack in formal human factors training, most make up for in their enthusiasm to bring strong user-centered design practices to speech projects. To get a sense of the VUID community, consider visiting the VUIDs Yahoo group at *http://tech.groups.yahoo.com/group/vuids*.

Strong support for human factors in speech technology systems has come from AVIOS, the Applied Voice Input Output Society. AVIOS is a nonprofit membership organization for speech technology professionals that has been influential for over 20 years. AVIOS often sponsors speech technology conferences and previously had editorial control of the *International Journal of Speech Technology*, an important source of rigorous technical and user interface information for speech systems.

More so than in the GUI world, VUIDs need a mix of both technical and design skills. GUI interactions occur using a mouse, keyboard, or some other precise input device, and GUI designers can reasonably assume that user input will be interpreted correctly by the system. In speech-enabled applications, well-formed user input will be misinterpreted in a significant proportion of cases. Most current speech recognition algorithms boast near 100 percent recognition accuracy, but in practice, error rates of 10 to 20 percent or more at a given recognition state are not uncommon. Therefore, the VUI designer must understand the limitations of the technology and design using strategies that minimize technology-level errors. VUIDs also serve as the voice of the user in speech projects, much as other human factors professionals do in the GUI world. On many projects, the VUI designer is working alone without any other human factors support, so it is vital to have firm guidelines for appropriate user-centered design processes, as well as guerrilla techniques to fall back on when the ideal is not possible. The ability to justify and evangelize for user-centered design (UCD) processes is vital for VUIDs who are often the first exposure project members have to human factors.

6.4.1 User-Centered Design for Speech Projects

User-centered design practice for speech technologies is similar but not identical to established UCD techniques. At a general level, the process is similar: Gather information from users and the business, design a preliminary interface, test with real users, modify the design, and iterate. There are different twists at each stage, however, because of the unique position of speech technologies.

During requirements gathering, VUI designers have unique sources of information relevant to the voice channel. Since most speech technology is deployed over the phone in call centers, call center agents and managers can provide valuable background on the user's state of mind when calling, terminology used and understood by callers, and nuances of calling patterns that may not be obvious from other sources (e.g., users who call during business hours tend to be more impatient than users who call later in the day). In call centers with existing IVRs, usage statistics can reveal frequency of use for various functions, patterns of use over time, and points at which users tend to abandon self-service in favor of live agents. Existing IVRs also offer the opportunity to listen to real calls so that designers may understand the current user experience and thus gain insight on how to improve it.

Another data point in the requirements process is the details of the particular speech recognition system being used on a project. While all recognition engines have the same basic capabilities, there are a multitude of differences in systems sold by different vendors that can affect the VUI. Understanding what speech technology is available and how to put it to best use is a foundation for designing an effective VUI.

The process of VUI design is similar to GUI design except, of course, for the modality of the presentation layer of the interface. In spite of being auditory, VUIs need to be documented visually. Most VUI designers rely on a combination of flowcharts to represent the logical flow of the application (in a program such as Visio) and text documents to specify the prompts and detailed behavior at each recognition state. This combined call-flow–plus–text-document arrangement is suboptimal and difficult to maintain, but most VUI designers have no better alternative. There is no tool in wide use that captures both call flow and detailed specification information in a single program.

6.4.2 How Much Speech Should We Use?

One of the most important questions for VUI designers is how to use speech technology smartly to achieve business goals and meet user needs. The aim is to avoid using speech technology for its own sake, and instead look for opportunities in which speech provides distinct value to the user and to the business. These opportunities arise under several conditions:

+ The current technology underaddresses user or business needs, such as typing city names using the numerical telephone keypad.
+ The typical context of use is an eyes-busy, hands-busy situation, such as any application where users are likely to be mobile.
+ Users have a disability that limits their ability to use their hands. (See the United States Access Board, Section 508, and the Web Accessibility Initiative of the W3C for further recommendations for speech and accessibility.)
+ The wait time to speak to a live representative is unacceptable to the users for their task.

Some factors weigh against using speech. Many users have concerns about privacy and may not be willing to speak certain information aloud (e.g., passwords, Social Security numbers, account numbers, medical data, other financial information). If you choose to speech-enable collecting this type of sensitive information, be sure to allow users an alternate method to enter their information. Speech is also a poor fit for collecting unconstrained alphanumeric sequences because of the difficulty of recognizing sequences of variable lengths composed of letters and numbers that are notoriously confusable. Various noisy acoustic environments also make speech a less desirable solution. In many other situations, speech works reasonably well for the task at hand, but may not offer any real benefits over another technology (like touch-tone input). In these scenarios, the case for using speech is more about the ability of speech interfaces to generate positive customer experience for users.

When it is clear that speech recognition technology is a good fit for the task, there is still the choice of the type of dialog and grammar to use. For telephone-based speech systems, a directed dialog approach is often appropriate. Directed dialog refers to the class of VUIs in which the system offers users the choice among a finite set of options, and those options are recognized against a standard grammar defined in advance. Directed dialog can provide effective and satisfying interactions for wide variety of routine information retrieval tasks (e.g., getting an account balance, checking flight status) and for many transactions such as placing an order or changing an address. In the speech industry press, directed dialog is sometimes portrayed as unnatural or inferior to so-called "natural-language" systems.

Natural language is an imprecise term that is used differently by each author. A common interpretation for "natural language" is the sort of minimally constrained responses possible when using a statistical language model. As explained above, SLMs are simply a more expansive grammar, not true language understanding, and can be costly to implement. And because most users are unaccustomed to speaking naturally to automated systems, they do not know how to best form their responses, and are thus uncomfortable with these more "natural" interfaces. To the contrary, my experience shows that well-designed directed dialog interfaces are often the most successful and comfortable for users.

6.5 TECHNIQUES FOR TESTING THE INTERFACE

Collecting data from end users of speech-enabled applications in usability tests is a vital step in any speech project, as it is for any automated system. Usability testing for speech-enabled applications should begin at the earliest stages of a project in order to be maximally effective. Early user input allows the designer to make changes when it is still relatively quick and inexpensive to make them. Moreover, subjective data from end users is an important factor in interpreting objective measures of the technical performance of a speech application. In isolation, measures such as recognition rate or call containment do not give organizations a meaningful way of measuring the overall success of a speech application. Only the combination of objective performance measures (such as recognition rate) coupled with subjective measures of performance (such as perceived frequency of being misunderstood by the system) allows us to know if a speech-enabled application is good enough (Hura, 2007a).

As in any usability project, there is a trade-off between how early you test and how representative the collected data are. There are direct correlates in speech to many of the testing techniques that are used in GUI projects. For formal VUI usability tests, there is a choice between using actual speech recognition during testing and mimicking it. This is a significant decision because fully functional speech prototypes are typically coded by developers (rather than usability specialists) and thus usually occur later in the project life cycle. The alternative is a so-called Wizard of Oz test in which no application has been developed at all. Instead, a usability specialist (the "wizard") supplies speech recognition by listening to user input and then playing sound files of prompts to mimic the behavior of the speech application (Hura, Polkosky, & Gilbert, 2006). There are benefits and downfalls to each type of testing.

Wizard of Oz (WOZ) testing is often discussed as being a "discount" usability method equivalent to prototype testing, but no coding is required. Although there is no coding, the benefit is exaggerated in relation to the overall amount of work to run a WOZ test which is equivalent to a prototype test. WOZ testing requires a complete VUI design, professionally recorded prompts, a system for playing prompts over the phone, and the full-time involvement of two human factors professionals for every test session (one to act as the wizard and one to facilitate test sessions—both demanding jobs).

It is certainly possible to run WOZ tests that compromise on one or more of these factors, but the test is then not a true substitute for prototype testing. For example, if you choose to have the wizard simply speak the prompts rather than using recorded prompts, there are multiple disadvantages: First, you lose the ability to gather feedback about the voice talent; second, the prompt will be spoken differently every time it is uttered, which introduces unwelcome variability into the testing process. Because so much of the user experience is bound up in the minute details of the prompts, this compromise is hardly worthwhile in many cases. In spite of the intensive preparation required by the usability team, WOZ tests are typically conducted earlier in the project than tests using functional speech recognition prototypes.

Testing with a functioning speech prototype is the gold standard for assessing VUI usability (Hura, 2003). Prototypes give a realistic view of the flow of the automated conversation, and the data that can be gathered in a prototype usability test are an order of magnitude more realistic than the data from a WOZ test. The timing of the interaction is also more realistic in prototype testing. It is notoriously difficult in a WOZ test to accurately simulate the response latency of a speech application, and wizards typically respond more quickly than the system in some situations and more slowly in others.

One later-stage technique that is underused in VUI testing is field tests, in which users interact with a VUI from their typical usage environment. Field tests are logistically complicated because telephone-based systems can be used anywhere and while in motion (as opposed to computers, which are stationary when used). Additionally, because many commercial speech applications are used rarely by a given individual, it can be difficult to arrange to be present at the time of use. Instead, most VUI projects rely on live call monitoring, or call recording, in which usability specialists can listen to users as they have real-life interactions with a VUI. These techniques do not allow the designer to give the user a set of tasks to complete (hence we can never be totally sure of a user's motivation in the interaction), and there is generally no ability to speak to users directly to understand reactions to their experience following an interaction (although surveys are sometimes administered at this time).

A final consideration for VUI usability testing is logistics: VUI usability testing can take place with the user and facilitator in the same location or remotely over the phone. The technology is readily available to allow users to interact with a speech application while usability specialists (and others) listen in, all over a standard telephone connection. The benefits of over-the-phone testing are that it saves the considerable expense of securing a testing facility, and the time and cost associated with having participants travel to this facility. The downfalls all relate to the fact that we can only listen to the interaction, not watch it, which deprives us of cues that users give through their facial expressions, postures, and body language.

In-person usability testing for VUIs allows us to watch and listen to interactions, and therefore record both audio and video of test sessions. Building rapport with a user is easier when you are physically in the same room with one another, which encourages users to give uncensored responses during interviews. The downfalls of in-person testing are the time and cost associated with bringing a group of participants to one location for testing. In-person testing also limits test participants to those who are willing to travel to the test location, as opposed to over-the-phone testing where remote users can be included easily.

One final VUI testing method is tuning. Tuning is the process of optimizing performance of a speech application overall. The focus in tuning is very broad: Everything from low-level recognition parameters to prompts is evaluated. Tuning involves recording live user interactions with a speech system—several hundred interactions are a typical analysis set for a round of tuning. Recorded calls are transcribed and transcriptions are used as a basis for determining recognition accuracy in the application overall and state by state. Utterances are categorized as in grammar or out of grammar and as correctly recognized, incorrectly recognized (or false accept in which an out-of-grammar utterance is mistakenly recognized as an in-grammar utterance that fails to be recognized), or correctly rejected. Along with computing statistics, tuning also involves listening to as many user calls as possible.

Tuning calls are sometimes the only opportunity that VUI designers have to observe their design in action and determine whether conclusions from usability testing hold in the real world. Based on the computed statistics and insights gained from monitoring calls, designers can identify issues and propose solutions. Tuning offers the opportunity to test some types of solutions before they are put into practice. It is often possible to modify a grammar, run the recorded speech samples through, and observe the results. If tuning data are interpreted narrowly (applying results only to specific dialog states rather than overall across the application) and used judiciously—without making too many assumptions about what users are thinking—it is possible to come to some conclusions about usability from tuning data.

There are few established metrics of VUI usability available or in wide use. Gupta and Gilbert (2005) propose the holistic usability metric, which allows the tester to apply weights to factors judged to be most important for individual interfaces. Because each tester may assign weights differently, this metric can be tricky to interpret universally, but can be quite valuable for testing one application over time or a set of similar applications using the same version of the metric. Polkosky (2005) has constructed an opinion metric specific to VUIs for collecting subjective user data. Her research shows that users perceive the behavior of a VUI as evidence of higher-order factors, such as the overall level of customer service orientation within an organization. This metric is valuable in that it addresses issues beyond pure usability, such as desirability and customer satisfaction, and it has wide implications for exactly which factors in a VUI influence user opinions.

Neither of these VUI metrics has been put to widescale use in the industry. Instead, organizations tend to rely on home-grown surveys or customer satisfaction surveys designed for Web or other media, with a few minor adjustments for speech, to evaluate the quality and usability of speech-enabled systems. The items on such surveys are rarely vetted for validity or reliability, and as such the data gathered from them are less than ideal. One notable exception to this pattern is the services offered by Vocal Laboratories, which offers professionally designed surveys and the ability to quickly collect data from very large populations of users.

6.6   DESIGN GUIDELINES

There are a number of references that provide an overall philosophy of VUI design, including Cohen, Giangola, and Balogh (2004), Ballentine (2001), Kotelly (2003), and Larson et al. (2005). The guidelines presented here are intended not as all-inclusive, but rather as a collection of practical tips for writing effective prompts. The data supporting these tips derive from observations of hundreds of users interacting with dozens of speech-enabled IVR applications during usability tests.

6.6.1   Do Not Worry So Much about the Number of Menu Options

George Miller's 50-year-old research on short-term memory (1956) is often portrayed as the final word on the permissible number of auditorily presented menu options. The thinking here is that it is primarily the limits of auditory short-term memory that determine the usability of a voice menu. There is a kernel of truth here—it is certainly possible to overload users with too much auditory information—but the strict limit of seven-plus-or-minus-two simply does not hold for VUI menus (Hura, 2007b). The flaw in applying Miller verbatim is that using a menu is not a recall task but rather a selection task. Users should never need to hold an entire VUI menu in memory in order to make their selection. The goal is to have them evaluate and accept menu options one by one or, at most, remember a few likely candidates.

6.6.2 Present Menu Options in a Way That Makes Them Comprehensible and Easy to Retain

Especially in menus with more than a small handful of options, the best way to enable successful selection is to ensure that the overall set of menu options is *descriptive* and *distinct*. *Descriptive* menu options give users the immediate sense of "I know what that is!" This enables users to evaluate and discard nondesired options without holding them in memory. Similarly, users will immediately know the differences between *distinct* menu options, which eliminates the need to remember the entire list while making their menu selection. Current short-term memory theorists agree that there are many factors beyond the number of items that affect auditory short-term memory, such as word length, frequency, familiarity, and inter-item discriminability (i.e., how distinct items on a list are from one another [Nairne, 2002]). The importance of descriptiveness is supported by recent research in human–computer interaction showing that meaningful menu option labeling is more important than information architecture for users (Resnick & Sanchez, 2004).

6.6.3 Do Not Let Persona Get in the Way of Effectiveness

Users will always attribute a persona to a VUI, but persona should always be subordinate to effectiveness. Some designers argue that a VUI should never say anything that a person would not say, but following this rule will often lead the VUI designer to write prompts that encourage the user to overestimate the abilities of the speech system. Instead, focus on crafting prompts that lead users to give appropriate responses. It is far better to be somewhat unnatural than to take a user into an error state for the sake of a nice-sounding prompt (Klie, 2007; Rolandi, 2007).

6.6.4 Consider Error Sources When Writing Error-Handling Prompts

Users find themselves in error conditions for a variety of reasons, many of them unrelated to the VUI itself. There are distractions in the environment that may take the user's attention away from the VUI, and background noise that may trigger misrecognitions—or there may truly be an issue with the interface. Error prompting must work for all users. A good practice for error prompts is to allow users to barge in quickly and self-correct (for users who know what they want and simply need a chance to repeat it), but to give additional information to help users who were unable to give a good response at the initial prompt.

6.6.5   Provide Landmarks

Because audio-only interfaces are nonpersistent, one of the most challenging things for users is staying oriented. Using landmark prompts can serve as a way to acknowledge a user's choice and reinforce a sense of place. For example:

> SYSTEM: You can say: account balances, last contribution, transfers, allocations, or life insurance.
>
> USER: Transfers.
>
> SYSTEM: *Transfers.* You can say: hear pending transfers or make a transfer.

6.6.6   Use Hints and Reminders

Presenting information to users before they are in a context in which they can make use of it is ineffective. The more usable alternative is to provide instruction to users in small chunks at times when they are likely to need the information (Hura, 2006). Hints are information presented to users to give them context-specific instruction for the current dialog state or to inform them of shortcuts or alternate methods of completing tasks in the system. For example, if a user has reached the "order status" state in the application by navigating through several menus, consider offering a hint at the order status module: "Next time, you can say 'order status' at the main menu to get here quicker." Reminders, not surprisingly, are brief prompts that remind the user of a relevant bit of information at a particular state in the dialog, such as "You can speak your account number or key it in."

6.6.7   Favor Models over Instructions

It is sometimes necessary to provide an example to users of how to form a response. In these cases, it is often easier and more effective to simply model the desired sort of response than to explain it. Consider two versions of a prompt requesting date of birth to be entered as an eight-digit number. The second version, using a model, is more concise and much easier to understand than the first version.

Version 1: Please enter your date of birth as an eight-digit number, with two digits for the month followed by two digits for the day and four digits for the year.

Version 2: Please enter your date of birth as an eight-digit number. So, March 14, 1986, would be 03141986.

6.6.8   Use Discourse Markers Liberally

Remember, this is a conversation! Discourse markers serve to connect utterances within a dialog. Discourse markers help establish a sequence, change the topic, provide acknowledgment, and convey the speaker's attitude toward what's being said. This does not mean that the VUI should respond to every user utterance with "All right" or "Okay"—this tends to make the VUI sound overly perky. Instead, vary the discourse markers with the conversational context as in the following conversation (discourse markers are in italics):

SYSTEM: Main menu. You can say: balances, transfers, transaction history, or loans.

USER: Balances.

SYSTEM: *All right*, balances. To get you that information, we need to get you logged in. *First*, please enter your account number.

USER: 123–45–678.

SYSTEM: *Next*, enter your password.

USER: 54321.

SYSTEM: *Thanks*. The balance on account 123-45-678 as of May 23, 2007 is . . .

6.6.9   Allow for Conversational Repair

Human conversation is full of small problems. Speakers hesitate, start an utterance and then back off and start again, and mispronounce words; listeners misunderstand and request clarification. We do all of this without thinking about it and without the conversation failing in a vast majority of cases. When there is a small problem in a VUI interaction, users naturally fall back on their techniques to fix conversations (recall the example of the user speaking more slowly and clearly after being misunderstood).

In crafting error prompts, be aware of this tendency and write prompts and grammars that accommodate the user. A common example that is easy to accommodate is prompting following an explicit confirmation. If what is being confirmed is wrong, users often want to correct the mistake in a single turn:

SYSTEM: To confirm, you're looking for flights from Los Angeles to Boston on May 23rd.

USER: No, to Austin.

To accommodate such corrections, build grammars that include "no" plus any correction the use might offer (here, "no" plus city names).

6.6.10   Be Careful with Terminology

Using terminology that is familiar and comfortable for users is a basic guideline for any automated system. In the VUI world, this guideline takes on more importance because the interface is nonpersistent. If users do not understand a term on a web page, they can rescan the options with minimal effort. In a VUI, if users miss a term, they have to spend the time to listen to the entire list of options again. Steer clear of jargon and branded or technical terms in a VUI unless users use these terms.

6.6.11   Give Users an Escape Hatch

Users detest being held captive in an automated system. Most users are willing to try automation to accomplish simple tasks, but when the automation fails, they just want a real person. Users also tend to pick up the phone when they have already established that they have a problem—meaning that a user may be entering his interaction with a call center VUI already certain that automation cannot solve his problem. Businesses deploying speech often want to deflect as many calls as possible away from live agents and want to completely prevent users from requesting a live agent. These efforts are often counterproductive in that users always find the escape hatch, and if you have made them work hard to find it, they will be frustrated and angry when they get there. Take the middle ground: Offer to transfer to agents after you have evidence that the user is having a problem such as repeated time-outs or no matches. It is helpful not only to count the number of errors at a single dialog state but also to use a global counter that keeps track of errors for the entire interaction.

6.7   CASE STUDY

A case study for a speech interface can be found at *www.beyondthegui.com*.

6.8   FUTURE TRENDS

What's the next big thing for VUIs? This is necessarily linked to the next big thing for speech technology overall. One trend that has emerged recently is voice search—using speech recognition to specify a search term and then receiving the output auditorily or displayed as text on some device. One highly anticipated voice search application is 1-800-GOOG-411, Google Voice Local Search, from Google Labs (*http://labs.google.com/goog411/*). Google is the search engine giant on the Web. The Google Labs decision to offer a speech option for specifying searches over the phone makes a portion of the Web accessible to mobile users in a way that it has not been until now. Searching the Web using even the most advanced smart phone is painful. Typing search terms on a tiny keyboard is overwhelmingly more difficult than simply speaking the term. Speech-enabling search is a logical step.

It is interesting to note that the user interface of 1-800-GOOG-411 is quite minimal. There is nothing flashy or chatty about it. The prompts are brief and to the point and are not necessarily phrased exactly the way a person would phrase things in a human-to-human conversation. The VUI designers at Google clearly conceive of 1-800-GOOG-411 as a tool, not as a live operator emulator. The prompts for 1-800-GOOG-411 are certainly pleasant and not at all terse, but they represent a stylistic shift away from the more verbose, more human-aspiring systems of the last five years.

As of this writing, 1-800-GOOG-411 is still described as an experimental service and is limited to Yellow Pages–type searches for local businesses. A number of the features of 1-800-GOOG-411 are novel and would be useful in broader searches as well. For example, the user can browse the results of a search in multiple ways; using simple voice or touch-tone commands, users can navigate the list of results and select the results they want. Alternately, users can choose to have the list of results sent via short message service (SMS, or text messaging) to their mobile phone. Google gives users the choice of modality (speech and touch-tone for input, audio and SMS for output) throughout the application. This decision to be modality agnostic should serve tomorrow's increasingly mobile users well. When the context of use is "anywhere, anytime," the user interface must allow users to select the best way to interact based on their current circumstances.

Speech technology is also poised to move into the home. Appliances, home security systems, and "smart home" systems will soon be supplemented with speech technology. In these cases, speech recognition algorithms are unlikely to reside on the devices themselves, but will be housed on a central server to which the local device communicates in real time. These developments will be propelled initially by the advantages that speech offers in terms of accessibility for the mobility impaired. Being able to speak commands in order to control lights, televisions, heating, cooling, and the like, would provide control and freedom to individuals who today cannot manage these tasks independently. It is likely that speech in the home will be a phenomenon (like curb cuts) in which an affordance designed to assist a small group ends up benefiting the larger community in unexpected ways.

Finally, there is the promise of speech for mobile and handheld devices. Multimodal speech, touch, and graphics devices have been promised by many companies large and small for more than 10 years, but there has yet to be a true multimodal breakthrough. The proliferation of mobile wireless devices with small screens and audio input and output makes the possibility of real multimodality more likely over the next 10 years. Like any technology, breakthroughs will come when multimodality enables easy access to an activity that is cumbersome or impossible today.

REFERENCES

Attwater, D., Ballentine, B., Graham, M., & Hura, S. (2005). Voice user interface design: What works, what doesn't. Paper presented at SpeechTEK West Conference.

Baber, C., & Noyes, J. (1993). *Interactive Speech Technology: Human Factors Issues in the Application of Speech Input/Output to Computers.* London: Taylor & Francis.

Ballentine, B., & Morgan, D. (2001). *How to Build a Speech Recognition Application.* San Ramon, CA: Enterprise Integration Group.

Blumstein, S., & Stevens, K. (1981). Phonetic features and acoustic invariance in speech. *Cognition* 10:25–32.

Campbell, J. (1997). Speaker recognition: A tutorial. *Proceedings of the IEEE* 85:1437–62.

Cohen, M., Giangola, J., & Balogh, J. (2004). *Voice User Interface Design.* Boston: Addison-Wesley.

Denes, P., & Pinson, E. (1993). *The Speech Chain.* New York: Worth Publishers.

Furui, S. (1996). An overview of speaker recognition technology. In Lee, C., & Soong, F., eds. *Automatic Speech and Speaker Recognition.* Boston: Kluwer Academic, 31-57.

Gupta, P., & Gilbert, J. (2005). Speech usability metric: Evaluating spoken language systems. *Proceedings 11th International Conference on Human–Computer Interaction, Las Vegas, NV* (CD-ROM).

Grice, H. P. (1975). Logic and conversation. In Cole, P., & Morgan, J., eds., *Syntax and Semantics*, Vol. 3: *Speech Acts.* New York: Academic Press, 41-58.

Heider, F. (1958). *The Psychology of Interpersonal Relations.* New York: John Wiley & Sons.

Hura, S. (2003). Passing the test. *Speech Technology* 8.5 (September/October).

Hura, S. (2005). Great expectations. *Speech Technology* 10.6 (November/December).

Hura, S. (2006). Give me a hint. In Meisel, W., ed., *VUI Visions: Expert Views on Effective Voice User Interface Design.* Victoria, BC: Trafford Publishing.

Hura, S. (2007a). How good is good enough? *Speech Technology* 12.1 (January/February).

Hura, S. (2007b). My big fat main menu. *Speech Recognition Update.*

Hura, S., Polkosky, M., & Gilbert, J. (2006). Point, counterpoint: What three experts say about speech usability. *User Experience* 5:2.

International Organization for Standardization. (1998). ISO 9241:11. Ergonomic requirements for office work with visual display terminals (VDTs)—Part 11: Guidance on usability. Geneva, Switzerland: ISO.

Klie, L. (2007). It's a persona, not a personality. *Speech Technology* 12.5 (June).

Kotelly, B. (2003). *The Art and Business of Speech Recognition: Creating the Noble Voice.* Boston: Addison-Wesley.

Larson, J., Applebaum, T., Byrne, B., Cohen, M., Giangola, J., Gilbert, J., Nowlin Green, R., Hebner, J., Houwing, T., Hura, S., Issar, S., Kaiser, L., Kaushansky, K., Kilgore, R., Lai, J., Leppik, D.,

Mailey, S., Margulies, E., McArtor, K., McTear, M., & R. Sachs. (2005). Ten guidelines for designing a successful voice user interface. *Speech Technology* 9.7 (January/February).

Lea, W. (1980). *Trends in Speech Recognition*. Englewood Cliffs, NJ: Prentice Hall.

Massaro, D. (2006). Speech perception and recognition: Theories and models. *Encyclopedia of Cognitive Science*. New York: John Wiley & Sons, 1-9.

McClelland, J., & Ellman, J. (1986). The TRACE model of speech perception. *Cognitive Psychology* 18:1–86.

Mori, M. (1970). Bukimi no tani. The uncanny valley. MacDorman, K. F., & Minato, T., trans. *Energy* 7(4):33–35.

Nairne, J. (2002). Remembering over the short-term: The case against the standard model. *Annual Review of Psychology* 53:53–81.

Nass, C., & Brave, S. (2005). *Wired for Speech: How Speech Activates and Advances the Human–Computer Relationship*. Cambridge, MA: MIT Press.

Nielsen, J. (1993). *Usability Engineering*. San Diego: Academic Press.

Pinker, S. (1994). *The Langauge Instinct*. New York: HarperCollins.

Pitt, I., & Edwards, E. (2003). *The Design of Speech-Based Devices*. London: Springer-Verlag.

Polkosky, M. (2005). Toward a Social Cognitive Psychology of Speech Technology: Affective Responses to Speech-Based e-Service. PhD diss., University of South Florida.

Pruitt, J., & Adlin, T. (2006). *The Persona Life Cycle: Keeping People in Mind Throughout Product Design*. San Francisco: Morgan Kaufmann.

Rabiner, L. (1989). A tutorial on Hidden Markov Models and selected applications in speech recognition. *Proceedings of IEEE* 77:257–86.

Reeves, B., & Nass, C. (1996) *The Media Equation: How People Treat Computers, Television and New Media Like Real People and Places*. Stanford, CA: Center for the Study of Language and Information.

Resnick, M., & Sanchez, J. (2004). Effects of organizational scheme and labeling on task performance in product-centered and user-centered web sites. *Human Factors* 46:104–17.

Rolandi, W. (2002). Do your users feel silly? *Speech Technology Magazine*, November/December—available at *http://www.speechtechmag.com/Articles/ReadArticle.aspx? ArticleID=29420*.

Rolandi, W. (2007). The persona craze nears an end. *Speech Technology* 12.5 (June).

Sacks, H., Schegloff, E., & Jefferson, G. (1974). A simplest systematics for the organization of turn-taking for conversation. *Language* 50:701.

Schegloff, E., Jefferson, G., & Sacks, H. (1977). The preference for self-correction in the organization of repair in conversation. *Language* 53:361–82.

United States Access Board. (2007). Section 508 Accessibility Standard—available at *http:// www.section508.gov/*.

Shiffrin, D. (1987). *Discourse Markers*. Cambridge: Cambridge University Press.

Stevens, K., & Blumstein, S. (1981). The search for invariant acoustic correlates of phonetic features. In Eimas, P., & Miller, J., eds, *Perspectives on the Study of Speech*. Hillsdale, NJ: Erlbaum, 1-38.

Weinshenk, S., & Barker, D. (2000). *Designing Effective Speech Interfaces*. New York: Wiley.

World Wide Web Consortium. (2007). Web Accessibility Initiative—available at *http://www. w3.org/WAI/*.

Interactive Voice Response Interfaces

Jeff Brandt

7.1   NATURE OF THE INTERFACE

Interactive voice response (IVR) interfaces are chiefly telephony interfaces. They are the much maligned systems that you reach when calling a business and do not connect to a live person, but instead hear a recording something like this: "Your call is very important to us.... To continue in English, press 1." Most people can readily recall instances of terrible experiences with IVRs. IVRs, however, are not inherently bad. More often than not, they are created that way by their designers. This chapter will explore the challenge of designing IVRs that are usable and useful, and it will offer techniques for testing IVRs that will give the designer confidence that, once fielded, the IVR will fulfill its design goals and leave its users smiling (or at least not cursing) (Figure 7.1).

The most common IVR application is voice mail. Another well-known example is a 24-hour banking application where account holders can check balances and transfer funds from one account to another using the phone. The user dials a phone number associated with the bank, which is answered by an automated system (the IVR), which immediately plays out a recorded message: "Welcome to the Big Bank. Please enter your account number." Using the keys on the telephone, the caller enters an account number. The IVR receives these key presses and can then check the entered account number against a database and continue with the interaction.

It is the playing of recorded messages or prompts and the collection of responses through key presses that define a user interface as an IVR. On a phone the key presses are made on the dual-tone multifrequency (DTMF) keypad (DTMF was also trademarked for a time as "TouchTone"). The principles of IVR design can also be applied to other interfaces that combine the elements of an

FIGURE

7.1
Most people expect terrible IVRs.
Poor designs have conditioned users to dislike IVR—especially when they are expecting a person. Users tend to be surprised when an IVR has been well designed to meet their needs. *Source:* From "Blondie," February 7, 2007; © King Features Syndicate.

IVR with others such as those in a self-service check-out system. These systems typically play prompts to communicate with the shopper, and often have a numeric keypad used for responses, just like an IVR. Additionally, self-service checkouts have barcode scanners and frequently a touch screen that can both display information and accept input.

In general, the psychophysical requirements from an IVR user are minimal. The buttons on phones are designed to be used by people with a broad range of physical abilities (see Chapter 2 for more detail on haptic interfaces). Individuals are free to select a phone with keys that match their physical needs for size, labeling, and push force, so long as the phone produces standard DTMF tones. Users who can carry on a conversation with another person on a phone are usually able to hear the prompts from an IVR. Some IVRs provide a means for a user to adjust the volume or pace of prompts. Users requiring amplification require it of all calls, not just calls reaching IVRs. Callers with these needs typically make use of a phone with adjustable volume, or use a hearing aid and a hearing aid–compatible phone (see Chapter 5 for more detail on psychophysical considerations of auditory interfaces).

While the user's physical capabilities must be taken into consideration, the most challenging human characteristic to consider in IVR design is the cognitive capabilities of the expected users of the system. Quite frequently the intended user population is the general population, with its inherent range of cognitive ability. Differences between the user's expected task flow, the language model she uses to describe the flow, and the vocabulary presented by the IVR can cause difficulties. The serial nature of recorded prompts also presents a unique cognitive challenge. The designer must consider the fact that a user must hear an option and the associated key to press for that option, one following another, and may have to try to hold in short-term memory one or more possible choices while simultaneously evaluating the continuing options on a menu. The fact that most

IVR systems have a finite waiting period for a response before timing out and hanging up creates a further cognitive load. Techniques to make IVRs more accessible are discussed in Section 7.3.1.

TECHNOLOGY OF THE INTERFACE

Historically, IVRs were built on telephone companies' (telcos') class-5 switches such as the Nortel DMS100 or the Lucent 5E. These switches were used to provide the IVRs that the telcos themselves might need, and in some circumstances to provide IVRs for large corporate customers. Class-5 switches have many limitations that must be considered when designing an IVR. Typically, recorded prompts have length limitations—often as short as 17 to 20 seconds. Prompts may have to be recorded over a phone handset as a live recording, or may have to be recorded by the switch manufacturer onto an announcement card. Some keys on the phone keypad may have reserved functions that cannot be changed, and the timing parameters may not be sufficiently adjustable. The ability to interface with external data sources (information not on the switch or in the telephony network) is limited and in some cases not possible or not in real time.

Later, IVR capabilities were built into smaller telephony platforms including CENTREX platforms and key systems, allowing medium-sized businesses to create their own IVRs. Many of the limitations imposed by the class-5 switches were reduced on these platforms. Currently, IVRs can be built on hardware as small as a consumer-grade PC. Outfitted with a telephony modem or a purpose-built telephony line card and software, a system to serve one or more lines can be created for less than $10,000. Line cards allow phone calls to connect to the computer that is hosting the IVR and playing the prompts. Line cards are manufactured for both analog (plain old telephone service [POTS]), and digital (ISDN or T1) phone lines. Voice over Internet Protocol (VoIP) calls can be terminated through the computer's Ethernet/Internet network connection.

Voice eXtensible Markup Language (VXML) now allows IVR service providers to host your IVR code on their hardware—eliminating the need for small- and mid-sized businesses to maintain telephony servers. Current IVR technology affords the designer virtually complete control over the interaction. The limitations present in IVRs today are mainly those that are inherent to the technology. First, information must be presented serially due to the auditory nature of prompts. Second, responses are limited to the 12 keys on the phone keypad, hanging up, or doing nothing, which eventually results in a time-out. Third, the keypad makes number entry easy, but alphanumeric entry is tedious, slow, and error prone. Fourth, although the phone supports full duplex audio (both sides can talk and hear), IVRs cannot "hear" anything other than the DTMF tones produced by key presses—all the unkind words shouted at IVRs fall on deaf "ears." Speech recognition interfaces, where the system can "hear" (and hopefully understand) what is said, were covered in Chapter 6.

7.3 CURRENT IMPLEMENTATIONS OF THE INTERFACE

Interactive voice response interfaces are ubiquitous. It is nearly impossible to call a phone number without the likelihood of reaching an IVR. This bold statement is true because voice mail is an IVR application, probably the most common of them all. Every sizeable business seems to have an IVR call router that attempts to direct your call to the proper person or department. Once your call has been routed, it may go not to a person but to another IVR, such as an information service that allows you to retrieve data—your bank balance, the status of an airline flight, or movie show times, for example. There are also IVR systems that allow you to perform actions. These vary from home automation IVRs that turn lights on and off to a stock-trading system that allows a caller to place buy and sell orders. The list goes on and on.

Users will grudgingly accept an IVR so long as the task they called for can be completed nearly as well with the IVR as the user expects it would be by a person. In many cases, however, IVRs offer better service than would otherwise be available. They are typically available 24/7 and are not affected by weather, illness, strikes, or new movie releases. Balance-checking IVRs at banks are a great example—bankers' hours are not when most of us need to know if that last deposit has been credited or not. Although online systems perform many of the same functions as IVRs do, there are many times when and places where people are without Internet access, although most carry a cell phone that can be used to call anywhere, anytime.

In some cases, an IVR is preferred over a person. IVRs offer inherent privacy (Tourangeau & Smith, 1996). When you use an IVR, no person is listening to your Social Security number, your PIN, or your account number. The IVR will not recognize your voice or your name and will not talk over lunch about your medical problems, financial issues, test results, or scandalous interests—and an IVR will not complain that you should have written all this down when you call a second or third time.

Creating an IVR can reduce the cost of offering a service, possibly resulting in a cost savings passed on to the customer. The personal touch, however, is lost. No automaton can replace a person who is genuinely happy or sad for you, and not one of us puts much stock in the prerecorded "Please's," "Thank you's," and "We're sorry's."

7.3.1 Applications of the Interface to Accessibility

For a person with a speech production deficit—a person who has difficulty speaking, difficulty being understood, or is completely unable to speak—IVRs are a blessing. In this case, IVRs create a means for accomplishing a task that does not require speaking. Good points notwithstanding, IVRs tend to create more accessibility problems than they solve. IVRs tend to increase the cognitive requirements for completing a task when compared with a customer service representative. For some users with cognitive deficits, the predictability of an IVR ("Dial this number, press 1 and then 3") may be easier than interacting with a person. However, short time-outs requiring quick responses and the inability to get additional explanations may make IVRs more difficult for others.

In a very, very few cases, IVRs have been built that can interact with the TTY/TDD machines used by the Deaf to communicate via phone. However, most IVRs are unusable by people with severe or total hearing loss. IVRs also interact poorly with TTY relay service and communications assistants (CAs), as the time-outs on most IVRs are too short to allow the CA to type the text of the prompt to the relay caller, receive a reply, and then press a key in response. TTY is a slow means of communication, and in the United States, by law, CAs are required to transcribe what they hear from the IVR verbatim.

It might be assumed that if a person can dial a phone, he can use the IVR reached. While there are many dialing aids for people with motor disabilities, most of these function as speed dialers and do not have the capability to stay online after the call completes, and do not provide a faculty with which to perform postdialing key presses. By and large, those using adaptive technology to dial phone calls have difficulties with IVRs.

7.4 HUMAN FACTORS DESIGN OF THE INTERFACE

The human factors discipline is what makes the difference between an IVR that people do not mind using and one they love to complain about. Advances in IVR software have made it possible for people with low levels of skill in computer programming, in human factors, and in user interface design to quickly and easily set up a terrible IVR and inflict it on the unsuspecting public. This section will describe the human factors involved in designing an IVR. While much of this section is based on 10 years of experience as a human factors engineer with AT&T, many others have published on this topic, such as Gardner-Bonneau and Blanchard (2007), Schwartz and Hardzinski (1993), Schumacher (1992), and Marics (1991). I owe a great deal of what I know to those who have labored in this area before me.

7.4.1 When to Select an IVR Interface

Rarely, if ever, will the IVR be the only interface aimed at caring for the user's needs. Most commonly, IVRs share the burden with customer service representatives (CSRs) with the goal of achieving some level of customer self-support through the IVR, while the challenging, unusual, or nonstandard issues drop out of the IVR for the CSRs to handle. In these cases, it is also common to have a website that is designed to fulfill the same goal—to enable customer self-service to reduce the need for and cost of CSRs. Both the IVR and the website tend to cost less than staffing a call center for the full load, 24/7. Full automation and 100 percent coverage of all issues is rarely a realistic goal, and in most cases not appropriate. This means that the IVR cannot be designed in a vacuum, but must be considered as part of a system. Chapter 11 discusses in greater detail the human factors of offering the same service over multiple interfaces.

Once it has been determined that users will want to call, the most common trade-off to be considered is whether the phone interface should be an IVR or a speech recognition system. Many people erroneously assume that "If only the callers could talk to the system, it would be easier." Although speech recognition interfaces are the subject of Chapter 6, a few comments here would be appropriate. In this author's experience, there are three key factors to consider when choosing between speech and IVR systems: accuracy, privacy, and the nature of the menu content.

First, although it will be argued by every speech recognition vendor you encounter, an IVR should be the default choice on the basis of accuracy. The DTMF keys of an IVR are recognized with near-perfect accuracy; speech, on the other hand, advertises recognition accuracies in the middle- to upper 90th percentile. In practice, this level of accuracy is difficult to achieve—even granting it, this level of accuracy pertains to the recognition of single utterances. The cumulative chance for misrecognition grows with each utterance required to advance the user through the menus to complete a task. The typical task of a caller requires several menus to be traversed, and thus his or her success and satisfaction with the system is dependent on multiple recognition events.

Given a theoretical speech recognition system with 99 percent recognition accuracy for connected digits, the cumulative probability of collecting a 10-digit phone number without a recognition error falls to 90 percent. If the application needs a 16-digit credit card number to be spoken and a 9-digit Social Security number (SSN) for confirmation, the probability of an error-free attempt falls to 82 percent. Error recovery is an important, difficult, and time-consuming part of designing an IVR or a speech system, but given the higher probability of errors in a speech system, more attention is required compared to an IVR.

This brings us to the second consideration—privacy. In the preceding example, the user of a speech interface would have had to speak his credit card number and SSN aloud. Frequently, calls to systems such as these are made from public places such as coffee shops or airports where it would be unwise to speak aloud personal and otherwise private data. IVRs are more secure in these cases, as it is a challenge to determine from watching someone what the keys they are pressing mean, whereas phone numbers, SSNs, and credit card numbers each have a unique cadence that when spoken makes them easy to identify. And while it would be a challenge to unobtrusively attempt to watch a person enter key preses, it is often difficult not to hear the conversations of others whether one wishes to eavesdrop or not. Give people a cell phone and they will talk loud. Give them a speech system on a cell phone and they will talk louder still. Let the system make an error and they will shout their SSN, loud and slow, so the "stupid" system (and everyone else in the room) can clearly hear them. Good speech designers will "fall back" to DTMF in these cases rather than require speech, creating a hybrid interface. At a certain point one has to ask how much, in what way, and at what cost is speech recognition helping the user?

Hopefully, the honest answer to the above question is that speech is allowing your interface to do something well that an IVR cannot. This is the third consideration—the nature of the menu content. IVR menus do best with things that are natural ordered lists, and preferably lists with fewer than 10 items. A good example would be choosing a day of the week: "For Monday, press 1. For Tuesday, press 2," and so on through 7 for Sunday. IVRs also do okay with short menus that can be ordered, although no natural order exists: "For vanilla, press 1; for chocolate, press 2; for strawberry, press 3." IVRs fall down when the categories are large, creating an artificial need to break them into multiple menus.

One could imagine having to select one of the 50 U.S. states from an IVR. How many times would you have to "press 9 for more options" to get to the menu that would let you select Texas? What would the order on the menu be? Should it be alphabetical, or would it make more sense to order by size, population, or frequency of selection? It sounds awful—and it would be, in an IVR. Choosing a state is the perfect task for a speech recognition system. The vocabulary is clearly defined and bounded, but is too large to be easily done in an IVR. If your design need is one that will cause the user to make choices from large, difficult-to-organize groups, speech should be considered. If the need can be completed with tasks that are limited to menus that are short and can be sensibly organized numerically, an IVR should be considered.

The preceding is not meant to scare designers away from speech recognition systems, as they have their place; rather, it is meant to allow for a reasoned choice when speech is inevitably presented as the cure-all that will make users happier and more successful than a stodgy old IVR. Newer and flashier almost always means more expensive, and "better" can only really be judged by users.

7.4.2   What Data Should Be Collected to Design an IVR

The single most important thing to know is that you are not representative of the users of your IVR. You will have to find out who the users are and as much as you can about them. Why did they call the number that got them connected to the IVR? What goals do they have and what tasks are they hoping to complete? What do they know? Have they done this before? What level of functional vocabulary might they possess? When you begin to look really hard at the answers to these questions, the implications can be staggering.

Who are the users of voice mail? There are two kinds of voice mail users: the subscriber who is retrieving messages, and callers who leave messages for the subscriber. There are not many good assumptions you can make about either user. Both could be of nearly any age, 6 to 106, of any vocabulary level, and neither may speak the primary language of your area. A useful trick is to use grammar checkers available on the Web or in many word processors to give you the reading level of your prompt script. In Microsoft Word, go to Tools > Options > Spelling & Grammar, and place check marks in the boxes for "Check grammar with spelling," and "Show readability statistics." In most cases, for general telephony applications we have found that prompts written at the *third-grade level* or lower perform well in usability testing. The subscriber could have received some materials from you such as a user guide—and they may or may not have read them.

Voice mail illustrates two broad categories of IVRs: subscribed to and "walk-up-and-use." In a subscribed-to interface, the user may have received training materials, and is likely to use the IVR repeatedly with some learning. In a walk-up-and-use interface, the user may have no idea why they are hearing a recording, no training, no experience, and no desire to participate in what is being asked of them. For these reasons, the user interface design bar is much higher for a walk-up-and-use interface.

In most IVR designs, there will more than one set of users, so you must be certain that the user interface gets designed for the real users. For example, consider an IVR built for the accounting department that keeps track of the hours that employees work and then feeds those data to a payroll system. Without considering the large number of employees who will have to use the system to enter time, the IVR will get designed by and for the few payroll personnel who need to use it to pull reports. It will contain jargon and acronyms specific to accountants that will be unfamiliar to the majority of users. The process of entering time will be constructed around the database used to store time data, not the way a "normal" person thinks of the work week.

It is paramount to understand that your IVR will likely have more than one type of user, and that each type has differing goals and characteristics. If one group of users controls the design, there is a high risk that while their needs will be considered, the needs of any other user population will be left unconsidered, or poorly cared for in the design. The typical end result is a system that "functions," that is, the task can be completed using the IVR, but with high training costs, low success, high error rates, and low satisfaction. Recognizing that there are multiple user types and allowing for their needs in the design will lead to a more successful design.

Even an IVR that is designed to simply route callers to an appropriate person or department requires human factors techniques to succeed. Determining what to call items on a menu, and what subitems fall under each menu heading, should be a human factors exercise, not simply what the designer thinks will work. Card sort techniques are quite useful here for determining menu terminology and mapping. We once had an incoming call router in one of our regions that was intended to allow callers to reach a person in one of our departments who could solve the caller's problem. The original call router had been developed without the aid of human factors, and had been created from the company's and not the customers' point of view. The main menu offered choices of Departments A, B, and C. Customers frequently had difficulty determining which of our departments should handle their problem; they made a choice but reached the wrong department. The CSR they reached had to determine the nature of the problem, the correct department, and then manually transfer the call. Everyone was frustrated: the customer, the CSR, and the corporation because of the wasted time.

Staff in our human factors group sampled approximately 20,000 calls to determine the "reason for calling." We brought in participants and asked them to sort the 30 or so most common reasons for calling into groups. We did not specify the number of groups, just that there be at least two. We then analyzed the groupings for common patterns and found a set of three groups that appeared to fit most participants' view of how to organize the reasons for calling. A second group was brought in to view the groups and the reasons in them and to provide suggested names for the groups. A third group was then brought in to validate both the names and the groupings. Given a reason for calling and the names of the three groups, we asked the participants to tell us under which group name they would expect to find each reason for calling.

This process may be repeated until a satisfactory result is obtained—that is, that a group of naïve users can reliably select a named menu option that will solve their given problem. The menu for this call router was recorded using the names resulting from the study in the form of "For a problem of type 1, press 1; for a problem of type 2, press 2; or for a problem of type 3, press 3." No new technology was applied; the routing did not even have to be changed. The improvement in efficiency resulted in a recurring $29 million annual cost savings. The cost to achieve this savings? About 6 to 8 weeks of time for one human factors engineer, and less than $10,000 in recruiting and incentives for participants. Customers were less frustrated—they routed themselves correctly more often. The CSRs were happy—they got credit for serving more customers rather than spending time transferring misdirected calls. The corporation was pleased to have happy customers and happy employees, and to have improved efficiency.

7.4.3 How to Specify an IVR

Small IVRs can be specified in a text outline or as a page or two of pseudocode. These methods are sufficient for a one- to two-level–deep menu. An example might be a small company call router where you have three departments and one to five employees in each department. For anything larger or more complex, flowcharting or state diagramming will be required to fully describe the desired behavior of the IVR. There are many different styles and conventions for how to lay out the diagrams that describe a user's progress through an IVR. Consider using a format that completely describes only one menu per page. Formats that spread logic across multiple pages often make it challenging to understand what the desired behavior is.

7.4.4   What the Flowchart Should Show

What is the phone number that must be dialed to reach the IVR? How many times will it ring before the IVR comes on the line? As soon as the IVR answers, what is the first prompt to be played? Typically, the first prompt identifies the organization that you have called: "Welcome to AAA Plumbing." Immediately following the first prompt is usually a choice of some type. In order to fully describe the actions on a menu, each of the 12 DTMF keys (0, 1 to 9, * and #) must be assigned a function or path. A value for how long the system will wait for a response (5 to 6 seconds is good) before "timing out" and doing something for the user is needed, as well as the action to be performed on a time-out. What the system should do if the user hangs up at this point in the process must be described as well.

In the example of Figure 7.2, the user has made a selection on the main menu indicating that she wishes to change her language selection (currently English). Upon entry, the error count is set to zero and the menu is played. The IVR then waits for a key press (hopefully 0, 1, 2, 3, or *). If the user presses * (star), she is given a status message that the language selection has not changed and then returned to the main menu. Upon pressing 1, the flow continues to another page where the language will be changed to English. Upon pressing 2, the flow continues to another page where the language will be changed to Spanish. Upon pressing 3, the flow continues to another page where the language will be changed to bilingual, which will allow a caller to choose either English or Spanish when she calls. If the user presses a key with no function (4, 5, 6, 7, 8, 9, or #), she is informed that the key she pressed is not one of the options. The error count is then incremented.

On the first error, the user is returned to the menu to try again. On the second and third errors, the user is played a context-specific help prompt before being returned to the menu to try again. On the fourth error, the user receives a status report that the language settings have not been changed and the call is disconnected. Calls are disconnected after multiple errors to protect the IVR platform from being overloaded by calls that are not able to use the system. Most of these calls are from autodialers trying to make sales calls, fax machines, or hang-up calls. It is the responsibility of the IVR designer to ensure that it is easy enough to use that few of the disconnected calls are real users who are frustrated. If the user presses 0 (zero) or times out with no response after 5 seconds, the error count is incremented and error handling follows. This illustrates how a "simple" three-choice menu is more complicated to specify than one might expect.

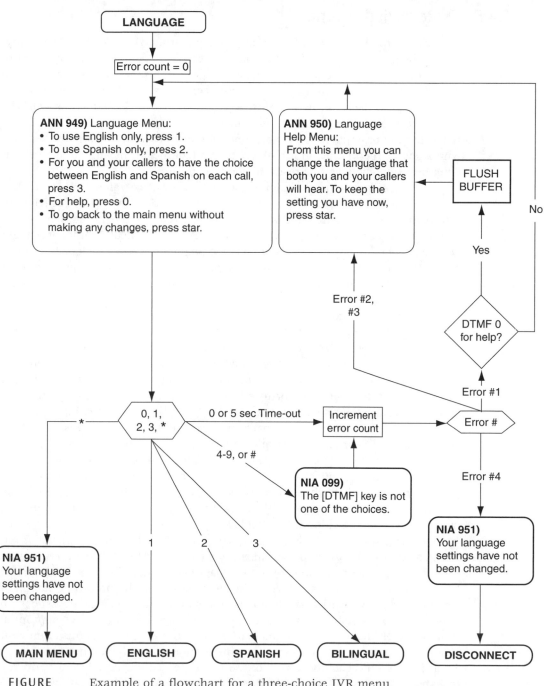

FIGURE

7.2

Example of a flowchart for a three-choice IVR menu.

The user has made a selection on the main menu indicating a desire to change the language selection (currently English).

7.4.5   Recorded Prompts

By the time the hard work of designing the logic of the IVR is complete, prototyping has been done, it has passed usability testing, and people are ready to get the IVR out the door. Consequently, one of the most important aspects of an IVR—the recorded prompts—is often given very little consideration. The prompts will have been "word-smithed" during usability testing, so that *what* is said works well. *How* it is said, though, is of equal importance. It is not uncommon for the task of recording the prompts to fall to someone on the development team, the secretary who answers the phones in the office, or the technician installing the hardware. Only occasionally does this result in good recordings.

The person reading the prompt script, your "voice talent" need not be a professional, but his or her voice must be pleasant, friendly, warm, engaging, relatively free from accents, and should give your callers the impression of your company or organization that you desire. Irritating voices; bored or haughty voices; speech impediments; and gum smacking ought to be avoided. The character of the voice should match your application. A chirpy, gushing cheerleader voice is probably not the best choice for a somber application like a hospital, church, or bank.

On the other hand, the voice talent for a health club, car dealer, or dance club may need to be engaging and upbeat. If your IVR may change over time, select a voice talent who is likely to be available in the future for additional recordings. It is best to have the entire IVR recorded by a single voice talent (Christodoulou, 1996). If you cannot use the previous talent again, you may incur the expense of rerecording the entire script to effect a simple addition or change. In some cases, the voice talent becomes a large part of the "brand" by creating a "persona" that comes to represent the company or organization that the IVR serves. Many celebrities lend their voices to corporations hoping to influence users of their IVRs through familiarity, recognition, and associations with the voice talent. Verizon has used James Earl Jones's distinctive voice (the voice of *Star Wars'* Darth Vader) primarily in advertising, but also in some of their IVRs. When creating a persona, it is important that the persona reinforces, rather than challenges, the politics and culture of the organization it is intended to represent.

One area that is particularly challenging for IVRs is handling bilingual or multilingual needs. Designers often flippantly suggest that "we'll just translate it and record the script in two, three, or ten languages, and then *all* of our customers will be able to use it." There are several fallacious points in the preceding statement. Even high-quality translations do not often result in usable IVRs. One issue is that concepts and metaphors that work in one language are not always appropriate in another. If English is the original language of the IVR, translations often result in longer prompts, in some cases exceeding limitations that were not a problem in English. Without usability testing of the translated script, how it will perform is unknown. The pool of human factors professionals is small to begin with; fewer still are able to perform testing in multiple languages. Suddenly our "easy" task is becoming more complicated.

The hardest part of a multilingual interface is simply getting users to stay with it long enough to use it. In order to tell the IVR the language that you want to use, you must wait until you hear—in a language you understand—the instruction to choose a language and the key to press to select it. A nice graduate thesis could be built around answering the question of how many options for a language a user will listen to in a language they do not speak before hanging up. It is therefore important to understand the expected user population before reaching for a multilingual solution. Be careful to evaluate your decision based on the number of non-native speakers rather than on ethnicity.

Experience tells us that most voice talents are female. For better or worse, our old cultural stereotypes are still with us. Unpublished research from our lab suggests that female voice talents are perceived as "friendlier" than male voices, while male voices are more "credible" than female voices. In one study, we varied the gender of the voice talent across two types of recordings: menus that offered choices and responses that provided information. Not surprisingly, the IVR that subjects in the study had the most confidence in and rated as the friendliest had a female voice talent offering choices and a male talent providing the information that the user was trying to retrieve. What was surprising was this—almost all subjects reported hearing only a single voice talent!

The script that the talent will follow to record your prompts is an important tool for increasing the quality of the IVR. In many cases, a simple list of phrases is provided to the voice talent. While this will get the job done, there is no context—that is, no means for the voice talent to try to understand what is happening or how he can make it easier for the user to understand by pausing or pacing the prose. It can help to provide some side notes to the talent, such as "This 'hello' is a statement of greeting, while this 'hello' is a question—are you there?" Scripts often break prompts into short phrases, and it is often not obvious to the voice talent how phrases will be combined. Without this knowledge, the combined phrase may sound awkward when played in sequence. If possible, it is productive to have someone from the design team who is intimately familiar with the flow and prompts of the application sit with the voice talent to coach her as prompts are recorded.

Regardless of the methods used to procure the prompts, it is necessary to perform a full test of the complete application to ensure that the right prompt plays at the right time, sounds good, and contains the correct verbiage. When producing a script, best practice is to spell out numbers as you want them read. Instead of "20," write out "twenty" or "two-zero" or "two-oh." In the United States, the number 0 is usually read as "zero" and in some cases as "oh," while in other countries it may be read as "zed." U.S. convention for the # key is "pound," and the * key is read as "star." There are international differences in these conventions. You must not only know your users so that a correct specification can be created, but with outsourcing and globalization, you must know your partners and vendors so that they do not unwittingly introduce inappropriate terms into the script.

Prompts need to be recorded in a suitably quiet environment. Background noise included in your prompts only makes the already challenging task of listening to the menu choices harder. From a recording point of view, no office is "quiet," and even the hush of the cooling fan on a PC will lower the quality of your recordings. "Hold music" is fine for being on hold, but it has no place as a background in an IVR prompt. Your prompts will be played over a telephone with about 3 kHz of dynamic range (Carey et al., 1984), which is of low audio quality when compared to a CD recording with a 44-kHz range. Production values should reflect this. IVR prompts are typically recorded at 8-kHz sampling in mono, and they do not need to be CD quality. Your prompts will not sound as good played over a telephone as they do in the studio or as .wav files on your PC, but this is not an excuse to accept poor recording quality. The IVR will never sound better than the source recordings, so the recordings must be as clean as practicable.

An important final step in producing prompt recordings for an IVR is to filter the recordings for DTMF frequencies. Some voice talents' speech naturally includes the same frequencies as a DTMF key press. Some recording tools such as Adobe Audition (formerly Cool Edit) have DTMF filters available as a function that can be applied to a sound file. Filtering the DTMF frequencies out of the prompts prevents the IVR from being tricked by its own prompts into thinking that the user has pressed a key.

7.5 TECHNIQUES FOR TESTING THE INTERFACE

The deceptively simple nature of IVRs requires that special attention be paid to the methods that are employed in the testing of the interface to ensure that valid, reliable, and useful results are obtained.

7.5.1 Task-Based Protocol

The best results in the AT&T Human Factors Lab have always been gained when a task-based testing protocol was used. Each participant is given a number of tasks that align with the reasons a person might call the number to reach the IVR under test. If, for example, the IVR under test is a call blocking service that allows you to block incoming calls from specified phone numbers, one task might be to add a phone number to the list of numbers that are blocked. A better task would be a narrative that gives less direction like the following: "Your daughter has an ex-boyfriend who keeps calling. See if you can set up the call-blocking service so that if he calls your daughter, the phone won't ring."

The least informative tasks are the ones that provide too much information and lead the test participant through the interface step by step. Create scenarios that will exercise the features of the IVR, and provide as little "instruction" as possible. It is fair and often informative to include a task that the IVR will not support. It allows you to see if a person can determine that the IVR will not do what she was trying to accomplish or if better feedback is required. Tasks that contain incorrect or partial information can be used to test the error recovery elements of the IVR. It is a good idea to test the error recovery paths in addition to the "sunny-day path" (in which the user does everything just as the IVR expects).

Our preferred test regimen is one or more iterative usability tests on prototypes of the IVR to improve usability, utility, and accessibility, followed by a summative usability test on the last prototype or preferably the production system to characterize the expected performance of the IVR. Iterative testing of a prototype can quickly improve a design. It is especially effective if a developer who can modify the prototype is on hand during testing. Using this method, weakness in the design can be corrected overnight between participants or, in some cases, as soon as between tasks. With IVRs, it is also helpful if the voice talent who recorded the prototype is available during the study. In many cases you may fill the roles of designer, tester, voice talent, and prototyper, giving you full control and responsibility to use the iterative process to get the IVR in final form.

7.5.2   Signal Detection Analysis Method

One technique that can be used to great effect in IVR testing borrows from signal detection theory, which is explained quite well in Wickens's *Engineering Psychology and Human Performance* (1984). By observation it can be determined if the user successfully completes a task. It can be quite instructive, however, to ask the user if he believes he has completed the task successfully. One hopes that each task will result in both successful task completion and perceived success, a "hit" in signal detection terms. The second best outcome is a "correct rejection," where the user fails the task and correctly believes that she has failed. In these two cases, the user has an accurate picture of her situation and can make an intelligent decision as to what to do next. The other two conditions, actual failure with perceived success (a false alarm) and actual success perceived as a failure (a miss), cause significant problems if they occur in deployed systems.

Given the task of making a car reservation at an airport, imagine what happens to the user. In a system that generates false alarm states, the user believes that he has rented a car when in fact he has not. Most often, some part of the interaction has given the user the impression that he has finished before having completed all necessary steps. Perhaps a case of too much feedback, too soon— or a poorly organized process that does not linearly drive the user to a successful conclusion. In any event, the user confidently hangs up, boards a flight, and lands thousands of miles from home without a car reserved. Systems that generate misses cause users to mistakenly repeat a process to a successful conclusion more times than they intend. The user is not getting the feedback that he is done, or that the system believes he is done. He repeats the steps looking for a positive confirmation that the reservation has been accepted, often multiple times. This person arrives at his destination only to find that he has reserved four cars, perhaps with nonrefundable deposits.

7.5.3   Prototyping

There are four methods of testing, each with increasing levels of fidelity, that can be employed to test an IVR: Wizard of Oz (WOZ), WOZ with sound files, a functional prototype, and beta code on a production platform.

WOZ

The lowest level of fidelity, and also the lowest cost to run is a WOZ study. A WOZ study is often done for speech recognition, but can also be employed for IVRs. Like the wizard in the movie, the experimenter takes on the role of the technology. You read from the prompt script and have the participant indicate the choice they would make on each menu. Following a flow or outline, you then read the next prompt and collect the next response until each task is complete. It is cheap, fast, and easy to make iterative changes. The face validity is a bit lacking and a wizard may introduce bias with or without intention or awareness through non-verbal cues, coaching, or giving a bit of the benefit of the doubt. Nonetheless, this method is useful if your means are limited, or as a first vetting of the interface before additional resources are spent on prototyping.

Note that the lower level of fidelity option of handing a participant the outline or flow and the prompt script was not offered as a viable method, as it is not. The first step in vetting prompts is to read them out loud. Many things that will pass by design teams and committees on paper sound as awful as they are when read aloud the first time. One discount method is to leave your prompt script as a voice mail to yourself and then listen to it. When you are forced to listen to and not allowed to read a prompt, its suitability can be better judged. Prompts with awkward structure, repeated words, or too many options will begin to make themselves apparent.

WOZ with Sound Files

The veracity of a WOZ can be improved by recording the prompts as sound files on a PC that can be played rather than read, as in the simplest WOZ. In this way, each participant hears the menus exactly the same as another participant with the same inflection, emphasis, and phrasing. The recordings provide some social distance between the experimenter and the participant and give the impression of working with an IVR system. A means of organizing the sound files so that they can be accessed easily and played correctly in response to the user's indicated key presses is needed. Adding recordings for the wizard to use is a relatively small step in the effort to increase test validity.

Functional Prototypes

Functional prototypes allow the test participant to experience your design without your intervention. A soft prototype could be built on a PC in a scripting language or even in HTML or VXML that ties your menu logic to sound files that play in succession as users click their way through. The next level of prototype is one that will answer phone calls. Now you have high fidelity in that users are interacting with a phone handset, just like they will with the real IVR. This prototype may be built on a small PC with a telephony modem, or with line cards, or through a website that will host VXML scripts and provide a call-in number. If your prompts are recorded well, your test participants may not know they are calling a prototype. One major headache in prompt scripting can now be evaluated.

When using a wireless or a cordless phone the keys cannot be used while listening. The handset must be held away from the ear to press a key and then returned to the ear to listen to the next prompt. If critical information resides in the first few words of the prompt, many users will not hear it as their phone will be on its way back to the ear. Testing with a cordless or princess-style handset instead of a desktop or speakerphone will help identify prompts that are vulnerable to this issue. In a functional prototype, connections to data sources may have to be faked. The PC in your test lab is unlikely to have access to the bank's account database. Fake databases or even static data are not usually a problem, and clever task design can often avoid exposing a prototype's limitations.

Production System Prototypes

The grand luxury in testing is to be able to run test code or beta designs on a production system. On the production system, you will likely have access to databases and live feeds. Everything is of the same quality and character that it will be once it goes live. Audio levels, delays to reach data, slowdowns due to traffic or loads can all be included as part of your test. You may even be able to use a load balancer to send every n-th call to your prototype instead of the current IVR. The downside is that your test may access live accounts, place real orders, and incur real charges. It can be difficult to obtain test accounts or test processes that require a real credit card or SSN on production systems. In some cases, testing is easier and safer on a prototype.

7.5.4   Testing Equipment

There are certain pieces of equipment, such as phone taps, DTMF keystroke loggers, and compressor/limiters, that make testing prototype or production systems easier and more informative. Phone taps can be legal or illegal depending on how they are used: Consult local law enforcement to determine the proper method in your area. There are three types of phone taps: contact microphone, handset cord, and line taps. Contact microphone taps are the least expensive and tend to have the lowest audio quality. As of this writing, Radio Shack carried

FIGURE

7.3

The THAT-1 phone tap.
Because the tap plugs into the handset cord, the audio quality is better than
contact taps. (Courtesy of JK Audio, Inc.—*www.jkaudio.com.*)

a model that adheres to the handset with a suction cup near the speaker. Better
audio quality can be obtained with a tap that plugs into the cord that goes between
the telephone base and the handset (e.g., from a manufacturer such as JK Audio)
(Figure 7.3). Both of these taps must be used on the phone that your caller is
using. The line tap goes on the phone line itself, and can be used on the caller's
line or the line that the IVR is connected to.

Line taps do a better job of separating the audio of the caller and the IVR. Line
taps such as Gentner's Hybrid Coupler or JK Audio's Innkeeper run from the low
hundreds to the low thousands of dollars; this is a piece of equipment where
higher cost tends to equate to higher quality. DTMF keystroke loggers can tie into
the phone line or the audio from a phone tap, and then log or display the key
corresponding to a DTMF tone. This makes testing IVRs simpler as you do not
have to watch what key the user presses. At one time during a period of heavy
IVR testing, I found that I had memorized the sound of most of the DTMF
keys—perhaps a sign that I needed more variety in my work.

DTMF key presses are loud, especially when played through headphones.
A compressor/limiter is a common piece of audio gear used in the recording
industry and on performance stages large and small. The compressor/limiter
takes an audio input and shapes the output to conform to limits that you set, such
as no matter what the input level nothing leaves the compressor over 65 dB. This
is done not by cutting off loud parts of the input but by compressing the audio to
fit within the limits you set. The benefit is that you can run microphones or phone

taps hot (at high volume levels) so that you can hear quiet inputs well, without the worry that when something loud (like a DTMF key press) comes along your ears will get blasted. The Alesis 3630 compressor/limiter is a workhorse in our lab that can often be sourced for as low as $100.

7.6   DESIGN GUIDELINES

IVR system interfaces need to be inherently easy to use. This section is intended to give IVR developers a set of tools that they can use during the design stage to help ensure that their product meets functional and usability expectations for a broad range of users. It should be noted that no set of rules or guidelines *alone* will lead to the development of an optimally usable and useful IVR. The goal in any interface design effort is to find the best "fit" between a system's specific requirements and the consistency and usability benefits that will result from the application of design guidelines. To achieve this goal, it is recommended that guidelines be used as part of a broader process, including usability testing.

7.6.1   General IVR Design Information

The guidelines that follow are drawn from several sources (HFES/ANSI, 2006; ANSI, 1992; ANSI/ISO/IEC, 1995; Davidson & Persons, 1992; Engelbeck & Roberts, 1989; Pacific Bell, 1992; Shovar, Workman, & Davidson, 1994; Simon & Davidson, 1990; Pacific Bell, 1990), and represent conclusions that in many cases were reached independently in multiple labs and that often cross-reference each other.

Do Not Lose Data When Transferring
Information entered by the user should not be lost in the transition between systems, nor should it be lost when routed from an automated system to a live representative. The user should never have to give the same information twice in the same session (except for required verification or other special circumstances).

Make Walk-up-and-Use Applications Simple
Walk-up-and-use applications should have simple interfaces accessible to the novice user. Menus should contain few choices, possibly containing only binary ones (e.g., yes/no or accept/cancel). System options should be explained fully in complete or nearly complete sentences. Users should not need to ask for help or for additional information, although it should be available.

Provide Shortcuts in Subscribed-to Applications
Subscribed-to applications can have more complex interfaces. While they should still be accessible to first-time users, shortcuts should be provided to allow experienced users to operate the interface more efficiently.

Use the Best Announcement Voice

+ Use human speech whenever possible.
+ Use trained voice talent whenever possible.
+ Use synthesized speech only for text that cannot be prerecorded or stored.

7.6.2   Opening Message

The opening message is not a throwaway greeting; it has work to do.

Opening Message Should Give and Gather Information

+ Present users with an opening message when they first dial into the system. Except in systems where subscribers are allowed to record their own greeting messages (e.g., voice mail).

+ The opening message should be worded to identify the relevant company (e.g., AT&T, Rice University) and the application.

+ For walk-up services that accept only touch-tone input, verify as early as possible in the script that the user is at a touch-tone phone, and if not, present a live-agent routing option for rotary phone users. The user should be instructed to press 1 to indicate touch-tone service ("If you are using a touch-tone phone, press 1; otherwise, please hold"), but the system should recognize any touch-tone input (not just the 1 key) as evidence of a touch-tone phone.

+ If a touch-tone phone is not detected in the time-out period at the opening message, it should be assumed that the user has rotary service (or is unwilling to use the IVR) and should be connected with a live agent.

7.6.3   User Control

Give users tools to allow them to succeed.

Give the User Control

+ The user should have control of the system wherever possible. This includes the ability to control the start and end of the system's actions, the ability to cancel transactions in progress, and the ability to exit from the system and/or speak with a service representative.

+ Present users with the option to choose between a self-service channel and a live-agent channel, with the options being presented as early as possible in the application consistent with organizational objectives.

+ Always prompt users with all of the available options.

+ There should be a common, universal set of functions that are always available within the system. These functions should be easy to use, and be consistent throughout the interface. An example is using the 0 (zero) key to transfer to a service representative or to hear a help message.

✦ Give the user immediate feedback for each action. An example of feedback is "Your order has been canceled." Beeps and tones by themselves are insufficient feedback and should be paired with an announcement when used.

Allow Dial-Through: Interrupting System Output

✦ Allow the user to override prompts, menus, and other statements at virtually any point before or during the time that they are offered. Allow the user to select a menu option immediately after it has been heard without having to wait for the entire menu to be presented.

✦ Some types of system output must not be interruptible. These include error tones and the noninterruptible portion of an error message.

✦ User input should terminate the prompt asking for user input within 0.3 seconds, and the system should then act on that input.

✦ Error messages have three parts. The first part of an error message states *what* went wrong. The second part states *why* it happened. The third part tells the user *how* to correct the problem if it is correctable, and what options they may have or whom to call for help. The *what* portion of the announcement should not be interruptible. It may be appropriate to combine the *what* and *why* in some cases.

Error: User enters an incorrect password.

What: "Access to your account has been denied."

Why: "The phone number and PIN you entered do not match our records."

How: "Please re-enter your area code and phone number."

Allow Dial-Ahead: Pre-empting System Output

✦ Allow users to jump ahead through several menus.

✦ User input should be queued for processing when the user is working ahead of the system.

✦ If an invalid input is encountered by the system while input remains queued for processing, the queue should be cleared at the time the user starts to receive error feedback.

Allow the User to Cancel Actions after Data Input

Always provide a way for users to cancel their input. This can be provided as either a prompted option ("If this is correct, press 1. If incorrect, press 2."), and/or as an implicit option (e.g., by enabling the * [star] key to cancel and/or start data entry over again).

Do Not Strand the User

For optimal user control, the user should be returned to a familiar and meaningful location after an activity is completed (e.g., to the system main menu or the current primary topic menu). The location should be titled and contain prompts for what to do next. When appropriate, present the user with an option to exit the system, or instructions on how to exit.

7.6.4   Menus

Menus must be purposefully designed to be easy. Rarely do good IVR menus organize themselves like existing business processes.

Make Menus Informative and Orderly

+ Prompted choices should be numbered sequentially, in ascending numerical order beginning with 1.

+ The user should not hear prompts for commands that are not currently available. It is preferable to skip numbers in the menu than to present a nonfunctional option. However, skipping numbers should be considered only for options that are state dependent and are available at other times. (For example, on a menu that has a delete function, the option to delete may only be offered when there are items to delete.)

+ When entering a new menu, the user should be given a title and a brief instruction with the appropriate inflection to separate title from instruction (e.g., "Forwarding Start Day. <pause> Choose the day to start forwarding on. For Monday, press 1...").

+ For walk-up-and-use services, tell the user how many entries to expect. This cues the user not to respond too soon if there is uncertainty (e.g., "Please make your selection from the following four choices.").

+ Present the next prompt or statement within 750 milliseconds after a user makes a menu selection.

Organize the Menu to Help the User

+ As a general rule, place the most frequently selected choices first. For example, if most incoming calls are for billing, the billing option should be the first one presented.

+ Follow a natural/logical order. For example, ask for the user's telephone number before asking whether there is trouble on the line. Ask for the area code and then the telephone number. Ask for the item and then the desired quantity. Days of the week should be in order rather than by most frequently selected.

+ Follow a functional order. Related menu items should be presented together in the same menu.

+ There may be cases where grouping options based on the user's knowledge of categories is more effective and may lead to better performance than grouping options by frequency alone.

+ The structure of menus can be largely dependent on application requirements. A hierarchical menu structure is generally the most appropriate for an interface with a large number of available options. On the other hand, a flat menu structure (i.e., just one menu) should be used when only a few options are available.

Limit the Number of Items on Menus

+ There is no universal optimum number of choices on a menu or menu levels in a hierarchy. In general, fewer options and fewer levels will be easier. Good design is a compromise between the number of options in a menu, the number of menus on a level, and the number of levels of menus. Given a set of requirements, create the smallest system that gives users access to the options that they want.

+ As the number of choices in a menu increases, expect increases in the number of errors and requests for help, and an increase in user response times. Design the menu structure to break up tasks or categories into subtasks or subcategories that can be accomplished with shorter menus. This structure should match with a representative user's model of the system.

+ No menu should ever require double-digit entry. Menus must be designed to use the 0 to 9 and * and # keys as single key press options. A menu design that has more than 12 options must be split up into separate menus, which may require adding levels of hierarchy or adding items to higher-level menus.

+ Avoid using "catchall" alternatives such as "For more options, press 3." Instead, provide a description of the option so that a user could select 3 on purpose, not just because 1 and 2 did not sound right.

Map Binary Menu Choices to the 1 and 2 Keys

+ The labels or choices on a binary menu should reflect the user's, not the system's, view of the choice.

+ Many yes/no decisions can be rewritten for clarity. Consider this example: "Do you have dial tone? For yes, press 1. For no, press 2." This prompt can be improved as follows: "If you have dial tone, press 1. If not, press 2."

+ When a yes/no binary choice is presented, "yes" must be assigned to the 1 key, and "no" must be assigned to the 2 key.

+ If possible, phrase the questions so that "yes" is the most frequent response.

+ For confirmations, star can be used in place of 2 to go back and fix the entry: "You entered 800-555-1212. If this is correct, press 1; if not, press *."

7.6.5   Statement Phrasing

Prompt statements need to be short and clear—the user will be hearing, not reading, them.

Keep Statements Brief

Prompts, messages, and menu selections should be as brief, simple, and unambiguous as possible (e.g., "To confirm your order, press 1"). Statements should be short enough not to discourage expert users, but contain enough information to be clear to novices.

Format Statements for Success

+ System prompts should be of the form "action-object (i.e., "To do X, press Y."). In other words, present the available action first, and the key needed to take the action second ("For call waiting, press 1.").

+ Avoid the passive voice. For example, do not say, "A telephone number must be entered now." Instead, say, "Enter the area code and phone number now."

+ Avoid negative conditionals. For example, do not say, "If these instructions were not clear, press 1." Instead, say, "To hear these instructions again, press 1."

+ System statements should present important information first to allow users who do not wish to listen to the full set of statements and prompts to quickly identify key information.

+ Critical information should not be the very first word in an announcement. Users with keypads embedded in the handset may not hear the first two words following a key press as the handset is brought back to the ear. Therefore, introduce critical information with phrases such as "You have five...," "There are five...," The number five..."

+ Terminology should be consistent throughout the system.

+ Terminology should match the intended audience of the interface. Avoid jargon related to telecommunications and computers when the intended audience is outside of these industries. For example, use "Enter the area code and the first three digits of your phone number," rather than something like "Enter your area code and prefix."

+ Avoid prompts, statements, and menu labels that sound too general, as users will often select an incorrect option before listening for one more appropriate. For example, when users are prompted with a statement that includes wording like "For all other billing questions..." many users will select the option for a variety of questions unrelated to billing needs.

✦ When referring to the system itself, or to the company or one of its representatives, use the pronoun "we." ("We"re sorry, we are unable to process your request at this time."). An exception is when referring to a specific individual or group within the company ("A repair technician will contact you within 24 hours.").

7.6.6 Wording Conventions

Following conventions and using standard terminology builds on what the user already knows.

Use Standard Terminology in Prompts

✦ Use the term "press" for single-digit entry ("If yes, press 1.").

✦ Use the term "enter" for multiple-digit entry of phone numbers ("Please enter the area code and then the telephone number for which you are reporting trouble.")

✦ Use "enter" to request information ("Enter your zip code now.").

✦ When prompting the user for a single key press (e.g., as in a menu prompt to press a key for a menu selection), use the "To do X, press Y" format ("For call waiting, press 1.").

✦ Refer to the DTMF keys with standard names.

✦ Refer to * as "star" ("To cancel this order, press star.").

✦ Refer to # as the "pound key" or simply "pound" ("Enter the amount of your payment in dollars and cents, then press pound.").

✦ Refer to 0 as "zero" ("To speak with a service representative, press zero.").

✦ Refer to the number keys (0 to 9) by their digit names ("For sales, press 3.").

7.6.7 Metaphors

Metaphors can be powerful. While a proper metaphor can aid the user, a poor metaphor will add confusion.

Use Metaphors Appropriately

✦ Choose an appropriate metaphor so that the users develop a reasonable conceptual model of the system (e.g., voice mail is based on a metaphor of the paper mail system). By using metaphors, users can develop a set of expectations that can then be applied to the system environment, and will help them to have a valid conceptual model.

✦ Reserve the use of a metaphor for places where it truly applies. Do not label functions with names that do not make sense in the context of the application simply to fit a given metaphor. This will confuse the user when the name does not represent the expected action.

✦ The use of a metaphor should be consistent within each interface. Do not mix metaphors within a single application.

✦ The use of metaphors should be consistent among applications. This is especially important in interfaces with interacting or identical functions. Otherwise, users who transfer between systems with different metaphors may become lost and confused.

Apply Directional Metaphors Consistently

✦ If a directional metaphor is used (e.g., backward, forward), use the numbers on the left side of the standard 3 × 4 keypad to indicate previous, lower, slower, backward, and so on, and the numbers on the right of the keypad to indicate next, higher, faster, forward, and so on. This provides the best possible match with learned expectations and should minimize any conflicts with other factors.

✦ This is a case where it may be desirable to violate the recommendation on ordinal numbering of options. For example, if 4 and 6 are used for previous and next, there is no requirement that 5 have an assigned function. In some cases, it may also be acceptable to voice these choices in order of most frequent use rather than numerically. For example, "For the next message, press 6; for the previous message, press 4."

7.6.8   Key Assignment

Key assignments should be chosen to simplify the interaction for the user.

Do Not Use Mnemonics

✦ Choices in a menu should be numbered consecutively in ascending order, starting with 1.

✦ Menu choices should be numbered to correspond with the numbers on the telephone keypad, rather than named with mnemonics, such as "To send a message, press 1," rather than "To send a message, press s." While using "w" for call waiting, "f" for call forwarding, or "s" to send a message might seem like a good way for users to remember menu commands, it will increase user input time and lead to errors. Some handsets do not have letters on the keys. In addition, persons with visual disabilities memorize the position of the numbers, not the letters on the keypad.

✦ The meaning of a key press must never depend on its duration. Calling cards that require the user to press and hold star for two seconds have violated this recommendation and should not be emulated. Using key press duration to change function creates accessibility issues for users with motor control difficulties and often for all users due to lack of clear instructions. Many users hold key presses longer when frustrated—having the function change when they do so will not improve satisfaction.

Create Functional Consistency

✦ Make key assignments consistent within applications, and wherever possible, across different applications. Keys should be consistently assigned to the same actions.

✦ Recognize that in new versions or upgrades to a service, reassigning a key associated with a core function to a different function may cause difficulties for users; they will have to then unlearn the original association and learn the new one, resulting in errors.

✦ If a core function is known to be coming in the future, but is not yet implemented, try to reserve the key for the future implementation of that function.

✦ Within a menu the function assigned to a key should not change as a result of time passing or background state changes. If the function assigned to a key needs to change, the system should play a new menu with the new key assignment.

Avoid Letter Entry on the Keypad

Whenever possible avoid letter entry on the keypad. If possible choose numeric rather than alphanumeric User IDs or passwords for IVR systems. If there is a database to match text entries against, the user should be able to enter text with a single keystroke per letter. For example, if a customer's name is "Pat," the customer should be required to only press the numbers 7, 2, and 8. Letters map onto keypad numbers, as shown in Table 7.1.

While most keypads begin lettering with the 2 key (A, B, C), some assign Q and Z to the 1 key. Therefore, although the preferred way is to map Q and Z onto the 7 and 9 keys, respectively, the system should also map the 1 key to Q and Z. When it is not possible to use a single keystroke per letter (e.g., when names are being entered for the first time and the system could interpret the input in several ways), the system should clearly describe the multikey method of text entry to users.

TABLE 7.1   Letters Mapped to DTMF Key Numbers

Key	Letter Mapping
1	None
2	A, B, C
3	D, E, F
4	G, H, I
5	J, K, L
6	M, N, O
7	P, Q, R, S
8	T, U, V
9	W, X, Y, Z

Avoid Terminator Keys When Possible

+ Do not require users to enter a terminator as part of a menu selection. For example, if the prompt is "For billing, press 1," then the user should only have to enter 1, not 1 followed by # (where # is the terminator). If the user mistakenly enters a terminator where one is not asked for, the system should ignore it and not flag it as an error.

+ If possible, do not require users to enter a terminator to signal the end of multiple-digit input. For example, the system should be able to detect the ninth and final digit of a nine-digit Social Security number entry.

7.6.9   The 0 Key

When help is provided, the 0 (zero) key should be used to access system help. Typically, system help takes the form of being transferred to a live attendant or of receiving recorded help information.

+ If transfers to a live attendant are available, pressing the 0 key should terminate the current action and take the user to a service representative. The specific facility to which the system transfers the user should be determined by the context of where in the system the user is currently located (e.g., if the user requests assistance while in the billing portion of an IVR flow, the user should be routed to a billing representative).

+ Avoid transferring the user to more than one representative.

+ When no representative is currently available (busy or after business office hours), the 0 key should lead to appropriate statements or messages.

+ If the system provides online help, the help message should be specific to the user's context in the IVR. Once finished with help, the user should be returned to the same system state that he or she was in prior to the help request. Incomplete actions may be cancelled, however. If they are, the user should be asked to start the action over again.

✦ If multidigit input is expected e.g. a phone number, and the user enters ZERO and then times out, the system should treat this as a "ZERO pressed for help" and transfer to a service representative or give context-specific help.

7.6.10   The * Key

The * (star) key is typically used to cancel user input, move backward in a sequence, or move upward in the menu structure.

✦ *Cancel input string:* If the user is entering a string such as a telephone number or pass code, pressing * should discard the digits that have been entered and return the user to the prompt preceding the entry of the digits.

✦ *Back up to previous action:* In functions that require many steps, pressing * should take the user back past the last completed step, and return to the prompt to allow the user to perform that step over again.

✦ *Exit submenu:* If the user is at a lower-level menu, pressing the * key should take the user back to a higher-level menu, perhaps even to the system's main menu.

✦ *Terminate call:* The * key can be used to exit system and terminate call.

For some walk-up applications where users are expected to be unfamiliar with IVR conventions, a context-specific help message tells users where the * key is located on the keypad the first time that it is mentioned or used.

Use the * Key to Its Best Effect

Since the * key cannot provide *all* of the functionality outlined in the above examples at any one point in time, use of the key in any particular instance should depend on where the user is in the application. In other words, the star key will have different results depending on whether the user is currently

✦ Entering a numeric string (the user will be taken back to the prompt for the numeric entry field)

✦ In the middle of a multistep function (the user would be taken back to the prompt preceding the last completed step)

✦ At a lower-level menu (the user would be returned to the higher-level menu that the lower-level one is nested within)

✦ At the highest-level menu in the application (the user would then exit the system)

Developers should carefully evaluate the entire application to determine where the * key should return the user at any given point, and assign the functionality of the key accordingly.

Special Uses of the * Key in Voice Messaging Systems

If the user is recording a message, greeting, or name, pressing the * key should stop the recording, discard what had been recorded, and return the user to the prompt preceding the record tone. In the process of logging in, if the user realizes that the system is assuming the wrong mailbox, he or she should be allowed to press * during the password prompt and be taken back to the mailbox prompt.

7.6.11   The # Key (Terminate, Skip Ahead)

The # (pound) key is typically used as a terminator for variable-length user input, to confirm user input, or to move forward in the system.

Use the # Key as a Terminator

+ When a terminator is needed, use the # key. Time-outs are always accepted as terminators in these cases as well.

+ Use # to indicate the end of a variable-length input string ("Enter the amount in dollars and cents, then press pound"). A terminator is required whenever the input can be of variable length. A terminator should not be used with fixed-length entries (e.g., phone numbers). When use of the pound key is required, the prompt must explicitly state this requirement, as in the example above.

+ Do not use the # key as a terminator as part of menu selection. For example, if an option is "To use our automated system, press 1," then the user should only have to enter 1, not 1 followed by #.

+ If a user enters the # key during the playing of an interruptible announcement, the announcement should immediately cease and the system should proceed with the next step. The # key is not in the dial-ahead buffer, as its action was to interrupt the announcement and it has acted.

+ The # key can also be used to do the "usual" or most common choice at a menu. An example would be selecting a day of the week, where "today" varies, but is the most common choice. "For today, press pound. For Monday, press 1. For Tuesday..."

Special Uses of the # Key in Voice Messaging Systems

+ The # key should be used when a terminator is required to end a recording (a time-out should also be accepted as the terminator).

✦ If the user is listening to a prompt that precedes a record tone, pressing the # key should skip the rest of the prompt and take the user to the record tone (e.g., when recording a message, greeting, or name). This also applies if the user is listening to a subscriber's greeting. The # key should skip the greeting and go straight to the record tone to leave a message for the subscriber.

✦ If the user is listening to a voice mail message, pressing the # key should stop playing the message and take the user to the next message.

7.6.12 Prompts and Feedback

The user should receive feedback from the system to confirm actions, warn about dangerous actions (e.g., request explicit confirmation from the user), echo user input, explain error conditions (e.g., an option is not available), and tell the user what the system is doing.

✦ If something has been added or deleted, confirm each addition or deletion as it is requested, then summarize overall changes at end of add/delete activities.

✦ Present the next prompt or statement within 0.75 of a second (750 milliseconds) after a user makes a menu selection.

✦ When possible, avoid using the words "to," "too" or "for" in menu prompts where they may be confused with the option keys 2 and 4.

Provide Appropriate Information about System Delays

✦ A *small* delay is defined as between 750 milliseconds and 3 seconds. This delay is noticeable and undesirable to customers, but no announcement explaining it *is required*.

✦ A *medium* delay is defined as between 3 and 8 seconds. This delay is noticeable and undesirable to customers; a short announcement explaining the delay *is required*.

✦ A *long* delay is defined as between 8 and 30 seconds. This delay is noticeable and undesirable to customers; a short announcement with an estimate of the delay *is required*. Music should be played during the wait. Music should not be a "canned" file that starts in the same place every time the customer experiences a delay.

✦ An *extreme* delay is defined as any delay longer than 30 seconds. This delay is noticeable and undesirable to customers; it *is required* that every 30 seconds the user hear an updated announcement with an estimate of the remaining delay, and music should be played between these announcements. Extreme delays will cause customer dissatisfaction. Music should not be a "canned" file that starts in the same place every time the customer experiences a delay.

Get Required Information Early

If an application requires customer information, collect it before completing steps that are dependent on correct entry of that information. For example, if the telephone number is required to access account information, get and confirm it before asking what type of information the customer desires.

Tell the User the Number of Digits to Enter

When the user is prompted to enter numeric input of a fixed length (e.g., a Social Security number), the prompt should state the number of digits that must be entered, as in "Please enter your nine-digit Social Security number." Exception: In cases where a fixed-digit number is being collected for security purposes, the number of digits required *should not* be stated.

Tell the User How to Indicate That Number Entry Is Finished

When the user is prompted to enter numeric input of a variable length, the prompt should ask the user to enter a terminator when data input is completed. However, a time-out will also be accepted to terminate input. The # key should be used as the terminator key.

Always Confirm User Input

 ✦ Always provide a way for the user to confirm (or cancel) input. This can be provided as an explicit, prompted option, such as "If this is correct, press 1. If not, press 2."

 ✦ The system should read back user input in proper phrasing. In many cases, the system should just read back the user's input exactly as it was entered. For example if the user enters 1-800-555-1212, the system may ignore the "1," but when confirming the entry it must ask the user to confirm all 11 digits entered, as in "You entered 1, 8, 0, 0..." However, if the system reports the time to the user, it should be read back as "eleven o'clock," not "1, 1, 0, 0."

 ✦ Repeat back to the user long data entries for confirmation with appropriate pauses. For instance, if the user is asked to enter a Social Security number, the system should respond with something like "You entered 123 <pause> 45 <pause> 6789. If that is correct, press 1. To re-enter, press 2." It is critical that phone numbers be read with proper phrasing and not as a 10-digit string.

Use a Tone to Signal the Start of a Recording Window

Voice messaging systems should have a distinct record tone, distinguishable from any other tones used in the system. The tone used to indicate that the recording is starting should be a single beep, of 440 Hz frequency, and 0.5 second (500 millisecond) duration.

7.6.13   Errors

Error statements and prompts should be of the following format:

✦ Optional error tone (standard double beep, as in the following example).

✦ What went wrong (and the current system status).

✦ Why the error happened.

✦ What the user should do next, possibly including more information than was presented in the original prompt.

> *Example*: <beep-beep> "Access to your account has been denied." (*what*) "The phone number and PIN you entered do not match our records." (*why*) "Please re-enter your area code and phone number." (*how*)

✦ After a particular number of consecutive errors (two or three), the system should take special action, such as transferring user to a live attendant. An error count on one prompt should generally not carry over to other parts of the interface, but should be cleared when user resolves the error. For example, in an ordering system, a user might enter a phone number incorrectly twice before getting it right. The system should not then immediately transfer the user if he or she makes a *subsequent* error in the ordering process, even though that would be the third error since entering the system.

✦ A single user error should not cause the system to disconnect, unless it is a security requirement.

Use the Standard Error Tone

If a tone is used to indicate an error, the tone should be a double beep, and should be 440-Hz frequency, 100 milliseconds on, 50 milliseconds off, and 100 milliseconds on.

Do Not Make the Error Message "Bad"

✦ Avoid negative words and phrases like "invalid response," "error," and so on.

✦ Error messages should be written so as to avoid compromising system security. In parts of the application where security is an issue (e.g., entering the system, changing account status), the system responses should not provide more information than is needed. If a phone number and PIN are required to authorize use of a system, do not prompt the user with the digit length of the PIN, and do not give individual confirmations on either the phone number or the PIN. If either or both are incorrect, use the error message "The phone number and PIN you entered do not match our records."

7.6.14   Time-outs

When the user has taken no action after hearing a prompt, the system should respond after the designated time-out period. The system response following an unanswered prompt should take the form of

+ First time-out or error: Repeat the prompt.

+ Second time-out or error: Play the help message and repeat the prompt.

+ Third time-out or error: Play the help message and repeat the prompt.

+ Fourth time-out or error: Transfer to a customer service representative if available; otherwise, play a polite message asking the user to try again later, and then terminate the call.

Time-outs can be of variable length; at present, there is no single standard for the amount of time to allow users to comprehend and act on a prompt before the system responds. Factors such as the context of the prompt, the frequency that the prompt will be used, and the complexity of the action should be considered in determining the length of the time-out. A postmenu time-out is the time that a user is given to initiate a response after a list of menu choices. The recommended value for postmenu time-outs is five to six seconds. An interdigit key press time-out is the time allowed between key presses in a multiple-digit entry. It also functions as the terminating time-out for variable-digit entry. The recommended value for interdigit time-outs is six seconds.

These are general guidelines and it is strongly suggested that time intervals used in IVR interfaces be carefully tuned for the intended audience, as well as for the situation in which they will be presented. Also, designers should be aware that users whose phones have keypads in the handset (cordless and cellular phones) may take longer to respond to prompts due to the time necessary to lower the handset from the ear to access the keys. Finally, individuals with perceptual or motor disabilities may also experience difficulty in responding to prompts, and their specific needs should be considered.

7.6.15   Announcement Voice

The sound of the digitized voice should be consistent within the application, as well as with all others offered by the company or organization. Whenever possible, the same person should be used for the recordings, and the recordings for a given application should take place in a single session. However, as it often becomes necessary to record over several sessions (e.g., as changes and updates to an application become necessary), it is understood that the same individual may not be available when needed. When a different individual will do subsequent recording, try to select a speaker in such a way as to minimize differences in pitch, accent, and other individual voice characteristics, as these differences can distract users from paying attention to the content of the statements or prompts.

When recording spoken digits such as telephone numbers, the digits must be recorded for each position in which that digit will occur. That is, any particular digit will have different intonations depending on whether it appears at the beginning, middle, or at the end of a spoken digit string.

Silence and pauses are just as important as the words being spoken. In the following example, pauses occur at different places. In the first case, the inappropriate location of the pause might cause some users to "press 1 to send." The appropriate pauses in the second case reduce the possibility of that error.

Inappropriate: "To listen <pause> press 1. To send <pause> press 2."

Appropriate: "To listen, press 1. <pause> To send, press 2. <pause>"

7.6.16   Techniques for Collecting Information

The techniques described below have been subject to multiple usability tests (Miller, 1996) and are widely and successfully deployed.

Collect Personal Identification Numbers Securely

For security reasons, do not tell the user how many digits they are to enter. The exception is when the user is changing a personal identification number (PIN)—then the system *must* tell the user how many digits to enter. If the entered PIN does not match the phone number and PIN pair maintained by the system for the user, inform the user and reprompt for both phone number and PIN. Do not voice back the erroneous data. Say instead, "The phone number and PIN you entered do not match our records."

Be Flexible When Collecting Phone Numbers

+ *Never* ask user for a "10-digit phone number." When a 10-digit phone number is needed, prompt user with "Enter just the area code and phone number" (or the regionally appropriate description of the phone number).

+ If the user enters a 1 or a 0 as the first digit, the system should ignore it and collect 10 more digits.

+ If the user enters the wrong number of digits, inform her: "You entered only 8 digits." Reprompt for the phone number. Do not repeat the short set of collected digits.

+ If the user leaves off the area code when asked for the area code and phone number, prompt him to add just the area code. However, the system should anticipate and accept either the 3-digit area code or the full 10-digit number as a response to this prompt.

Collect Days of the Week in Order

Prompt the user by reading the days, with each day followed by a number corresponding to the day's position in the week. Monday is day 1 and Sunday is day 7. If "today" is a likely response, assign it to the pound key and offer it as the first option; otherwise, start with Monday. "For today, press pound. For Monday, press 1. For Tuesday, press 2."

Be Flexible in Collecting Times

+ The system should accept input in military time (24-hour clock) format. (However, the user should not be asked to enter military time.)
+ Ask for hours and minutes, followed by 1 for A.M. or 2 for P.M.
+ If the user leaves out the A.M./P.M. indicator, prompt the user for it again while confirming the hours/minutes part of the entry. For example, if the user entered "2, 3, 0" corresponding to 2:30, prompt the user with "For 2:30 A.M., press 1. For 2:30 P.M., press 2. For a different time, press star."
+ The 0 (zero) key can still provide assistance, but only if it is the first and only digit entered after the prompt for the time entry. After a time-out, the user is either routed to a live agent or receives some other form of assistance.

Define Time Periods with a Start Time and a Stop Time

+ A start time and a stop time, *not* a start time and duration, define time periods.
+ The stop day is the same as the start day if the start time is earlier than the stop time. On the other hand, if the start time is later than the stop time, the stop day is the following day.
+ Identify overlapping time periods and provide options to correct them.

7.7   CASE STUDY

A case study for an IVR can be found at *www.beyondthegui.com*.

7.8   FUTURE TRENDS

It is unlikely that people will be willing to live without voice mail in the future, and so IVRs will continue so long as phones exist. In the next 10 to 20 years, I would expect little to change in the field, as IVR technology is already relatively mature. VXML has opened many doors, making it easier than ever to create and operate an IVR. However, that has been a mixed blessing. Many easily created

IVRs are poorly conceived and create headaches for their users. Promoters of speech recognition systems have been predicting the demise of IVRs and the rise of speech for the last two decades.

Currently there is little evidence that speech is about to take over the world, or that IVRs are about to go away. IVRs, once built, tested, and put into operation tend to be quite robust, and are fairly straightforward to update. Speech recognition systems require professional tuning and careful selection of grammar models to be successful, and seemingly insignificant changes often require these steps to be repeated.

One area for growth in the IVR field is standards. Currently there are few, ANSI/ISO/IEC 13714-1995 being one, and in the experience of this author, existing standards occasionally promulgate suboptimal solutions. What would be of great value to the designers of IVRs is a "tool kit" of proven solutions that could be adopted and copied. The open-source movement and VXML have the possibility of creating a public library of usability-tested solutions to common IVR tasks such as collecting a phone number, collecting a date and time, and entering a dollar amount. With such a library in widespread use, users may be able to count on some similarity across various IVRs rather then having to learn for each one "how we do it here."

Acknowledgment

The author is indebted to the many human factors professionals who have advanced the field of IVRs. Researchers from Southwestern Bell/SBC, Pacific Bell, and Ameritech have all made contributions to the body of knowledge that is carried forward in the AT&T Human Factors Lab of today.

REFERENCES

American National Standards Institute (ANSI). (1992). *Standard User Interface to Voice Messaging*. ANSI Working Document, Project 0976-D.

ANSI/International Standards Organization (ISO)/International Electrotechnical Commission (IEC). (1995). *User Interface to Telephone-Based Services—Voice Messaging Applications*. ANSI/ISO/IEC 13714-1995.

Carey, M. B., Chen, H. T., Descloux, A., Ingle, J. F., & Park, K. I. (1984). 1982/83 End office connection study: Analog voice and voiceband data transmission performance characterization of the public switched network. *AT&T Bell Laboratories Technical Journal* 63(9): 2059-2119.

Christodoulou, J. (1996). Strategies for Effective Voice Response Scripting. Draft. Austin, TX: Southwestern Bell Technology Resources, Inc.

Davidson, J., & Persons, K. S. (1992). *Quickservice Billed Toll Inquiry: Technology Assessment and Recommendations*. San Ramon, CA: Pacific Bell Human Factors Engineering Laboratory.

Engelbeck, G., & Roberts, T. L. (1989). *The Effects of Several Voice-Menu Characteristics on Menu-Selection Performance.* Boulder, CO: U.S. West Advanced Technologies.

Gardner-Bonneau, D. (1999). *Human Factors and Voice Interactive Systems.* New York: Springer.

Gardner-Bonneau, D., & Blanchard, H. E. (2007). *Human Factors and Voice Interactive Systems*, 2nd ed. New York: Springer.

Human Factors and Ergonomics Society (HFES)/ANSI. (2006). *Human Factors Engineering of Software User Interfaces.* HFES/ANSI 200 (Canvass draft 2006). Santa Monica: HFES.

Marics, M. A. (1991). *Interactive Telephone Services: Designing for the User.* C3F.91.008. Austin, TX: Southwestern Bell Technology Resources, Inc.

Miller, J. T. (1996). *An Interim Summary and Proposed Studies for Standardizing Interactive Voice Response and Speech-Based User Interfaces for Advanced Intelligent Network Services.* A3.126.96.79. Austin, TX: Southwestern Bell Technology Resources, Inc.

Pacific Bell. (1990). *Voice Messaging User Interface Forum: Specification Document.* San Ramon, CA: Pacific Bell Human Factors Engineering Laboratory.

Pacific Bell. (1992). *Guidelines for Interface Design of Automated Response Units.* San Ramon, CA: Pacific Bell Human Factors Engineering Laboratory.

Schumacher, R. M. (1992). Phone-based interfaces: Research and guidelines. *Proceedings of Human Factors Society, 36th Annual Meeting.* Santa Monica, CA: Human Factors Society, 1050–56.

Schwartz, A. L., & Hardzinski, M. L. (1993). *Ameritech Phone-Based User Interface Standards and Design Guidelines.* Hoffman Estates, IL: Ameritech Services.

Shovar, N., Workman, L., & Davidson, J. (1994). *Usability Test of Alternative Call Director Prototypes.* San Ramon, CA: Pacific Bell Human Factors Engineering Laboratory.

Simon, S., & Davidson, J. (1990). *Los Angeles CSO Audio Response Unit (ARU) Implementation: Customer Impact Assessment and Recommendations.* San Ramon, CA: Pacific Bell Human Factors Engineering Laboratory.

Tourangeau, R., & Smith, T. W. (1996). Asking sensitive questions: The impact of data collection mode, question format, and question context. *Public Opinion Quarterly* 60:275–304.

Wickens, C. (1984). *Engineering Psychology and Human Performance.* Columbus, OH: Merrill.

Olfactory Interfaces

Yasuyuki Yanagida

8.1   NATURE OF THE INTERFACE

Olfaction is the sense related to smell, and olfactory interfaces refer to devices and/or systems that provide users with information through smells. Olfaction is categorized as a kind of "special sensation," for humans have a dedicated sensing organ to detect the stimuli. Humans perceive smell by detecting certain molecules in inspired air via the olfactory organ in the nose.

The olfactory interface is a relatively novel device. In modern computer systems, visual and auditory interfaces are highly developed, and there are numerous ways for users to obtain information through visual and auditory channels. Recently, haptic interfaces have been actively developed such that, in certain environments, users are able to touch and feel virtual objects or environments. Among the so-called "five senses," only olfaction and gustation (sense of taste) have been left unexamined. These unexploited sensations are chemical senses, whereas the relatively well-developed sensory interfaces (visual, auditory, and haptic) are related to physical stimuli. This fact makes it difficult to introduce olfactory and gustatory interfaces in a similar manner to sensory channels based on physical stimuli. However, researchers have recently begun to actively discuss and develop the use of olfactory interfaces, including the introduction of an olfactory display in virtual-reality (VR) systems (Barfield & Danas, 1996), and transmitting olfactory information over telecommunication lines (Davide et al., 2001).

Figure 8.1(a) shows the structure of the human nose. In the nasal cavity, there are three nasal turbinates separating the path of the air flow in the cavity. There is an organ called the olfactory epithelium at the top of cavity, where smells are detected. Figure 8.1(b) shows the detailed structure of the olfactory epithelium. Odor molecules are captured by the mucus layer and then detected by olfactory cells.

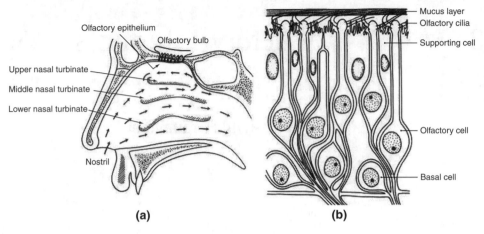

FIGURE

8.1

Nose anatomy.
(a) Structure of the nose. (b) Detailed structure of the olfactory epithelium.
Source: From Ooyama, *Handbook of Sense and Perception* (1994).

Materials that have smells are quite limited. Carbon (C), hydrogen (H), oxygen (O), nitrogen (N), phosphorus (P), sulfur (S), chlorine (Cl), bromine (Br), and iodine (I) are the elements related to smell, but only the last three (halogens Cl, Br, and I) have smells as single elements. Most molecules with smells are organic compounds and have a limited range of molecular weight. Only molecules whose weight is approximately 30 to 300 have smells; heavier and lighter molecules do not.

From the perspective of human–computer interaction, it may be interesting to examine how many kinds of smell are required to synthesize arbitrary smells. However, before discussing this important issue, let us review the case of vision. It is said that we can "synthesize" arbitrary color by mixing three primary colors: red, green, and blue. Physical light consists of electromagnetic waves whose power is described as a continuous distribution function of wavelength or frequency. This distribution is called a spectrum. Our visual system, however, might regard light with different spectrums as a single color if the balance of power of the three representative wavelengths (red, green, and blue) is the same. Hence, the phenomenon of the three primary colors is not the nature of physical color itself, but derives from the mechanism of visual sensation. This mechanism is now used by various visual recording and display systems, including color television sets, to transmit color information among remote places by coding color as a three-dimensional vector instead of a continuous distribution function that can be considered to have infinite dimensions. In other words, we can compress color information by making use of our visual mechanism.

The question now is whether olfaction has a mechanism comparable to the three primary colors—that is, are there primary odors? The answer is, unfortunately, no. Humans can detect the smells of approximately 0.4 million chemical compounds, and distinguish thousands (in some cases, tens of thousands) of smells. Amoore (1970), in his theory of stereochemistry, sophisticatedly categorized various odors into seven major groups of smells. Some researchers considered that these seven groups correspond to the three primary colors, but recent progress in olfaction research shows that this was not sufficient to explain human olfaction. Later, Amoore increased the number of "primary odors" to approximately 20 to 30, but this number has not been clearly tested.

A breakthrough in olfactory research occurred in 1991. Using an approach from the field of genetics, Buck and Axel (1991) estimated the number of receptor proteins in mice to be approximately 1,000. For this discovery, they received the 2004 Nobel Prize in physiology or medicine. Using recent data obtained in human genome analysis, this number is estimated to be 350 in humans. Each receptor protein responds to multiple molecules, resulting in the very complex functions of human olfaction. Currently, this number (350) is considered to be an index that shows the order of required odor components for obtaining arbitrary smells.

8.2 TECHNOLOGY OF THE INTERFACE

There are several technologies applicable to the development of olfactory interfaces. Because people smell by using their nose, odorants should be vaporized and delivered to the nose. Thus, the major technical fields involved in creating olfactory interfaces are the generation and delivery of smells.

8.2.1 Generating Smells

Most extracted smell sources are in liquid form (e.g., essential oils) or, occasionally, solid substances soaked in liquid. However, humans can detect airborne smells. This means that olfactory interfaces should include a device to vaporize sources of smells of which there are several methods.

Vaporization

A straightforward way of vaporizing smells is natural evaporation. This method can be used for liquid sources with relatively high volatility. By storing odor sources in a small vessel, such as an essential oil adhered to cotton or porous ceramics, and letting air flow through the vessel, we can generally obtain scented air in sufficient concentrations. If such natural evaporation is not sufficient, the material can be heated to accelerate the evaporation. Certain types of odor molecules, however, are easily destroyed by high temperatures, so heating of materials should be avoided whenever possible.

Another way of accelerating the vaporization of a liquid source is by making a fine mist of the liquid. One commonly used device is a sprayer. By continuously pushing the liquid out of a fine nozzle, the liquid becomes a fine mist that is easily vaporized. Electromechanical actuators can also be used to make a fine mist from liquids. Piezoelectric actuators are often used to push a small liquid drop out of the nozzle. With this method, the number of drops per second to be generated can be controlled so that the precise control of the aroma intensity can be achieved.

Blending

There are two ways to blend component odors at a specified ratio: in liquid status and after vaporization. To blend odors while they are in liquid form, dilution fluids (e.g., alcohol) are typically used, because aromatic sources (essential oils) are condensed and blending granularity would not be ideal if the undiluted solutions are directly mixed. Diluted solutions containing essential oils are mixed in drop units to make a mixed solution containing component odors at the desired ratio.

Another way of blending component odors is to blend them after each odor is vaporized. Air flowing through a vessel containing a liquid odor source becomes saturated with vapor, which is then mixed with air at desired mass flow rate. This mixing is usually performed by valves. Valvular systems that can continuously control the amount of air flow are called mass flow controllers (MFCs). As commercially available MFCs are often large and expensive, it may be difficult to integrate many MFCs into odor-blending systems. Instead, a method to blend multiple component odors by digitally controlling solenoid valves has been proposed (Yamanaka et al., 2002).

8.2.2 Delivering Smells

Once an odor source is vaporized, it should be delivered to the nose. The method of delivery depends on the application style, which involves whether users move or walk around, and how quickly the various smells should be delivered to the user. The traditional way of enjoying smells involves diffusing aromas throughout a room, where they are continuously enjoyed by users for a sufficient period of time. This is the reason why traditional aroma generators are generally called "diffusers." Widespread, long-term aromas are sufficient for communicating environmental or ambient information. Such smells do not disturb users explicitly, but can still subtly communicate what is happening in the surrounding world. Most of the aroma generators described in the previous section are categorized into this type because they, by themselves, do not have an explicit function for controlling the spatiotemporal distribution of smells.

This traditional way of enjoying aromas, however, may not be sufficient if we want to make use of smells as a part of media technologies. If we consider enjoying smells in conjunction with audiovisual entertainment such as movies, TV programs, and interactive video games, the smell corresponding to each scene might need to be replaced immediately. The problem with traditional "scattering-type" aroma generators is that they cannot erase the smell once it is diffused within the space. Hence, a mechanism that localizes scent delivery to a limited range (in time and/or in space) is required to use smell as a part of multimodal media.

There are two approaches to spatiotemporal control of smells. One is to emit the minimum amount of smell being delivered to the nose. Another is to incorporate smell elimination equipment into the system. The former approach has several variations (see Section 8.3). The most direct way of delivering small amounts of smell to the nose is to use tubes that carry scented gas from the scent generator. If the scent generator is placed on the desktop, the user's workspace (movable range) is limited by the length of the tube. Therefore, making scent generators compact enough to be wearable is an effective solution (Yamada et al., 2006). If the scent generators become extremely compact, they can be directly attached to the nose to configure a direct-injection type of olfactory display (Yamada et al., 2006).

Compact scent generators can also be mounted at the shoulders or chest (Morie et al., 2003). Mounting scent generators to these positions reduces the inconvenience of wearing devices and can result in natural human–computer interactions. There have also been attempts to deliver small clumps of scented air through free space without using tubes. By using the principle of a vortex ring, launched from the aperture of a box, scented air can be carried through free space (Watkins, 2002; Yanagida et al., 2004). This is an untethered approach for spatio-temporally localized olfactory displays.

The second option for achieving spatiotemporally localized olfactory display systems is to incorporate smell elimination equipment into the system. Without such equipment, the emitted scent will be gradually diffused and reduced in density. Such natural diffusion is adequate if the amount of emitted scent is sufficiently small. For relatively large amounts of scent, system designers should also design smell-eliminating equipment. If the amount of the smell is moderate, suction pumps and filters (such as a charcoal filter) can be used. For systems that emit a massive amount of scent, ventilating functions should be incorporated into the system. If these suction or ventilation functions are included in the system, the scent-emitting subsystem is less critical in delivering smells to the user.

8.3 CURRENT IMPLEMENTATIONS OF THE INTERFACE

Although olfactory interfaces are still at an early stage of development, several successful implementations can be found. Systems are focused on generation of aroma or spatiotemporal control of smells.

FIGURE

8.2

Sensorama simulator.
The Sensorama was developed by Morton Heilig around 1960. (Photo from *http://www.telepresence.org/*.)

8.3.1 Historical Background

An early approach to incorporating smells with other kinds of sensory displays can be found in Heilig's Sensorama (Figure 8.2), developed around 1960 (Heilig, 1962, 1992). Users could enjoy multimodal movies incorporating breezes and smells, although the displays were not interactive. There have also been some entertainment attractions using scent; for example, McCarthy (1984) developed a scent-emitting system called a "Smellitzer" (Figure 8.3) that could emit a selected scent and produce a sequence of various smells.

8.3.2 Systems for Scent Generation and Blending

An olfactometer is a reliable instrument used in experiments on human olfaction. Experimenters construct olfactometer systems very carefully to ensure that the

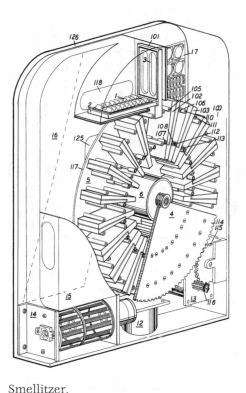

FIGURE

8.3

Smellitzer.

This scent-emitting system was developed by McCarthy (1984). *Source:* From McCarthy (1984); U.S. Patent 4,603,030.

experiment is both safe and accurate. Air flow is divided into several channels, each of which consists of a pair of tubes equipped with solenoid valves. A pair of solenoid valves is controlled to open exclusively so that the total air flow of the channel is kept constant. One of the tubes in each channel is connected to an odor vessel to produce scented air. The intensity of each odor is controlled by the timing ratio according to which the valve connected to the odorant vessel is opened. All valves in the system are controlled simultaneously by a host computer to produce blended smells that consist of the desired density of selected odorant(s). Figure 8.4 shows an example of an olfactometer developed by Lorig et al. (1999).

Although carefully designed and implemented olfactometers have successfully performed strict control of smell type, concentration (intensity), and temporal profile, these systems may be cumbersome for ordinary users. Recently, several trials have been made to construct easy-to-use, computer-controlled odor-emitting systems.

In 1999, DigiScents announced plans to release a scent diffuser, called iSmell, that can emit multiple smells. Shaped like a shark fin, the diffuser functioned as a

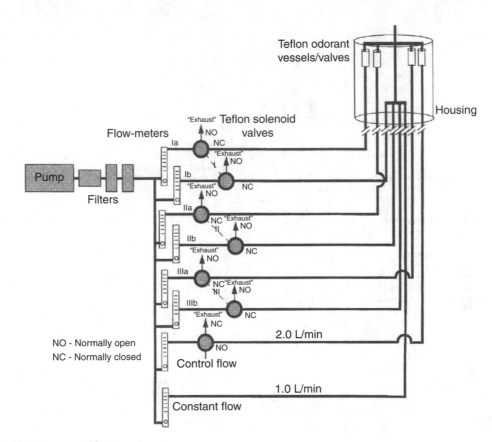

FIGURE   Olfactometer.
                A schematic diagram of an olfactometer. *Source:* From Lorig et al. (1999).
8.4

peripheral device for personal computers (Figure 8.5). As of this writing, iSmell is still not commercially available.

Kaye (2001, 2004) has developed various systems to emit scents in the context of ambient media (Figure 8.6). Air brushes are used in the system named "inStink" to emit 12 kinds of smells. Sprays driven by solenoids are used in the Dollars & Scents, Scent Reminder, and Honey, I'm Home systems, which emit two smells, five smells, and a single smell, respectively.

Scent Dome, developed by TriSenx (2007), is a computer-controlled scent diffuser that can blend 20 different scents. TriSenx also provides exchangeable cartridges, so that users can customize the aromas (Figure 8.7).

France Telecom developed multichannel scent diffusers that consist of vessels containing aromatic gels and a fan controlled via a USB communication line.

FIGURE

8.5

iSmell scent diffuser.
Developed by DigiScents, iSmell emits multiple smells and functions as a PC peripheral. (Courtesy of DigiScents.)

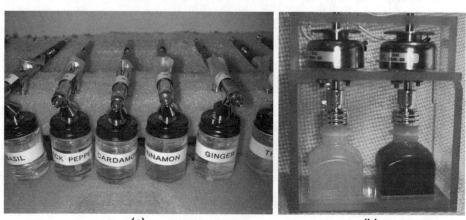

(a)   (b)

FIGURE

8.6

Computer-controlled scent emitters.
(a) inStink. (b) Dollars & Scents. *Source:* From Kaye (2001).

FIGURE

8.7

Scent Dome.
Developed by TriSenx, this scent diffuser blends 20 different scents. (Image courtesy TriSenx Holdings, Inc, from *http://www.trisenx.com/*.)

The system configuration is simple and straightforward, and this system was incorporated with practical audiovisual programs and demonstrated at various sites. In a typical program, called Kaori-Web ("Kaori" is a Japanese word meaning aroma), the cooking process is displayed by images, sounds, and smells of the food being cooked. These multimodal stimuli are synchronously controlled by a PC (Figure 8.8).

Nakamoto et al. (2007) developed a real-time odor-blending system as a part of an odor-recording system (Figure 8.9). These authors used solenoid valves that could open and close very rapidly, and succeeded in controlling the density of multiple odors in real time. Although the valves only had a digital function (open and close), they introduced the concept of delta–sigma modulation, which has been used in digital-to-analog conversions for audio players, to their real-time blender (Yamanaka et al., 2002). The high-speed switching of the valves made it possible to quantitatively control odor density by compact digital devices and computer interfaces, providing integration of up to 32-channel blenders in a compact enclosure (Nakamoto et al., 2007).

8.3.3   Systems for Spatiotemporal Control of Scents

Although spatiotemporally controllable olfactory interfaces have only recently been developed, there are a variety of potential applications.

Cater (1992, 1994) constructed a firefighting simulator that emits fire-related odors, and embedded all necessary parts in the set of equipment used by firefighters (Figure 8.10).

FIGURE 8.8

Kaori-Web (Scent-Web).
Developed by the Tsuji Wellness Corporation and France Telecom R & D, this system is a multimodal program with audio, images, and scents. (Courtesy of Exhalia Japan.)

水晶振動子匂いセンサ (20MHz, AT-cut)

ブレンダ

要素臭

(a)

(b)

FIGURE 8.9

Real-time odor blender.
The blender uses high-speed switching in solenoid valves. (a) Prototype, part of the odor recorder system. (b) Commercial version of the real-time odor blender. (Photo (a) courtesy of Nakamoto Lab, Tokyo Institute of Technology.)

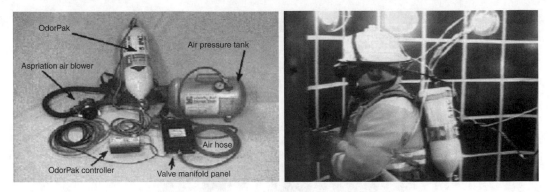

FIGURE
8.10

Firefighting simulator.
The D.I.V.E. firefighter training system by Cater. *Source:* From Kaye (2001).

Tanikawa and colleagues (Yamada et al., 2006) have developed a wearable olfactory display system. By focusing on the spatial distribution of virtual olfactory space, and by making the entire system (including scent generators) compact and wearable, their system allows users to move around the environment and actively explore virtual olfactory space. This olfactory display system has four odor vessels, and a micropump (driven by a DC motor) is used to produce air flow for each vessel (Figure 8.11).

Wearable olfactory displays under development are direct-injection–type displays. These use piezoelectric actuators (inkjet head) to produce small drops of essential oil that are directly injected into the user's nostril (Yamada et al., 2006). Although this research is ongoing, the goal is to ultimately create a compact olfactory display that will make a user feel unencumbered and able to move freely around the environment. If this system is developed, olfactory-augmented reality will be realized and various types of information can then be communicated by odor.

Mochizuki et al. (2004) developed an arm-mounted olfactory display system, focusing on the human action of holding objects in the hand, bringing them close to the nose, and then sniffing. They configured the system so that the odor vessels of the olfactometer are mounted on the arm and the odor-emitting end of the tube is positioned on the user's palm (Figure 8.12). This arrangement enabled quick switching among different smells so that an interactive game interface using smell was achieved.

At the University of Southern California Institute for Creative Technologies (ICT), Morie and coworkers (2003) designed and implemented a necklace-shaped unit equipped with four small wireless-controlled scent emitters. This system is called the Scent Collar (Figure 8.13, page 281). By emitting scents close to the

<table>
<tr><td>FIGURE

8.11</td><td>Wearable olfactory display.
Developed by Tanikawa and colleagues (Yamada et al., 2006) this system allows users to move around the environment and explore virtual olfactory space. (Courtesy of Hirose Tanikawa Lab, Graduate School of Information Science and Technology.)</td></tr>
</table>

nose, the collar enables location-specific olfactory stimuli without forcing users to wear encumbering tubes or devices near the face.

MicroScent developed a scent generator that can deliver a small amount of scent locally by making use of the vortex ring launched from the aperture of an enclosed space (Figure 8.14). This principle (also known as an "air cannon") is the basis of a popular scientific demonstration for children. In this system, multiple diffusers are embedded in the enclosure and multiple smells can be emitted. With this configuration, the scented air in the enclosed space has to be cleaned by a suction filter before each scent is emitted.

Yanagida et al. (2004) proposed a method of scent delivery that also uses the principle of the vortex ring. Their system has a tracking function that is aimed at the user's nose and a scent-switching mechanism that can provide a different smell for each launch of the vortex ring (Figure 8.15, page 282). The user's nose is tracked by detecting its position through image processing, and the two-degrees-of-freedom platform carrying the body of the air cannon is maneuvered to direct the air cannon at the nose. The scent-switching mechanism is implemented by placing a small

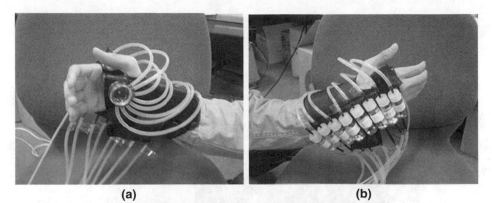

(a)   (b)

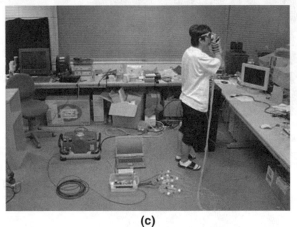

(c)

FIGURE   Arm-mounted type of olfactory display for active sniffing.
8.12   This display was developed by Mochizuki and colleagues (2004). (Courtesy of Nara
              Institute of Science and Technology, Japan.)

cylinder chamber in front of the aperture of the air cannon. The scented air is then injected into this chamber instead of into the air cannon body so that the scented air is completely pushed out of the chamber for each shot and no scented air remains in the air cannon body. This system was named the Scent Projector.

During the development of scent projectors, several issues emerged. For instance, the feeling of suddenly being wind-blasted impairs the natural olfactory experience of the user. To solve this problem, a method has been proposed to use two air cannons and let the vortex rings collide with each other so that they collapse in front of the user and the high-speed air flow composing the vortex rings is reduced (Figure 8.16, page 282).

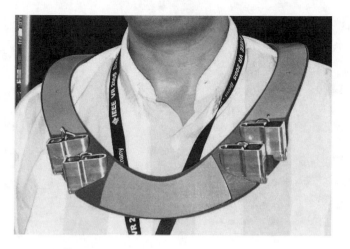

FIGURE

8.13

Scent Collar.
Developed by Morie and colleagues (2003) at the Institute for Creative
Technologies, this device permits location-specific stimuli without clumsy
tubes or devices near the face.

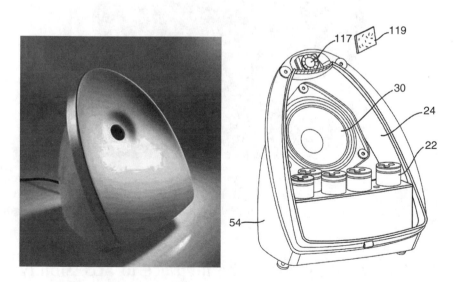

FIGURE

8.14

MicroScent's scent generator.
This device delivers scent locally. (Courtesy of MicroScent; U.S. Patent
6,357,726.)

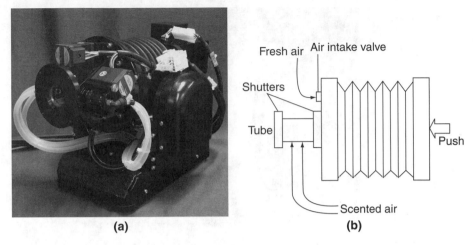

(a)   **(b)**

FIGURE

8.15

Scent Projector, with nose-tracking and scent-switching functions.
(a) System overview. (b) Scent-switching mechanism. The system was developed
at the Advanced Telecommunications Research Institute International.

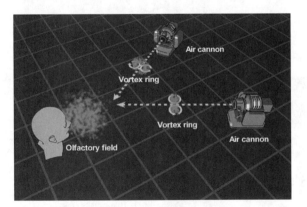

FIGURE

8.16

Generating an olfactory field in free space with two air cannons.
The vortex rings collapse in front of the user and the high-speed air flow is
reduced. *Source:* From Nakaizumi et al. (2006); courtesy of Joseph "Jofish" Kay.

8.3.4 Applications of the Interface to Accessibility

Olfactory interfaces can be an intuitive means to communicate danger to others.
For example, people notice leaks of inflammable (but invisible) gas by smell.
Usually, methyl-mercaptan, to which humans are highly sensitive, is blended

with methane or propane gas. The olfactory communication function is effective for both sighted and visually impaired people. If computer-controlled olfactory interfaces become common, a wide variety of physiologically intuitive signals can be provided, especially for communicating dangerous situations.

8.4 HUMAN FACTORS DESIGN OF THE INTERFACE

Although humans are said to be able to distinguish thousands of odors, there is currently no method, as mentioned previously, to synthesize arbitrary smells by mixing a small set of "primary" odors. In this section, various aspects of olfaction are described.

8.4.1 Odor Intensity

The subjective intensity of a smell is related to the concentration of odor molecules, but there is a wide variety in subjective intensity for various odor materials at identical concentrations. When multiple odor molecules coexist, odors at higher concentrations are not always subjectively dominant. Physical aspects that are related to subjective odor intensity follow:

Concentration: The most basic value, described in grams, molar concentration, percentages, or parts per million (ppm).

Diffusivity: This parameter refers to how fast the odor diffuses into the atmosphere, and is defined as the distance that the odor molecule can reach within a specified time or the time required for the odor molecule to travel a certain distance. The higher the concentration and the lower the molecular weight, the larger the diffusivity obtained. However, air flow has a significant effect on odor diffusion when compared to static diffusion of odor material.

Boiling point: The temperature at which vapor pressure becomes equal to environmental pressure. The boiling point for most odor compounds is within the range of 150° to 300° C. Odors with a significantly low boiling point tend to have an intense smell, and odors with a higher boiling point have a weaker smell.

Volatility: The amount of odor material that is consumed in a given time period.

Generally, Weber-Fechner's law can be applied to the relationship between the concentration of odor material and subjective intensity.

8.4.2 Threshold Value

The minimal concentration at which humans can "detect" a smell is called the odor threshold value. There are multiple definitions of threshold value: detection, recognition, and differential.

Detection threshold value: Smells cannot be detected at very low odor concentrations. If the concentration is gradually increased, subjects can detect "some kind of smell," even though they cannot distinguish the type of smell. This concentration is called the detection threshold value.

Recognition threshold value: By increasing the odor concentration beyond the detection threshold, the quality or impression of the smell at a certain concentration can be described. The minimal concentration at which the quality or impression of a smell can be described is called the recognition threshold value.

Differential threshold value: The minimum difference in concentration that can be detected is called the differential threshold value. Usually, it is described in percentages as the ratio of the minimal detectable difference to the original intensity (Weber ratio). This ratio varies depending on the type of odor. The differential threshold value generally ranges from 10 to 30 percent for olfaction, as opposed to 1 to 2 percent for differences in light intensity.

8.4.3   Adaptation

After continuous exposure to an odor, the subjective intensity of the odor gradually decreases; after a certain period (typically several minutes), the odor is no longer consciously detected. This phenomenon is called adaptation, and is considered to be a mechanism of the sensing organ for automatically adjusting sensitivity to an odor. Through adaptation, smells that continuously exist, such as body odor, are not detected, while odors that occasionally exist are clearly noticed. The level and the temporal aspect of adaptation vary depending on the odor material. The adaptation caused by a single odor material is called *self-adaptation*. The recovery phase is generally faster than the adaptation phase.

Cross-adaptation is the phenomenon of decreased sensitivity of one odor through exposure to another. This adaptation occurs for some combination of odors but not others.

Another mechanism is called *habitation*, which is often used in a similar context to adaptation. Habitation is considered to occur within the brain, while adaptation is a function of the olfactory cells. However, the definition of habitation appears to vary among researchers, and its definition has not been standardized.

8.4.4   Multiplier Effect and Masking Effect

When two different odors are mixed, the intensity is enhanced in some cases and suppressed in others. Enhancement is called the multiplier effect and suppression is called the masking effect. The masking effect in particular is used in daily life. When an aroma is introduced into a space where an unpleasant smell exists, the unpleasant smell is perceived to decrease. This is a subjective phenomenon known as sensory deodorization, as the unpleasant smell is not actually removed. Typical examples of sensory deodorization are perfumes and deodorants in lavatories.

8.4.5   Nonlinearity

Changes in concentration sometimes result in qualitative rather than quantitative changes, especially for very low concentrations. For example, a type of fecal odor (indole) changes to a flowery smell at very low concentrations. In addition, a mixture of smells sometimes results in a qualitative change of smell. This phenomenon is called *modification*. Usually, surrounding odors consist of multiple components, and these odors are modified. The blending of aromas is considered to be a sophisticated technique in creating desired aromas via modification.

8.4.6   Spatial Perception

The question of how humans perceive the surrounding environment through olfaction is an interesting topic for interactive human–computer interfaces, especially in virtual reality. Finding the location of a smell source is an important task related to spatial perception by olfaction. By moving and comparing the trajectory and the temporal profile of the intensity, the smell source can be located (Yamada et al., 2006). This might be considered to be a counterpart to the motion parallax cue in visual perception. Another interesting question is whether there are olfactory counterparts to binocular cues through the use of two nostrils.

In his historic research, Békésy (1964) reported that the time difference of the olfactory stimuli for each nostril contributes to the directional localization of the smell source, just as the direction of auditory sources can be determined by the difference in sounds perceived by each ear. More recently, Porter et al. (2007) reported that scent tracking is aided by internostril comparisons. Considering that the internostril distance is only 2 to 3 cm, which is less than the interpupillary (6 to 7 cm) and interaural (approximately 15 cm) distances, the "stereo effect" in olfaction may not be as strong as the "motion effect." However, it is very interesting to note that humans make use of two nostrils to help understand the environment by olfaction.

8.5   INTERFACE-TESTING TECHNIQUES

Historically, the evaluation of odor has long relied on sensory analysis by experienced human testers. This means that it was difficult to evaluate a smell objectively. Even though experienced testers were trained to provide objective results rather than subjective impressions, these results are assumed to be affected by environmental conditions, the tester's physical condition, and so on.

However, recent progress in chemical sensors has resolved this problem. The technical field related to smell-sensing machinery is called electronic nose or e-nose (Pearce et al., 2003). The electronic nose is a system to identify the components of odors. Multiple gas sensors (chemosensors) are used in the system to detect, analyze, and categorize multiple component odors. Among the various types of chemosensors, metal oxide semiconductor sensors, conducting polymer sensors, and quartz microbalance sensors have been frequently used for electronic nose systems.

Metaloxide gas sensors have been used to detect flammable gases such as methane and propane or poisonous gas such as hydrogen sulfide, and are now used as odor-sensing elements as well. When odor molecules are absorbed onto the surface of a semiconductor, the electric conductivity of the element is changed according to the amount of absorbed molecules. Conducting polymer sensors also use the principle of electric conductivity variation, while quartz microbalance sensors have a resonance frequency output that varies when odor molecules are absorbed in the sensitive membrane on the surface of a quartz resonator.

One problem with odor-sensing equipment is that the temporal response is still insufficient for interactive use. The response time for a typical gas sensor is dozens of seconds. However, some recent sensors have improved response times; if the response of gas sensors in particular becomes faster than human olfaction, these sensors could be incorporated into the interaction loop of an olfactory display system.

8.6   DESIGN GUIDELINES

Although the development of computer-controlled olfactory interfaces is still at an early stage, some guidelines can be indicated when designing and implementing them.

8.6.1   Expected Effect of Appending an Olfactory Interface

First, the designer should take into account whether it is meaningful or effective for the entire system to append an olfactory channel. If the olfactory interface is inherent to the designer's intention, such as creating odor as the main content of the interface, this decision is reasonable. However, there may be issues when the olfactory interface is not the primary mechanism of interaction. In this case, designers should decide whether the olfactory effect is worth the additional cost and complexity to the system.

8.6.2   Number of Odors

As mentioned previously, there is no method for synthesizing arbitrary smells from a small set of component odors. Hence, determining the number of odors required for the application is important. Even though humans can distinguish thousands of smells, the number of smells required for a single application may be relatively small. For example, if we want to emit smells according to the user's behavior in an interactive game, several to a few dozen odors may be sufficient. In this case, it is sufficient to blend a relatively small number of smells or select pre-blended aromas; blending hundreds or thousands of odors is not necessary. The number of simultaneously required odors directly affects the complexity of the system; thus, careful selection of the odors is essential for creating cost-effective systems.

8.6.3   Number of Users and Spatial Dimensions

The next point to consider is the number of target users for simultaneous smell deployment. If the application is for a single user, a device of personal size, such as a small scent generator located close to the user, can be applied. If the application is for many simultaneous users, odors should be delivered to a larger space. There are two approaches to achieve multiuser olfactory interfaces: providing a personal device for each user or controlling a large-volume olfactory space at any one time. The effects of the volume of space on the temporal aspects of odor are discussed in the next subsection.

8.6.4   Temporal Aspects of Odor Control

When applying olfactory interfaces with audiovisual programs or interactive applications, the speed of changes or replacements in olfactory stimuli is an important issue. If the odor is to be used as background or ambient media, the time constant is relatively large with constant or slow/gradual changes in odor. This is an effective way of using odor to provide information in a subtle manner.

However, the use of olfactory interfaces together with audiovisual programs or applications with rapid interaction requires short-term changing or switching of odors. In this case, designers face the problem that emitted smells cannot be erased instantaneously. Once an odor is diffused in the air, it remains unless it is moved by air currents. Possible solutions are (1) design the entire facility to include ventilation systems and (2) emit as small an amount of scent as possible.

The first approach is effective when the olfactory interfaces are used in dedicated, specially designed facilities, such as amusement parks. The olfactory interfaces used at such locations can simply emit a large amount of odor materials,

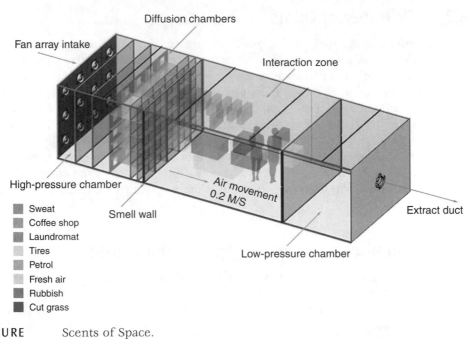

Fan array intake

Diffusion chambers

Interaction zone

High-pressure chamber

Air movement
0.2 M/S

Smell wall

Extract duct

Low-pressure chamber

 Sweat
Coffee shop
Laundromat
Tires
Petrol
Fresh air
Rubbish
Cut grass

FIGURE

8.17

Scents of Space.
An interactive smell system. *Source:* From Haque (2004), Haque Design and
Research/Josephine Pletts.

which ensures the olfactory experience. The emitted odors are then removed by
the ventilation system and fresh air or another odor is provided. This approach
can also be applied to a closed-booth configuration. An example of this type is
the interactive smell system shown in Figure 8.17, which makes use of a slow,
undetectable air flow to produce spatiotemporal patterns of smell in a rectangular
parallel-piped booth.

If one cannot make dedicated facilities, such as for home use, then the second approach would be effective. Because the amount of the emitted smell is
small, the odor emission device should be placed close to the nose. Although
the emitted smell is diffused into the surrounding atmosphere, the amount is
sufficiently small that the concentration quickly diffuses below the detection
threshold and the smell is unnoticed. There are several approaches of this type,
depending on the location of the odor-emitting device. Nose-mounted and face-
mounted types are possible solutions (Yamada et al., 2006). Odor emission
points can include other parts of the body, such as the chest (Morie et al.,
2003) or arm (Mochizuki et al., 2004). For users who do not want to wear
devices, a local scent delivery technique through free space can be adopted
(Watkins, 2002; Yanagida et al., 2004).

8.7   CASE STUDIES

Two case studies exploring olfactory interfaces can be found at *www.beyond-thegui.com*.

8.8   FUTURE TRENDS

As mentioned previously, olfactory interfaces are still in an early stage of development compared to visual, auditory, and haptic interfaces. In recent years, however, research papers and/or demonstrations have been presented in almost every academic conference related to human–computer interaction, including the field of virtual reality. This implies that many researchers in various countries are now interested in olfactory interfaces and that researchers in this field are currently networking to establish a research community.

The future of olfactory interfaces depends on at least two factors (Wasburn & Jones, 2004): the development of practical applications to make olfactory interfaces more common, and fundamental improvements in olfactory interfaces, including the elucidation of human olfaction at both the physiological and psychological levels. From a technical viewpoint, it is very important to know how many types of odors are necessary for a certain application and for producing arbitrary smells. If this number is less than a hundred, constructing practical interfaces will not be difficult.

REFERENCES

Amoore, J. E. (1970). *Molecular Basis of Odor*. Springfield, IL: Charles C. Thomas.

Barfield, E., & Danas, E. (1996). Comments on the use of olfactory displays for virtual environments. *Presence* 5(1):109–21.

Békésy, G. (1964). Olfactory analogue to directional hearing. *Journal of Applied Physiology* 19(3):363–73.

Buck L., & Axel, R. (1991). A novel multigene family may encode odorant receptors: A molecular basis for odor recognition. *Cell* 65:175–87.

Cater, J. P. (1992). The nose have it! *Presence* 1(4):493–94.

Cater, J. P. (1994). Smell/taste: Odors in virtual reality. *Proceedings of IEEE International Conference on Systems, Man and Cybernetics*, San Antonio, TX.

Davide, F., Holmberg, M., & Lundström, I. (2001). Virtual olfactory interfaces: Electronic noses and olfactory displays. In Riva, G., & Davide, F., eds., *Communications Through Virtual Technologies: Identity, Community and Technology in the Communication Age*. Amsterdam: IOS Press, 193–219.

Dinh, H. Q., Walker, N., Song, C., Kobayashi, A., & Hodges, L. (1999). Evaluating the importance of multi-sensory input on memory and the sense of presence in virtual environments. *Proceedings of IEEE Virtual Reality*, Houston.

Haque, U. (2004). Scents of Space: An interactive smell system. *Proceedings of ACM SIG-GRAPH 2004 Sketches*, Los Angeles.

Heilig, M. L. (1962). Sensorama Simulator, U.S. Patent 3,050,870.

Heilig, M. L. (1992). El cine del futuro: The cinema of the future. *Presence* 1(3):279–94.

Kaye, J. N. (2001). Symbolic Olfactory Display. Master's thesis, Massachusetts Institute of Technology.

Kaye, J. N. (2004). Making scents. *ACM Interactions* 11(1):49–61.

Lorig, T. S., Elmes, D. G., Zald, D. H., & Pardo, J. V. (1999). A computer-controlled olfactometer for fMRI and electrophysiological studies of olfaction. *Behavior Research Methods, Instruments & Computers* 31(2):370–75.

McCarthy, R. E. (1984). Scent-Emitting Systems, U. S. Patent 4,603,030.

Mochizuki, A, Amada, T., Sawa, S., Takeda, T., Motoyashiki, S., Kohyama, K., Imura, M., & Chihara, K. (2004). Fragra: A visual-olfactory VR game. *Proceedings of ACM SIGGRAPH 2004 Sketches*, Los Angeles.

Morie, J. F., Iyer, K., Valanejad, K., Sadek, R., Miraglia, D., Milam, D., Williams, J., Luigi, D–P., & Leshin, J. (2003). Sensory design for virtual environments. *Proceedings of ACM SIGGRAPH 2003 Sketches*, San Diego.

Nakaizumi, F, Yanagida, Y, Noma, H., & Hosaka, K. (2006). SpotScents: A novel method of natural scent delivery using multiple scent projectors. *Proceedings of IEEE Virtual Reality 2006*, Arlington, VA, 207–12.

Nakamoto, T., & Hiramatsu, H. (2002). Study of odor recorder for dynamical change of odor using QCM sensors and neural network. *Sensors and Actuators B* 85(3):263–69.

Nakamoto, T., & Minh, H.P.D. (2007). Improvement of olfactory display using solenoid valves. *Proceedings of IEEE Virtual Reality*, Charlotte, NC.

Ooyama, A. (1994). *Handbook of Sensation and Perception.* Tokyo: Seishin Shobor.

Pearce, T. C., Schiffman, S. S., Nagle, H. T., & Gardner, J. W., eds. (2003). *Handbook of Machine Olfaction.* Weinheim: Wiley-VCH.

Plattig, K. H. (1995). *Spürnasen unt Feinschmecker.* Berlin: Springer-Verlag.

Porter, J., Craven, B., Khan R. M., Chang S. J., Kang, I., Judkewitz, B., Volpe, J., Settles, G., & Sobel, N. (2007). Mechanism of scent-tracking in humans. *Nature Neuroscience* 10(1):27–29.

Tonoike, M., ed. (2007). [*Information and Communication Technology of Olfaction*] (in Japanese). Tokyo: Fragrance Journal Ltd.

TriSenx. (2007). Home page—available at *http://www.trisenx.com/.*

Yamada, T., Yokoyama, S., Tanikawa, T., Hirota, K., & Hirose, M. (2006). Wearable olfactory display: Using odor in outdoor environments. *Proceedings of IEEE Virtual Reality 2006*, Charlotte, NC, 199–206.

Yamanaka, T., Matsumoto, R., & Nakamoto, T. (2002). Study of odor blender using solenoid valves controlled by delta–sigma modulation method. *Sensors and Actuators B* 87:457–63.

Yanagida, Y., Kawato, S., Noma, H., Tomono, A., & Tetsutani, N. (2004). Projection-based olfactory display with nose tracking. *Proceedings of IEEE Virtual Reality 2004*, Chicago.

Yatagai, M., ed. (2005). *Kaori no hyakka-jiten* [Encyclopedia of aromas]. Tokyo: Maruzen.

Washburn D. A., & Jones, L. M. (2004). Could olfactory displays improve data visualization? *Computing in Science & Engineering* 6(6):80–83.

Watkins, C. J. (2002). Methods and apparatus for localized delivery of scented aerosols, U.S. Patent 6,357,726.

9

Taste Interfaces

Hiroo Iwata

9.1 NATURE OF THE INTERFACE

Taste is an important sense that is seldom used in traditional displays. It is the fundamental nature of the underlying sense and the difficulty of implementing unobtrusive devices that have made taste interfaces so difficult.

9.1.1 The Sense of Taste

Humans detect chemical features of food with taste receptor cells. They are assembled into taste buds, which are distributed across the tongue. Although humans can taste a vast array of chemical entities, they evoke few distinct taste sensations: sweet, bitter, sour, salty, and "umami." Distinct taste receptor cells detect each of the five basic tastes (Chandrashekar et al., 2006). Examples of these basic tastes follow:

Sweet: A common sweet substance is sucrose, known as table sugar.

Bitter: A strong bitter taste prevents ingestion of toxic compounds. A small amount of a bitter substance contributes to perceived interesting or good taste/flavor. Caffeine in coffee is an example.

Sour: Sour substances are mostly acids. Vinegar is an example.

Salty: A common salty substance is NaCl, known as table salt.

Umami: This is a Japanese word meaning "savory," and thus applies to the sensation of savoriness. Typical umami substances are glutamates, which are especially common in meat, cheese, and other protein-rich foods (Kawamura & Kare, 1987).

9.1.2   Why Is a Taste Interface Difficult?

Taste is the last frontier of virtual reality. Taste is very difficult to display because it is a multimodal sensation comprising chemical substance, sound, smell, and haptic sensations. The literature on visual and auditory displays is extensive. Smell display is not common, but smell can easily be displayed using a vaporizer. Although not as extensive as visual and auditory displays, the literature on olfactory displays is growing (e.g., Nakamoto et al., 1994; Davide et al., 2001).

Taste perceived by the tongue can be measured using a biological membrane sensor (Toko et al., 1994, 1998). The sensor measures the chemical substance of the five basic tastes. Any arbitrary taste can easily be synthesized from the five tastes based on sensor data.

Another important element in taste is food texture. Measurement of biting force has been studied. A multipoint force sensor has been used to measure force distribution on the teeth (Khoyama et al., 2001, 2002). The sensory properties of the texture of real food have also been studied (Szczesniak, 1963, 2002).

Dental schools have been working on the training of patients to chew properly. A master–slave manipulator for mastication has been developed (Takanobu et al., 2002). A robot manipulated by the doctor applies appropriate force to the patient's teeth.

Haptic interfaces for hands or fingers are the focus of major research efforts in virtual reality. In contrast, research into haptic displays for biting is very rare.

9.2   TECHNOLOGY OF THE INTERFACE

There have been no interactive techniques developed to date that can display taste. The Food Simulator Project was an effort to develop just such an interface.

9.2.1   Existing Techniques for Displaying Taste

The traditional method for displaying taste is the use of filter paper discs, such as for patients with taste disorders (Tsuruoka, 2003). A small filter paper disc is soaked with a chemical (taste) substance and put on the patient's tongue. Filter paper discs with various density of the chemical substance are used to test taste disorders. However, there has been no technique to display taste for human–computer interaction.

9.2.2   Food Simulator Project: A Taste Interface Challenge

In 2001, I launched a project called the "Food Simulator" to develop an interface device that presents food texture as well as taste. The unsolved problem in taste display is a haptics issue. The first goal of the project was developing a haptic interface for biting. The device should be suitably shaped for placing in the mouth, and be effectively controlled to simulate food texture. The second goal was to present a multimodal sensation to the user. To this end, biting sounds and chemical tastes should be integrated with the haptic interface.

To achieve these goals, we developed a haptic interface to present the biting force. The Food Simulator generates a force simulating the previously captured force profiles of an individual biting real food. A film-like force sensor is used to measure biting force associated with real food. A force sensor is installed in the Food Simulator and the device is actuated using force control methods.

The Food Simulator is integrated with auditory and chemical sensations associated with taste. The sound of biting is captured by a bone vibration microphone. The sound is then replayed using a bone vibration speaker synchronized with the biting action. The chemical sensation of taste is produced using an injection pump, with a tube installed at the end effecter.

9.3 CURRENT IMPLEMENTATIONS OF THE INTERFACE

The Food Simulator employs mechanical linkages to apply biting force to the teeth. It is integrated with sound, vibration, and chemical taste.

9.3.1 Haptic Device in a Food Simulator

The haptic device is composed of a one degree-of-freedom (DOF) mechanism that is designed to fit in the user's mouth. The device is aimed at applying a force representing the first bite. Chewing is not addressed in this prototype. Thus, the device is designed to apply force in a direction normal to the teeth. The configuration of the mechanical linkage takes into consideration the jaw structure. The shape of the linkage enables the application of force to the back teeth. The width of the end effecter is 12 mm, and the device applies force to two or three teeth.

Regarding hygiene, the end effecter includes a disposable cover of cloth and rubber. Figure 9.1 shows the overall view of the apparatus.

The haptic device is composed of a ones DOF mechanism that employs four linkages. Figure 9.2 illustrates its mechanical configuration. The linkages are driven by a DC servo motor (MAXON Motor, RE25). The user bites the end effecter of the device. The working angle of the end effecter is 35 degrees, and the maximum force applied to the teeth is 135 N. A force sensor that detects force applied by the user's teeth is attached to the end effecter. The device is controlled by a PC (Pentium 4, 2 GHz). The update rate of force control is 1,700 Hz, which is sufficient for controlling a haptic interface.

FIGURE

9.1

Overall view of the apparatus.
The 1-DOF haptic device is a core element of the system.

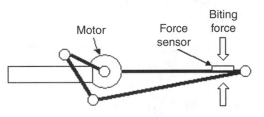

FIGURE

9.2

Mechanical configuration of the haptic device.
Four linkages are employed in the device, which is designed to fit into
the mouth.

9.3.2   Multisensory Display

The second goal of the Food Simulator project was to present multimodal sensation. We also tried to integrate a multimodal display with the haptic interface.

Sound

Sound is closely related to biting action. Mastication of a hard food generates
sound. This sound is perceived by vibration, mostly in the jawbone. Sounds of
biting real food were recorded using a bone vibration microphone. Figure 9.3
shows a recording session. The bone vibration microphone was inserted into
the ear. An accelerometer, installed in the microphone, picked up vibration of
the jawbone.

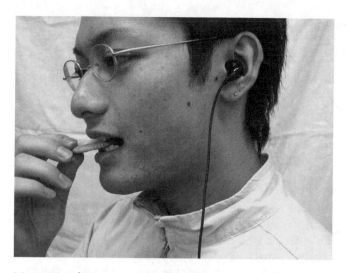

FIGURE
9.3

Measuring biting sound.
An accelerometer picks up vibration of the jawbone while biting real food.

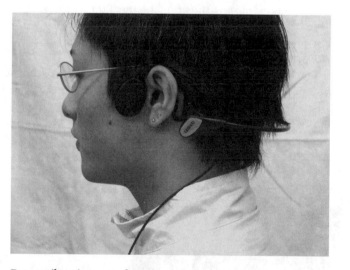

FIGURE
9.4

Bone vibration speaker.
The recorded sound is synchronized with the biting action.

The recorded sound was then displayed using a bone vibration speaker. Figure 9.4 shows the speaker that generated vibrations in the jawbone. The sound was synchronized with the biting action. For example, the sound of a virtual cracker was displayed at the beginning of the second stage of the force control shown in Figure 9.7 (see page 298).

TABLE 9.1 Chemical Substances Used in Synthesis of Five Basic Tastes

Basic Taste	Chemical Substance
Sweet	Sucrose
Sour	Tartaric acid
Salty	Sodium chloride
Bitter	Quinine sulfate
Umami	Sodium glutamate

Chemical Taste

Chemical sensations perceived by the tongue contribute greatly to the sense of taste. An arbitrary taste can be synthesized from the five basic tastes. Table 9.1 summarizes common chemical substances for the basic tastes.

The chemical sensation of taste was presented by injecting a small amount of liquid into the mouth. A tube was attached to the end effecter of the haptic interface. Figure 9.5 shows the tube at the top end of the linkage. The liquid was transferred using an injection pump (Nichiryo, DDU-5000). Figure 9.6 shows an overall view of the injection pump. Injection of the liquid was synchronized with the biting action. The pump provided 0.5 ml of solution for each bite.

Vision and Smell

Visual and olfactory sensations of food occur independently of the biting action. Thus, the Food Simulator prototype did not support these sensations. However, head-mounted displays can provide visualization of food. Also, smell can be displayed using a vaporizer. These displays could easily be integrated with the haptic interface.

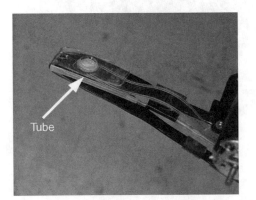

Tube

FIGURE

9.5

Tube for injection of chemical taste.
A small amount of liquid is injected from the top end of the tube into the mouth.

FIGURE

9.6
Injection pump.
The injection of the liquid is synchronized with the biting action. The pump provides 0.5 ml of solution for each bite.

9.4 HUMAN FACTORS DESIGN OF THE INTERFACE

The Food Simulator is designed to generate biting force according to that of real food. There are a number of human factors issues that must be considered when designing a taste interface.

9.4.1 Measurement of Biting Force of Real Food

The Food Simulator generates force, simulating the force recorded from an individual biting real food. A film-like force sensor (FlexiForce, Kamata Industries) is used to measure this biting force. The sensor has a thickness of 0.1 mm and a circular sensitive area of 9.5 mm in diameter. The sensing range is 0 to 110, with the maximum sampling rate of 5,760 Hz. Figure 9.7 shows an overall view of the sensor. Figure 9.8 shows a view of the participant in the experiment, who is biting a cracker and the FlexiForce sensor.

Figure 9.9 shows measured biting force associated with a real cracker. Two peaks appeared in the profile of the measured force. The first peak represents destruction of the hard surface of the cracker. The second peak represents destruction of its internal

FIGURE

Film-like force sensor.
The sensor is placed in the mouth with real food to measure biting force.

9.7

FIGURE

Force sensor and real food.
The participant is biting real food along with the force sensor.

9.8

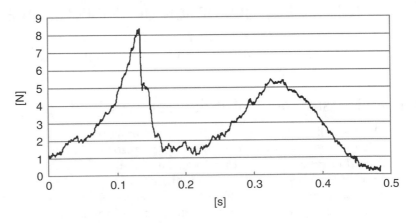

FIGURE

Measured force of a cracker.
The first peak represents destruction of the hard surface of the cracker.
The second peak represents destruction of its internal structure.

9.9

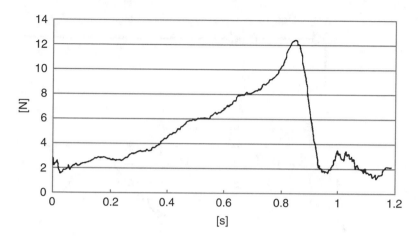

FIGURE

9.10

Measured force of cheese.
The slope represents the elastic deformation of the cheese.

structure. Figure 9.10 shows the measured biting force associated with real cheese. The slope observed in the figure represents the elastic deformation of the cheese.

9.4.2   Method of Generating Biting Force

Two food types are simulated: crackers and cheese.

Crackers

The force control of the device has two stages (Figure 9.11). First, the device applies force to maintain its position. This process represents the hard surface of the virtual cracker. When the biting force exceeds the first peak, the force control moves to the second stage. The device generates the same force as that measured from the real cracker. This stage is subject to open-loop control. Destruction of the internal structure of the virtual cracker is simulated by the second stage.

Figure 9.12 shows the measured force of the virtual cracker displayed by the Food Simulator. The profile of the measured force has two peaks, as in the real cracker. However, the shape of the profile is different from that associated with the real cracker. This difference seems to be caused by the open-loop control.

Cheese

Figure 9.13 shows two stages of force control for biting real cheese. In order to simulate elastic deformation of the cheese, the spring constant of the cheese was estimated from the collected data. The device generates force according to this spring constant to display the elasticity of the virtual cheese. When the biting force exceeds the peak force, the control enters the second stage. The device generates the same

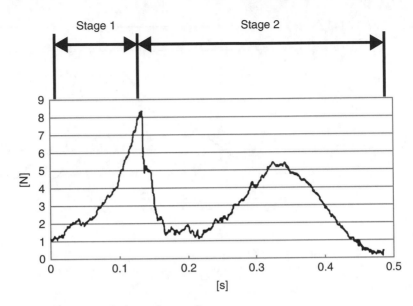

FIGURE

9.11

Method of simulation of a cracker.

First, the device applies force to maintain its position. When the biting force exceeds the first peak, the device generates the same force as that measured from the real cracker.

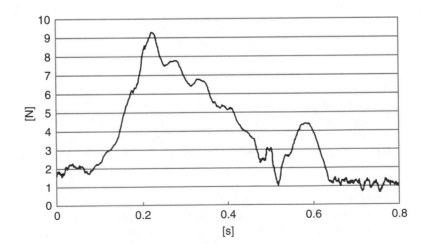

FIGURE

9.12

Measured force of a virtual cracker.

Two peaks are observed.

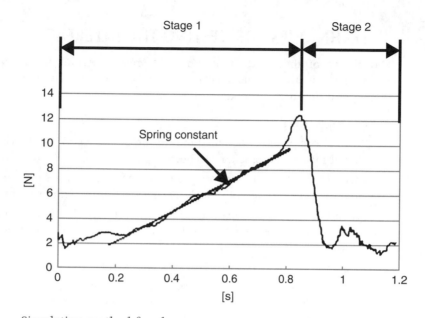

FIGURE

9.13
Simulation method for cheese.
The device generates force according to the spring constant of the real cheese.

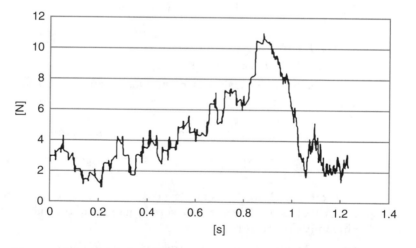

FIGURE

9.14
Measured force of virtual cheese.
The profile resembles the measurements for real cheese.

force as that measured from the real cheese. This stage is subject to open-loop control. Destruction of the virtual cheese is simulated by the second stage.

Figure 9.14 shows the measured force of the virtual cheese displayed by the Food Simulator. The profile of the measured force is similar to the real one.

9.5   TECHNIQUES FOR TESTING THE INTERFACE

The ability of the Food Simulator to represent virtual foods was evaluated. The major objective of this experiment was to discover whether the device could represent differences in food texture.

Experiment 1

Task: Participants were asked to bite the device and answer whether the virtual food was a cracker or cheese.

Participants: The participants (22 to 24 years of age) were six university students (male) who voluntarily participated in the experiment.

Procedure: Each participant tried 10 virtual foods. Crackers and cheeses were randomly displayed, with the subjects being requested to distinguish between the two foods.

Results: The answers were 98.3 percent correct.

Discussion: All of the participants were perfectly able to distinguish between the two virtual foods. The device thus succeeded in representing hardness of the food, perceived by the first bite.

Experiment 2

Task: The following three foods were displayed:

- ✦ *Food A:* Cracker (same as in the previous experiment)
- ✦ *Food B:* Biscuit ("Calorie Mate")
- ✦ *Food C: Japanese crispy snack ("Ebisen")*

These foods possess a similar texture. Figures 9.15 and 9.16 show the force profiles of Foods B and C, respectively. The same force control method as shown in Figure 9.11 was used to simulate Foods B and C. The participants were asked to bite the device and select one of the three foods.

Participants: The experiment was performed using the same participants as in Experiment 1.

Procedure: Each food was displayed 60 times; thus, each participant sampled 180 virtual foods. The subjects were requested to distinguish among three randomly displayed foods.

Results: The percentages of correct answers followed:

- ✦ *Food A:* 69.1 percent
- ✦ *Food B:* 59.1 percent
- ✦ *Food C:* 28.2 percent

Discussion: Force profiles of the cracker and the Ebisen were very similar. The peak force of the Ebisen was greater than that of the cracker, but most

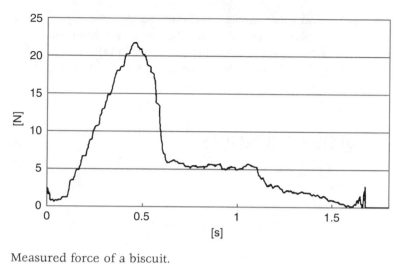
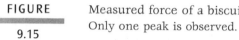

FIGURE

9.15

Measured force of a biscuit.
Only one peak is observed.

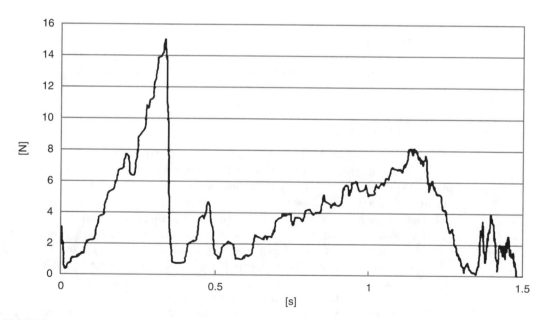

FIGURE

9.16

Measured force of a Japanese snack.
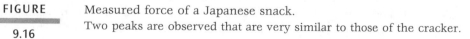
Two peaks are observed that are very similar to those of the cracker.

participants were unable to distinguish the Ebisen from the cracker. The peak force of the Calorie Mate was the greatest of the three foods. The internal structure of the Calorie Mate is soft so that the force profile did not exhibit a second peak. This difference contributed to the 60 percent of correct answers.

9.6 DESIGN GUIDELINES

Design of the Food Simulator is in a very preliminary state and it has much room for improvement. However, the following items should be considered for designing the device.

+ Linkages of the force feedback device should be fit to the structure of the user's jaw.
+ The mouthpiece of the force feedback device should be thin, but it has to support large biting force. The material of the linkage should be carefully chosen.
+ Due to sanitary issues, the mouthpiece should be covered by disposable material. The material should not infringe taste perceived by the tongue.
+ The injector tube for the chemical taste should be set at an adequate position so that the liquid drops on the tongue.

9.7 CASE STUDY

A case study of a taste interface can be found at *www.beyondthegui.com*.

9.8 FUTURE TRENDS

A major advantage of the Food Simulator is that it can display biting force by the use of a simple mechanism. Although the apparatus is inexpensive, it can effectively display food texture. The method of force control requires improvement, but most of the participants in the Special Interest Group on Graphics and Interactive Techniques (SIGGRAPH) greatly enjoyed the demonstration.

On the other hand, major limitations of the current prototype are concerned with its thickness and the weight of the linkages. The linkages are 2 mm thick. Because great force must be supported, the device must be fabricated in steel. When users finish biting, they can feel the unnatural sensation of the linkage thickness. The weight of the linkage causes unwanted vibration.

The user's teeth contact the flat surface of the linkage. This flat surface degrades food texture. A person feels the force on independent teeth while biting real food. Moreover, the current device applies force only to the teeth. The texture of real food is perceived, in part, by the tongue. However, displaying food texture to the tongue is very difficult.

Biting exercises contribute to human health. Some of the many application areas for the Food Simulator follow:

Training: The Food Simulator can be programmed to generate various forces other than that of real food. Elderly people can practice biting with reduced resistance to the teeth. On the other hand, increased resistance enables younger people to perceive the difficulty in biting experienced by elderly people.

Entertainment: The Food Simulator can change the properties of food while chewing. A cracker can be suddenly changed to a gel. The user can enjoy a novel experience while chewing. This kind of entertainment contributes to the chewing capability of children.

Food design: Preferred resistance by the teeth can be found using the Food Simulator. Such findings could contribute to the design of new foods.

This chapter has demonstrated a haptic device that simulates the sense of biting. The device presents a physical property of food. The Food Simulator has been integrated with auditory and chemical taste sensations. The chemical sensation was successfully displayed by a tube and an injection pump.

The device can be used in experiments of human taste perception. Taste is a multimodal sensation, making it very difficult to control modality by using real food. The device can display food texture without chemical taste, unlike real food. This characteristic contributes to experiments in modality integration. Future work will include the psychological study of sensory integration regarding taste.

REFERENCES

Chandrashekar, J., Hoon, M. A., Ryba, N. J., & Zuker, C. S. (2006). The receptors and cells for mammalian taste. *Nature* 444:288–94.

Davide, F., Holmberg, M., & Lundström, I. (2001). Virtual olfactory interfaces: Electronic noses and olfactory. In Riva, G., & Davide, F., eds., *Communications through Virtual Technologies: Identity, Community and Technology in the Communication Age.* Amsterdam: IOS Press, 160–72.

Kawamura,Y., & Kare, M. R. (1987). *Umami: A Basic Taste.* New York: Marcel Dekker.

Kohyama, K., Ioche, L., & Martin, J. F. (2002). Chewing patterns of various textured foods studied by electromyography in young and elderly populations. *Journal of Texture Studies* 33(4):269–83.

Kohyama, K., Sakai, T., & Azuma, T. (2001). Patterns observed in the first chew of foods with various textures. *Food Science and Technology Research* 7(4):290–96.

Nakamoto, T., Ustumi, S., Yamashita, N., Moriizumi, T., & Sonoda, Y. (1994). Active gas sensing system using automatically controlled gas blender and numerical optimization technique. *Sensors and Actuators* B 20:131–37.

Szczesniak, A. S. (1963). Classification of textural characteristics. *Journal of Food Science* 28:385–89.

Szczesniak, A. S. (2002). Texture is a sensory property. *Food Quality and Preference* 13:215–25.

Takanobu, H., Takanishi, A., Ozawa, D., Ohtsuki, K., Ohnishi, M., & Okino, A. (2002). Integrated dental robot system for mouth opening and closing training. *Proceedings of ICRA*, 1428–33.

Toko, K., Matsuno, T., Yamafuji, K., Hayashi, K., Ikezaki, H., Sato, K., & Kawarai, S. (1994). Multichannel taste sensor using electric potential changes in lipid membranes. *Biosensors and Bioelectronics* 9:359–64.

Toko, K. (1998). Electronic tongue. *Biosensors & Bioelectronics* 13:701–709.

Tsuruoka, S., Wakaumi, M., Nishiki, K., Araki, N., Harada, K., Sugimoto, K., & Fujimura, A. (2003). Subclinical alteration of taste sensitivity induced by candesartan in healthy subjects. *British Journal of Clinical Pharmacology* 305:807–12.

10 Small-Screen Interfaces

CHAPTER

Daniel W. Mauney, Christopher Masterton

10.1 NATURE OF THE INTERFACE

The use of small screens as a tool for displaying dynamic information is becoming ubiquitous. Displays range from very simple screens as seen on clocks, microwaves, alarm systems, and so on, to highly capable graphical displays as seen on mobile phones, medical devices, handheld gaming devices, and personal digital assistants (PDAs). The number of products sold with small screens is staggering. Mobile phone sales, just one category of small-screen products, were estimated at 825 million units in 2005 (Gohring, 2006) and have over 2.6 billion subscribers worldwide (Nystedt, 2006). In comparison, personal computer (PC) sales were estimated at 208.6 million units in 2005 (Williams & Cowley, 2006).

Small-screen design primarily makes use of the visual system. Questions in small-screen design often center on how small data elements can be displayed so that they are properly seen or recognized. To gain a deeper understanding of the answers to this question, a brief background of light, eye anatomy, and the sensitivity and acuity of the eye are discussed.

10.1.1 Light

Light is generally described by its wavelength and intensity. The subjective or psychological correlate to wavelength is color (more accurately known as hue). Thus, the color that people see is their perception of the light's wavelength. Wavelength is measured in nanometers (nm). The subjective or psychological correlate to intensity is brightness. Brightness is the impression produced by the intensity of light striking the eye and visual system.

People do not normally look directly at a light source. Most of the light seen is reflected from surrounding surfaces. The measurement of the intensity of light focuses on two aspects: the amount of light falling on an object, called *illuminance*, and the amount of light emitted from or reflected from a surface, called *luminance*. The common English unit of illuminance is the foot-candle (ft-c), while the metric unit of illuminance is the meter-candle (m-c). The common English unit of luminance is the foot-Lambert (ft-L), while the metric unit of luminance is the millilambert (mL). The amount of light emanating from a source is called *radiance*. To put all of this together, consider the act of reading a book. The amount of light generated by the light bulb is called radiance, the amount of light falling on the book's pages is called illuminance, the amount of light reflected from those pages is the luminance, and the amount of light perceived by the visual system is called *brightness*.

Contrast is a measurement of the luminance difference between a target and its background. It is often expressed as a ratio, such as 10:1.

10.1.2   The Eye

As light rays enter the eye, they first pass through the cornea, the pupil, and the lens to ultimately fall on the retina. The cornea and the lens refract (or bend) the light so that images come into focus on the retina. The iris controls the amount of light entering the eye by dilating or constricting the size of the pupil (the round black opening surrounded by the iris).

The retina is composed of nerve cells and photoreceptors that are sensitive to light. There are two types of photoreceptors: rods and cones. Rods are heavily concentrated in the peripheral region of the retina, while cones are concentrated primarily in a small pit called the fovea. Many rods share a common optic nerve fiber, which pools their stimulation and aids sensitivity to lower levels of light. In contrast, cones have a more or less one-to-one relationship with optic nerve fibers, which aids their image resolution, or acuity. Vision accomplished primarily with cones is called photopic vision, and vision accomplished primarily with rods is called scotopic vision. Only in photopic vision do people actually perceive colors. In scotopic vision, the weak lights are visible, but not as colors. Instead, all wavelengths are seen as a series of greys.

10.1.3   Sensitivity

In poorly lit environments, scotopic vision dominates because rods are far more sensitive to light. Since rods are located primarily in the periphery of the retina, sources of low light are best seen in the visual periphery. This explains why it is easier to see a faint star at night when not fixating directly on it.

The sensitivity of the eye to light is heavily dependent on the wavelength of the light. Figure 10.1 shows the spectral threshold curves for photopic and scotopic

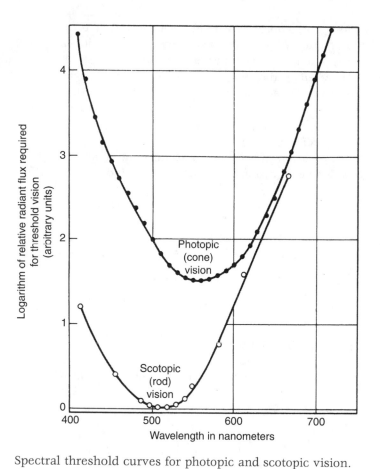

FIGURE

10.1

Spectral threshold curves for photopic and scotopic vision.
Rods are maximally sensitive to wavelengths of around 500 nm, compared to
550 nm for cones. Scotopic vision is generally more sensitive to light than
photopic vision. *Source:* From Schiffman (1990), as adapted from Chapanis (1949);
courtesy John Wiley and Sons, Ltd.

vision. These curves show that the rods are maximally sensitive to wavelengths of
around 500 nm (yellow–green), while the cones are maximally sensitive to wavelengths of around 550 nm (green). It also shows, as noted before, that scotopic
vision is generally more sensitive to light than photopic vision.

10.1.4 Visual Acuity

The smallest detail the eye is capable of resolving at a particular distance is known
as the minimum visual acuity (Grether & Baker, 1972). Visual acuity of an object
depends on both its size and its distance from the viewer. A small object that is

very close may appear larger than a large object that is far away. The distinguishing feature of an object's size, therefore, is the size the object projects on the retina, also known as its visual angle (Figure 10.2). The probability of an object being detected is directly related to the visual angle, expressed in seconds, minutes, and degrees, of that object subtended at the eye. Figure 10.3 graphs the probability of detecting an object at different visual angles.

Factors Affecting Acuity

A number of factors affect acuity, including illumination, contrast, time, and wavelength. In general, as illumination, contrast, and time spent focusing on a target

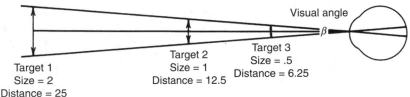

FIGURE

10.2

Impact of visual angle.

All three targets appear to be the same size because their visual angle is the same. *Source:* From Schiffman (1990); courtesy John Wiley and Sons, Ltd.

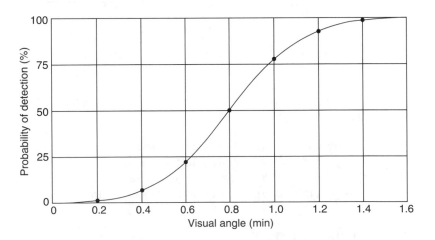

FIGURE

10.3

Probability of detection for objects of varying visual angles.

This probability is directly related to the visual angle of an object subtended at the eye. *Source:* From Grether and Baker (1972), as adapted from Blackwell (1946); courtesy John Wiley and Sons, Ltd.

increase, acuity improves. Wavelength has little impact on acuity if there is high luminance contrast. However, if there is low contrast, color contrast will improve acuity. A red signal is most easily seen in low-contrast conditions, followed by green, yellow, and white in that order (Sanders & McCormick, 1993).

10.1.5   Color

Color perception is the ability to discriminate among different wavelengths of light. It is useful to note that the light itself is not colored. Instead, different wavelengths of light produce the sensation of different colors. Thus, color is a psychological experience that different wavelengths of light have on the nervous system. There are three psychological dimensions of color: hue, brightness, and saturation. Hue varies with changes in wavelength, and the term is often used interchangeably with the word *color*. Brightness varies with the physical intensity of the light. Saturation refers to the spectral purity of the wavelength. The addition of other wavelengths, white light, or grey to a single wavelength will desaturate the color (Schiffman, 1990).

Approximately 8 percent of men and 0.5 percent of women have some form of color vision deficiency (Sanders & McCormick, 1993). The most common form of color weakness involves the inability to differentiate between red and green. As such, color should never be used as a primary cue (the primary distinguishing feature between two items), and requiring users to distinguish between the colors of red and green should be avoided.

10.2   TECHNOLOGY OF THE INTERFACE

Until 1970, the primary display technology available to electronics designers was the cathode ray tube (CRT). Although the CRT is still in use today and still outperforms competing technologies in some areas (especially manufacturing cost), it has two major drawbacks: size and power draw. Cathode ray technology requires a great deal of power to fire electrons at the screen and a relatively great distance in order to display those electrons on a reasonably sized screen. Despite continued improvements in the CRT engineering process, these two important flaws have not been overcome.

In 1970, the Swiss team of Schadt and Helfrich built the world's first liquid crystal display (LCD). While original LCD designs had their own shortcomings (namely cost, response time, and contrast), the technology overcame the limitations of the CRT by requiring much less power and a vastly decreased depth.

The continued improvement of the LCD has led to the emergence of an enormous range of small-screen products, from the simple digital watch to powerful laptop computers. In this section, we explore the various technologies used in current small-screen design.

10.2.1   Display Technologies

Today, LCD technology can be broken into two major categories: active- and passive-matrix displays. Passive-matrix displays work by rapidly activating corresponding rows or columns of the display sequentially to update the screen with new images. The technology is called "passive" because each pixel must retain its state between refreshes without an electrical charge. Passive-matrix screens require less power and are less costly to manufacture than active-matrix displays, but as the resolution of the display increases, the response time and the contrast of the passive display become worse.

Active-matrix LCDs employ an individual transistor for each screen pixel. Active-matrix displays use more power because each transistor requires a steady electric charge to maintain or update its state. However, because each pixel is controlled separately, active-matrix displays are typically brighter, sharper, and have a faster response time than passive screens.

Both active and passive LCD technologies can be used in color or monochrome displays. Monochrome displays use less power, have higher contrast, and cost less to manufacture than color displays (Wright & Samei, 2004). Obviously, color LCDs provide a richer and more compelling graphical user interface (GUI) and are typically more desired by users.

While LCD screens are the primary display technology in use today in small-screen devices, other technologies are in common use depending on the device's application. The simplest display technology is the vacuum fluorescent display (VFD). VFDs are used in microwave ovens, home theater components, and other consumer electronic devices that do not require complex user interfaces. VFDs emit very bright light, are very high contrast, and are inexpensive to manufacture. The downside is that a single VFD screen element can only display a single color at a time and is too slow for more complex computer-like interfaces. VFD displays are frequently designed in a segmented fashion with relatively large segments (compared to LCD or CRT pixels), thus limiting the range of shapes that can be created. Figure 10.4 shows a close-up of a VCR's segmented VFD display.

FIGURE

10.4

Close-up of a vacuum fluorescent display.
VFDs have been adapted to display dynamic data by breaking down the segments into individual pixels. *Source:* From Wikipedia.org, 2007.

Another straightforward technology that has blossomed into a viable display modality is the light-emitting diode (LED). From the single LED light that tells you that your smoke alarm is receiving power to fiber-optic light pipes that allow for high-resolution LED displays, the LED has evolved into an extremely useful display technology. LEDs use a very small amount of power, and have a long life span and quick response times. Figure 10.5 shows a car-parking assistant that tells the user how far away he or she is from backing into objects by displaying the distance on an LED display.

One of the most exciting developments in display technology is the organic LED (OLED) display. Organic dyes are deposited on a layer of glass or plastic that emits light when excited by electrons. OLED displays have great potential as a display technology because they require little power, produce a bright picture, and can be made very thin. The major disadvantage of current implementations of OLED displays is the short life span of the organic materials used (current implementations are limited to a maximum of 10,000 hours). If this limitation can be overcome, OLED displays may quickly overtake the LCD as the primary display technology. Commercial implementations of OLED technology are already becoming available; Figure 10.6 shows a Samsung OLED watch.

An additional benefit of OLED displays is that the entire screen may be made flexible. Electronic paper (also called e-paper) is an OLED display technology designed to look and act like printed paper but with the ability to display dynamically changing data. The key difference between e-paper technology and other displays is that no power is needed to keep pixels on the screen (a technique known as image stability); power is only needed to change the picture on the display. The image stability properties of e-paper mean that these displays require very little power to operate over long periods of time.

FIGURE

10.5

Car-parking assistant with an LED display.
The number on the display is the distance the user is from any objects behind him. *Source:* From Alibaba.com, 2007.

FIGURE

10.6

Commercial implementation of OLED.
The Samsung GPRS Class 10 Watch Phone uses an OLED display. *Source:* From
Engadget.com, 2004; courtesy Samsung.

10.2.2   Screen Size, Resolution, and Dots per Inch

Screens are defined by both their physical size and their resolution (the number of pixels they can display). Physical screen size is measured in inches by diagonal length. Typical displays use a ratio of horizontal to vertical size of 4:3. Newer, so-called "wide-screen" displays, use a 16:9 ratio, and other less common ratios (e.g., 3:2, 5:4, or 16:10) are employed depending on the application or device (Figure 10.7).

The resolution of the display also plays an important role in the amount of information that can be presented on screen. Dots per inch (DPI) (also called pixels per inch) is a measure of the number of pixels in each square inch of the display. The higher the DPI value, the denser the pixels are in the display and thus the better the visual quality of the display. A typical DPI value for a computer screen is 72, but newer screens are already exceeding 200 in density.

Another more common measurement of resolution is the number of horizontal and vertical pixels the screen is capable of displaying. Small-screen device resolutions can range from as low as 16×16 pixels for watch displays and other

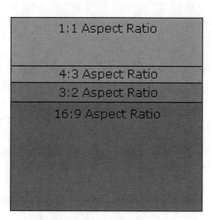

FIGURE

10.7

Four common aspect ratios.
Typical displays use a ratio of 4:3. *Source:* From Cambridgeincolour.com, 2007.

TABLE 10.1   Standard Small-Screen Resolutions

Name	Aspect Ratio	Resolution
QQVGA	4:3	160 × 120
QCIF	1.22:1	176 × 144
QCIF+	4:5	176 × 220
QVGA	4:3	320 × 240
CGA	16:5	640 × 200
EGA	64:35	640 × 350
VGA	4:3	640 × 480
SVGA	4:3	800 × 600
XGA	4:3	1024 × 768

similar very small screens, to 800 × 600 pixels (SVGA) found in some high-end devices (like the Motion LS800 Tablet PC). Newer devices are continually pushing this limit for higher-resolution displays. Various resolution standards are shown in Table 10.1.

10.2.3   Input Technologies

It is difficult to design for small-screen devices without considering how information is inputted to the device. The design of the display interface is heavily dependent on the data entry method. In fact, human data input for small-screen devices is arguably the biggest challenge facing device designers. Today, the QWERTY keyboard and mouse are the ubiquitous input devices for the PC; however, the small size of today's mobile devices requires different input techniques.

For mobile phones the standard telephone keypad (known as the dual-tone multifrequency, or DTMF, keypad) is a standard input device. For entering numbers, the DTMF keypad is fast and effective, but for alphabetic textual input, it is slow and inefficient. The most common method of textual input is called multitap entry. It requires the user to press the number key that represents a particular letter a number of times in order to produce the desired character. For example, pressing the 2 key three times would produce the letter C. Clearly this method requires several times more effort than a typical computer keyboard. Figure 10.8 shows the edgy DTMF keypad on a Motorola RAZR phone.

Several methods of predictive text input have been developed to overcome the limitations of the DTMF keyboard, including the T9, SureType, and iTap (to name only a few). Generally these technologies rely on dictionary completion, meaning that words are guessed from a dictionary based on the keys pressed. In the T9 system the user presses the number key representing the letters in the word (once for each letter). The software then guesses the most appropriate word based on dictionary matches of the letters entered. If the word is not correct, the user is given the option of cycling through other words that match the keys

FIGURE

10.8

DTMF keypad for the Motorola RAZR.
This model has a five-way rocker, two softkeys, and four hardkeys. *Source:* From Wikimedia.org, 2007; courtesy Motorola.

pressed. While predictive techniques are faster than multitap input, they are still slow in comparison with a standard computer QWERTY keyboard. A recent contest between a champion predictive-text phone user and a 93-year-old telegraph operator found that Morse code was a faster text entry method than predictive text (Engadget.com, 2005).

Even the small QWERTY keyboards employed in some phones tend to make the phone volume too large to be desirable for some consumers. The need for smaller form factors has led to the exploration of different input technologies. Techniques such as the Palm's handwriting recognition software and virtual soft keyboards (which display a QWERTY keyboard on a touch-sensitive screen) have been implemented. However, Western consumers have not taken to these new input styles en masse due to their limited input speed and poor accuracy rates (Bohan, 2000).

It is clear that a full QWERTY keyboard is a quick and usable input device, so companies like Lumio have developed projection keyboards. The projection keyboard (like the one shown in Figure 10.9) is a device that projects light in the shape of a regular keyboard onto a flat surface. The user interacts with it by virtually pressing the keys on the surface. Based on which light beams are broken, the device can tell which keys are being "pressed." So far, few commercial devices use the projection keyboard, but as the technology matures it will no doubt be integrated into more devices.

Clearly, textual input is not the only way in which users input data to their devices. For many years, hardkeys have been used in devices from microwaves

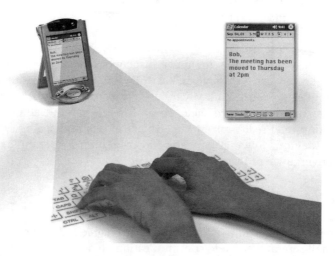

FIGURE
10.9

Simulation of a projection keyboard.
This one is in use on a Windows Mobile device. *Source:* From Roeder-Johnson. com, 2007; courtesy Cayerta, Inc.

and watches to phones and media devices. A hardkey is a dedicated button that when pressed activates one or more functions within the device. A similar technology is the softkey, which is a dedicated button with a software-determined label. The softkey performs context-sensitive actions based on the state of the software, and is very common in mobile phones. A popular trend in small-screen devices is the use of touch-sensitive screens. Recently the touch screen has become a more viable input option because of both the reduction in production cost and the improved sensitivity of the screens themselves. There is a good reason why keyboards have not been replaced by touch screens in larger devices; tactile feedback is an important part of the input experience. This limitation of feedback has hampered the development of touch screens until recently, when improved sensitivity and responsive interfaces have made the touch screen a reasonable input technology.

Other common input devices include the click wheel used in the popular iPod, the four- and five-way rocker used in a large number of phones, the track wheel used in BlackBerries, and the joystick used in phones and media devices. As yet no ideal all-around input device has been discovered. When determining a device's input technology, the hardware should be designed in cooperation with the software so that the primary use cases of the device are made easy and efficient.

10.3 CURRENT IMPLEMENTATIONS OF THE INTERFACE

Small-screen interfaces have become a pervasive part of today's society. From the simple digital watch to in-car displays, almost everyone interacts with small-screen devices on a regular basis. Despite the prevalence of these devices, their complexity and usability vary widely across the spectrum of implementations. Diversity of hardware aside, several trends can be seen throughout the industry: a continued reduction in the size of devices, increasing computing power within the device, growing software and user interface complexity and the enhanced convergence of features.

10.3.1 Mobile Phones

In many ways, mobile phones are responsible for today's explosive proliferation of small-screen, handheld communication devices. In 2005, worldwide mobile phone sales were estimated at more than 825 million units sold (Gohring, 2006). This massive market has driven manufacturing costs of small-screen devices down to a level where they are accessible to new user groups. The reduction in the cost of parts has also made other small devices more cost-effective to manufacture.

Typically mobile phones are smaller, less feature filled, and less costly than their larger PDA cousins. Mobile phones in the past usually performed a limited number of key functionalities such as making phone calls, writing short message service (SMS) messages, and managing contacts. Screens on mobile phones are found in both monochrome and color but tend to be lower resolution and tend to have simpler and more straightforward software interaction design than PCs. Newer phones follow the trend toward more features: camera integration, video playback, high-resolution screens, and multiple displays, to name a few.

10.3.2   Smart Phones and Personal Digital Assistants

Not only do users want smaller devices, they also want devices that have advanced technological and networking capabilities (Sarker & Wells, 2003). Nokia's Series 60–based devices allow users to perform a nearly unlimited number of tasks previously only available in full-blown PCs. According to the promotional material for the Nokia N95, "It's GPS. It's a photo studio. It's a mobile disco. It's the World Wide Web. It's anything you want it to be" (s60.com, 2007). The combination of improved battery technology, miniaturization of components, and a seemingly endless number of new hardware and software features are making small-screen devices into true portable PCs.

The downside of increased technological complexity, generally, is a reduction in usability. This is true of many of the cutting-edge small-screen devices. However, improved integration of features is making it easier for users to understand and even manage the complexities of multimedia data. The Research In Motion (RIM) BlackBerry Pearl keeps the user in control of his or her busy life by providing simple but effective filters for e-mail, integration of instant messaging, and the ability to easily synchronize the device with a PC. Despite the complexity of the underlying technology, usability-focused PDAs like the Blackberry can provide an easy-to-use integrated computing experience.

10.3.3   Digital Audio Players

In a 2003 study, Sarker and Wells found that reduced device size played a major role in the adoption of mobile devices. Developments in hardware miniaturization have allowed manufacturers to continue to reduce the form factors of small-screen devices while increasing their computing power. The iPod is currently one of the most popular mobile devices on the market with an estimated install base of 40 million units (AppleInsider.com, 2006). The iPod Nano uses a touch-sensitive click wheel, holds up to 8 gigabytes of music (or around 2,000 songs), provides a vibrant color LCD display, and offers many different ways to enjoy music. The Nano is barely bigger than a credit card at 3.5 in × 1.6 in × 0.26 in. The iPod's success makes it clear that users have a strong desire for small devices, and the number of similar devices has skyrocketed.

10.3.4   Mobile Gaming Devices

Mobile gaming devices are another small-screen information appliance that has found wide acceptance in the marketplace. While mobile gaming devices have been around since the 1980s, newer devices play multiple games, have bigger screens, and are packed with advanced features. The Nintendo DS Lite with its two 3-in LCD screens, touch-screen input, four-way rocker, and WiFi connectivity is a perfect example of a powerful computing device designed to be used by children and adults. DS games seamlessly integrate the touch-screen interface into play, encouraging an immersive interaction experience.

10.3.5   Scientific and Medical Equipment

Scientific equipment also employs small-screen technology to convey complex information. Progress in screen technology has vastly improved the reliability and accuracy of visual display–reliant scientific equipment. Small screens are found in a myriad of scientific devices, including oximeters, signal analyzers, infusion pumps, and oscilloscopes. For example, modern digital storage oscilloscopes (DSOs) employ small color CRTs or LCDs to display detailed test data. The Tektronix TDS 3052B shown in Figure 10.10 has an active-matrix VGA (640 × 480 pixel resolution) LCD screen, numerous hardkeys and softkeys, and an Ethernet network interface, and can even connect to larger monitors.

Another area of prolific small-screen usage is consumer medical devices. Modern thermometers, insulin pumps, glucose meters, and medical-alert devices all make use of small-screen device technology. For example, glucose meters for diabetics have advanced significantly in the past 10 years into digital devices that

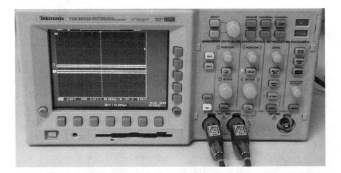

FIGURE
——————
10.10

Tektronix TDS 3052B LCD-based digital storage oscilloscope.
This device has an active-matrix VGA LCD screen and many other sophisticated features. (Courtesy Tektronix, Inc.)

can display glucose levels within seconds of testing. The OneTouch Ultra meter not only displays the detected glucose levels on a high-contrast monochrome LCD screen, but also allows the user to keep track of exercise, health, and medication by downloading the data from the device to a PC (Onetouch.ca, 2007).

10.3.6 Information Appliances

Devices like watches, microwaves, alarm clocks, cameras, stereo equipment, digital thermostats, and even exercise machines all make use of small screens. Donald Norman, a leading expert in cognitive psychology and interface design, calls these devices "information appliances," devices that perform a limited number of functions but carry out those functions in a simple and usable way (Norman, 1999). The Logitech Harmony series of remote controls employs a small screen accompanied by a series of softkeys and hardkeys to perform task-based control over even the most complicated home theater setups. The design of the Harmony Remote mirrors how users actually interact with their home theater, allowing the user to request simple tasks such as "Watch a movie" or "Watch TV." Hidden from the user, the Harmony intelligently performs a number of different actions such as changing TV inputs and receiver inputs and turning devices on or off (Logitech.com, 2007).

10.3.7 Media Center Interfaces

Small-screen devices share many design similarities with one particular large-screen interface paradigm: the media center. Media center technology has evolved out of the user's need to control the various types of home theater media. Products like the Tivo Series 3, the Motorola 6200 series, and Microsoft's Windows XP Media Center Edition allow the user to watch television, watch recorded shows, watch movies, and listen to music through a single integrated interface. Even though televisions are far from "small screens," the design guidelines used for these devices are nearly identical to those used for small-screen devices because the user sits a great distance from the display and the interface must be designed to accommodate the lower resolution of the majority of televisions in today's homes.

10.3.8 Hybrid Devices

The increasing demand by users for simplicity has driven manufacturers of small-screen devices to attempt combining various small-screen technologies into hybrids. Typically these devices are not as robust or simple as dedicated small-screen devices. However, due to their convenience, this is a rapidly

FIGURE

10.11

Example TTYs.
Left: Uniphone 1140 TTY. *Right:* Compact/C cellular telephone TTY.
Source: From Enablemart.com, 2007.

growing market. The Apple iPhone is a good example of a hybrid device. It contains the functionality of a phone, an iPod, and a PDA in a single small package. The iPhone employs a large multitouch screen interface, WiFi and cellular connectivity, and a built-in hard drive. The iPhone even has a gyroscope that rotates the screen based on which direction the phone is being held. Additionally, like a small PC, external programs can be added to the iPhone to expand its functionality.

10.3.9 Implications of the Interface for Accessibility

While small screens tend to have poor accessibility themselves (due to the fact that designers look to make things small in order to display as much information as possible on the screen), they can be used to aid accessibility to other products. Two examples are the small screens employed in Teletypewriters (TTYs) (Figure 10.11) and the use of the bottom of the television screen for closed captioning. The use of small screens in these examples aids those who have difficulty hearing the telephone or television audio by providing alternative modalities of feedback.

10.4 HUMAN FACTORS DESIGN OF THE INTERFACE

Before beginning a design, it is important to establish guiding principles that will be used to focus the design in a direction that will ultimately yield a successful product. There are many sources for guiding principles. Guiding principles can be style guide–like in nature (i.e., all text should be in the 8-point Arial font because research has determined that this size is readable at the distance the device will be used), or they can be interaction guideline–like in nature (i.e., menu structures should favor depth, not breadth; or the user interface should make primary functions easy and secondary functions should be removed). A complete set of guiding principles should be a combination of both.

This section, together with Section 10.6, will provide information to help the designer develop the principles that guide small-screen interface design. This will be accomplished by presenting important information every practitioner should know about small-screen design. This section will focus largely on the scholarly work in the area of small-screen interface design, covering topics such as character size, contrast, and font to enable readability on small screens, and on organizational structures for accessing the available functionality. It will also cover considerations for reading and comprehension on small displays, and include a discussion on the adaptation of common GUI elements for small screens. Section 10.6 will focus on providing guidelines for creating successful user interface designs on small screens. Together, this information can be used to develop the guiding principles that drive the design of the small-screen interface.

10.4.1   Information Gathering

One of the first steps in small-screen design is to determine the characteristics of the screen itself. To define those characteristics, it is wise to first collect the background information needed to make an informed decision. This includes collecting details about the following, which is adapted in part from Muto and Maddox (in press).

User characteristics: Information about users, such as their visual acuity, color deficiencies, and anthropometric characteristics (particularly information relevant to eye position) should be collected. These characteristics will be a source of requirements used to define screen characteristics. A practical way to collect this information is to note how users differ from the general population and use published measurements of the general population for instances in which there are no differences.

Position(s) of display: Information about possible positions and locations of the display.

Position(s) of eye: Information about the user's eye positions relative to the display is very important. Knowledge of the farthest distance from the display that should be accommodated in the design is very important for computing the appropriate size of characters and objects presented on the display. Knowledge of the widest angle of view that should be accommodated in the design is also valuable for ensuring the proper choice of display technology.

Lighting and environmental conditions: Information about lighting in the expected environment is useful. If a range of environments is expected, identify the extremes and ensure that the display properties are satisfactory in those extremes.

Type of information to be displayed: The type of information to be displayed should drive the characteristics of the display, such as resolution, size, and color range. For example, if fine detail needs to be displayed, a screen with high resolution will be required. But if purely alphanumeric characters need to be displayed, a low-cost segmented display may suffice.

10.4.2   Information Format

For small-screen interfaces, human factors designers face the challenge of displaying all the information they want to present, while making sure that what they present is not too small to be seen by the user. As a result, they often ask questions about the minimum size of objects displayed on the screen, particularly text.

As described earlier, the minimum size of any object displayed on a screen is directly dependent on the visual angle, contrast, and luminance. The impact these have on determining the size of text on small-screen displays is discussed in the following subsections. Note that viewing distance is a key component in many of the size calculations demonstrated in this section. The viewing distance used should be the maximum distance utilized in a typical installation. Specifying a typical installation is important because devices can be placed in locations not anticipated by designers or in locations that may not be optimal. Nevertheless, these locations need to be accounted for when specifying the maximum distance in the following calculations.

Visual Angle

Many of the recommendations for minimum size of an object are based on the angle that the object subtends at the eye, measured in minutes of arc. The formula for computing the visual angle, in minutes of arc, is shown in this equation:

$$Visual\ Angle\ (\text{min}) = \left(\arctan\frac{L}{D}\right) * 60 \qquad (10.1)$$

where L is the size of the object measured perpendicular to the line of sight, and D is the distance from the eye to the object (Figure 10.12).

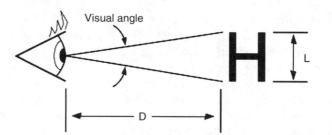

FIGURE

10.12

Computing the visual angle of an object subtended at the eye. Relationship between visual angle, size of the object, and distance from the object to the eye.

For angles less than 600 minutes, the following simpler form of the previous equation can be used (Grether & Baker, 1972):

$$Visual\ Angle\ (\min) = \frac{(57 \cdot 3)(60)L}{D} \tag{10.2}$$

The 57.3 and 60 in the formula are constants for angles less than 600 minutes. Solving for L, Equation 10.2 can be rearranged as follows:

$$L = \frac{(Visual\ Angle)(D)}{(57.3)(60)} \tag{10.3}$$

L, or character height, is a function of visual angle and distance. This formula will enable the designer to determine the object size when the distance from the eye to the object and the visual angle are known. For text, however, a conversion of this object size to font size measured in points is desirable. The following formula makes this conversion:

$$F = \frac{L}{0.0139} \tag{10.4}$$

where F is the font size in points and L is the size of the object measured perpendicular to the line of sight in inches.

Viewing Distance

Viewing distances can vary greatly, but the minimum viewing distance should be greater than 12 inches. It can be assumed that most office workers have a corrected, or uncorrected, visual ability for comfortable reading at 12 to 16 inches (ANSI/HFS 100–1988, 1988).

Character Height

Text is one of the most important object types displayed on a screen, and the size of that text affects the identification of characters and the readability of words. The minimum character height should be 16 minutes of arc to support the rapid identification of characters, and preferably 22 to 30 minutes of arc at the maximum viewing distance. For characters in which speed of recognition is not important (such as footnotes, superscripts, or subscripts), the character height should be at least 10 arc minutes (BSR/HFES 100, 2006).

Character heights of 20 to 22 arc minutes are preferred for readability, while the threshold for readability is 16 to 18 arc minutes. Large character heights, more than 24 arc minutes, may inhibit readability by reducing the number of characters that may be viewed per fixation (ANSI/HFS 100-1988, 1988).

For discrimination of the color of an alphanumeric string, 20 arc minutes are needed, and for discrimination of the color of a single character or symbol, 30 arc minutes are required (BSR/HFES 100, 2006).

In many situations, the designer has determined the viewing distance and, from the recommended visual angles for character height just presented, the designer can choose a preferred visual angle. With these two pieces of information, the desired character height and the resultant font size can be calculated with Equations 10.3 and 10.4, respectively. Table 10.2 shows character heights and font sizes for a sample of visual angles and viewing distances.

Character Width-to-Height Ratio

Based on the uppercase letter H without serifs (the small lines extending from the main strokes of a letter), the width-to-height ratio should be between 0.5:1 and 1:1. For optimal legibility and readability, it should be between 0.6:1 and 0.9:1 (BSR/HFES 100, 2006).

Stroke Width

Characters are made up of lines, called strokes. The stroke width of characters should be from $\frac{1}{6}$ to $\frac{1}{12}$ of the maximal character height (BSR/HFES 100, 2006).

Spacing between Characters

The spacing between characters without serifs should be at least equal to the stroke width, and preferably 25 percent to 60 percent of the width of the uppercase letter H. The spacing between characters with serifs should be at least 1 pixel (BSR/HFES 100, 2006).

Spacing between Lines

The space between lines of text, including diacritics, should be at least 1 pixel and preferably at least 15 percent of the maximal character height. For users with partial vision, use larger spacing, or 25 percent to 30 percent of the character height (BSR/HFES 100, 2006).

TABLE 10.2 Character Height and Font Size as Function of Viewing Distance and Visual Angle

Viewing Distance (inches)	Visual Angle (minutes)	Character Height[1] (inches)	Approximate Font Size[2] (points)
16	16	0.0745	5
	18	0.0838	6
	20	0.0931	7
	22	0.1024	7
	24	0.1117	8
24	16	0.1117	8
	18	0.1257	9
	20	0.1396	10
	22	0.1536	11
	24	0.1675	12
36	16	0.1675	12
	18	0.1885	14
	20	0.2094	15
	22	0.2304	17
	24	0.2513	18

[1] Calculated with Equation 10.3.
[2] Calculated with Equation 10.4.

Spacing between Words

The spacing between words should exceed the spacing between characters, and preferably be at least half the width of an uppercase letter *H* without serifs (BSR/HFES 100, 2006).

Luminance

In the case of light-emitting visual displays, luminance is a measure of the intensity of the light emitted from the display. It is an objective measurement of what most people think of as "brightness," and is usually measured as candelas per square meter (cd/m^2). The display should be capable of producing a luminance of at least 35 cd/m^2, and preferably, 100 cd/m^2. Users often prefer a high luminance $(>100\ cd/m^2)$, as reading speed and accuracy increase with increasing luminance, and legibility decreases when luminance falls below 35 cd/m^2 (BSR/HFES 100, 2006).

Luminance Contrast

Two of the most important components that influence recognition are object size and contrast. For contrast, the display should exhibit a minimum contrast ratio of 3 to 1 under all illumination conditions (BSR/HFES 100, 2006). A luminance contrast of 7 to 1 is preferred (ANSI/HFES 100-1988, 1988). Note that reflections of external light sources from the display reduce contrast and therefore legibility, so it is important to measure contrast under the proper environmental conditions.

Font

The determination of a font should take into consideration the available pixels to display a character. BSR/HFES 100 (2006) recommends a minimum pixel height of 9 pixels and a minimum pixel width of 7 pixels for Latin alphanumeric characters used in tasks requiring continuous reading or identification of individual characters (Figure 10.13). For purely uppercase letters and numerals, characters should be 7 pixels high and 5 pixels wide, while superscripts, subscripts, and the numerals in fractions should be at least 5 pixels high and 4 pixels wide. BSR/HFES 100 (2006) also states that the minimum height should be increased by 2 pixels if diacritic marks are used.

However, many fonts are designed for printed materials, and the legibility of these fonts suffers when font size falls below 14 pixels in height because these typefaces cannot be displayed fully on a pixel grid smaller than this size (Zwick et al., 2005). The typeface will otherwise appear jagged. As a result, pixel fonts have been designed for display on a small pixel grid. These are often optimized for a particular pixel grid and are not scalable. Table 10.3 presents a few examples of pixel fonts that are optimized for a particular pixel grid.

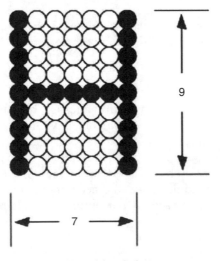

FIGURE
10.13

Height and width of characters.
Each circle represents 1 pixel.

TABLE 10.3   Example Pixel Fonts

Font	Example	Uppercase Pixel Height	Lowercase Pixel Height
Aldebra Regular	ABCDEabcde1234	9	7
Argon	ABCDabcdSTUVW	6	5
Axxell	ABCDabcdSTUVW	7	5
Foxley 816	ABCDEabcede1234	8	6
Mini 7	ABCDEABCDE1234	5	Not applicable
MiniMicra	ABCDabcd1234	6	4
Pico12	ABCDabcd1234	7	5

Source: MiniFonts.com (2006).

Pixel height differs from a font size in points. Font size in points is a measurement of the physical height of a character (1 point = 0.0139 in). For example, a 12-point font is 0.1668 in tall. Pixel height is the number of vertical pixels on a display used to render the character. The actual height of a pixel font is dependent on the number of pixels per inch the display is capable of rendering, also known as DPI. To calculate the actual height of a pixel font, use

$$L = \frac{P}{A} \qquad (10.5)$$

where L is the height of the character in inches, P is the height of the character in pixels, and A is the number of pixels per inch the display is capable of rendering (DPI). At 72 DPI, a 12-pixel font is 0.1667 in, approximately the same height of a 12-point font. Increasing the DPI of a screen will result in decreasing the height of a pixel font.

Another attribute of fonts is whether the font is a serif (has the small lines extending from the main strokes of a character) or sans serif (does not have those lines). Research on the use of serif fonts for displays is somewhat contradictory. Muto and Maddox (in press), in their review and summarization of that research, recommend the following:

+ For displays of high resolution (>150 DPI), serif fonts are more readable for tasks that require extended reading.
+ For displays of lower resolution (<80 DPI), sans serif fonts are often preferred by users.

Display Placement

For handheld devices, display placement is at the discretion of the user. However, for mounted displays in a fixed location, displays should be placed in the optimum visual zone (Figure 10.14).

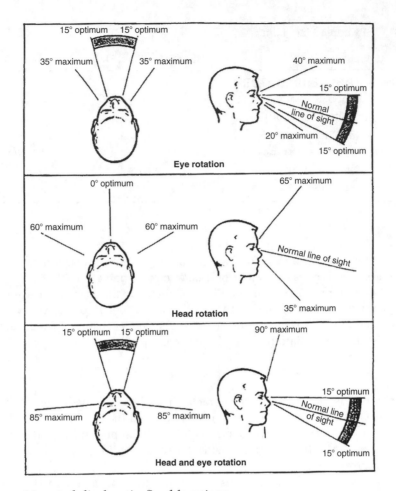

FIGURE

10.14

Mounted displays in fixed locations.
Vertical and horizontal visual fields. *Source:* Adapted from U.S. Department of Defense (1999).

10.4.3   Accessing Functionality

The computing power of devices today enables the creation of highly functional software. Even devices with small screens have the capability of performing a large number of features and functions. The challenge, then, becomes how the user accesses these features.

Menus are currently the dominant design for giving users access to a large number of features for both desktop computers as well as for handheld devices. For both applications, menus enable users to recognize the function they need and select it from a list (rather than recall it and type it in). The avoidance of text entry is especially beneficial on devices that do not have full keyboards.

However, even though menus are the dominant design choice, many software designs poorly implement menus, particularly on handheld devices, leading some researchers to investigate alternative methods. This section will present research on menu design and present factors that contribute to a usable menu design, as well as introduce research on alternatives to menus.

Menu Organization

The primary goal of menu design is to organize the menu items such that users are able to find the items they need quickly and with few errors and few keystrokes. Lee and Raymond (1993) reference a body of research showing that menu selection errors are most often caused by miscategorization, overlapping categories, and/or vague category titles. Focusing on avoiding these pitfalls will significantly improve the overall menu design.

There are two levels of menu organization: overall menu organization (often takes the form of hierarchical, branching tree structures) and within-menu placement (organizes items within a menu list). Studies have shown that the best way to avoid the common pitfalls of menu organization mentioned previously is to include end users in the design process. To help avoid overlapping categories and miscategorization of the overall menu structure, studies have shown that menu organizations generated by user data are superior to those generated by designer intuition (Hayhoe, 1990; Roske-Hofstrand & Paap, 1986). Capturing the users' perceived organization and organizing the menus in this way can improve the overall usability of the menu.

An in-depth discussion of techniques for collecting user data is beyond the scope of this chapter, but can be found in Paap and Cooke (1997), Cooke (1994), and Rugg and McGeorge (1997), and a discussion of scaling techniques for reducing user data into meaningful categories can be found in Paap and Cooke (1997) and Hayhoe (1990). One goal to strive for in the hierarchical structure is to develop orthogonal categories (mutually exclusive categories that have no common characteristics) that are clearly worded so that items under each category cannot be perceived as belonging to another category and so that every item fits well into a category.

To avoid errors due to vague category titles, menu titles should be clearly worded and accurately convey their contents. According to research referenced by Lee and Raymond (1993), clearly worded menu items can improve user performance and the best menu titles are derived from user data combined with designer expertise.

For within-menu placement, items can be arranged in several ways (Marics & Engelbeck, 1997; Paap & Cooke, 1997):

+ Alphabetically
+ Temporally (days of week, months of year)

- ✦ According to magnitude (small, medium, large)
- ✦ Consistently (if items appear in multiple places, keep the order the same)
- ✦ Categorically
- ✦ According to frequency of use (place the most frequently accessed items at the top)

Frequency of use is often the preferred choice. Organizing menus items in this way can save time and errors, as items located deep within a menu structure are often difficult to find. In our experience from viewing hundreds of users navigating menus, users will often select the first category that could contain the item for which they are looking. Thus, ordering items according to frequency of use will aid this search technique, as well as provide a more efficient interface for continued use.

This guideline can be adapted if the items have a natural order to them (such as temporal or magnitude), or to maintain consistency with other menus. For example, if Save commonly appears above Delete in other menus in the interface, then even in a menu in which Delete may be more frequently chosen, it may be best to again place Save above Delete from a consistency standpoint. Use alphabetical ordering only if those particular items are often presented in this way (e.g., a list of states).

Menu Enhancements

One promising approach to enhancing menus is to enable keyword entry. Lee and Raymond (1993) note two factors that cause difficulty for users of menu systems:

- ✦ A disproportionate number of errors occur at the top menu levels.
- ✦ Menu systems can be tedious, inefficient, and boring for expert users.

Menu keywords can assist in both situations. These systems associate a keyword with each menu, enabling expert users to bypass the problematic upper levels and reduce keystrokes to immediately access lower levels. Lee and Raymond (1993) present a series of studies showing a reduction in search time, failures, and number of frames accessed, with a marked improvement in user preferences.

Menu Depth versus Menu Breadth

Menus can be *broad*, with many options per menu category, or *narrow*, with few options per category. They can also be *deep*, with many levels of subcategories, or *shallow*, with few levels of subcategories (Figure 10.15). For a given set of features, the following question is often raised: Is it better to have a broad and shallow menu hierarchy, or a narrow and deep menu hierarchy?

This question has been heavily studied in the literature. Early studies on desktop computer menus show that broad hierarchies are usually the most efficient for users, especially when the options are unambiguous. Broad hierarchies also seem

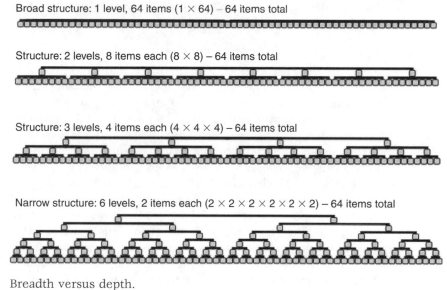

Broad structure: 1 level, 64 items (1 × 64) – 64 items total

Structure: 2 levels, 8 items each (8 × 8) – 64 items total

Structure: 3 levels, 4 items each (4 × 4 × 4) – 64 items total

Narrow structure: 6 levels, 2 items each (2 × 2 × 2 × 2 × 2 × 2) – 64 items total

FIGURE

10.15

Breadth versus depth.
Examples of broad, narrow, deep, and shallow menu structures.

best for expert users. However, if errors are costly or the number of errors is high, a narrow hierarchy that funnels users into a choice can be best. (A summary of these studies can be found in Bernard, 2002a, and Geven et al., 2006.)

The majority of studies conducted on this issue, however, have focused primarily on evaluating structures of relatively constant shapes (the same number of items per level). Few have examined the shape of the structure. Norman and Chin (1988) and Bernard (2002b) examined differently shaped structures, including a constant breadth (same number of items per level), a decreasing structure (more items in upper levels than in lower levels), an increasing structure (more items in lower levels than in upper levels), a concave structure (more items in top and bottom levels, fewer items in middle levels), and a convex structure (more items in middle levels, fewer items in top and bottom levels). Both studies found that concave structures are more navigationally efficient than relatively constant shapes of the same size and depth. Thus, they assert that broad menus at the middle levels will increase menu selection errors.

The majority of the above studies were performed on desktop computers where more information can be presented on the screen. If all items cannot be displayed on the screen at a time, the depth-versus-breadth trade-off shifts. Geven et al. (2006) re-evaluated this issue with menus on mobile phones and PDAs. They found that with mobile phones and PDAs, the most effective hierarchy has only four to eight items per level, and if a large number of items are needed, it is better to increase the number of levels than to increase the number of items per level.

Thus, a deeper hierarchy is more effective for users of products with small screens than a broad hierarchy, for expert users and novice users alike.

Icons

Icons are often used in menu design, particularly at the top levels of a hierarchy. Research into the value of the icons, however, is mixed. MacGregor (1992) found that when icons show an example of an item in that category, they reduce errors by 40 to 50 percent. The generalizability of this result, however, is dependent on the icon design and its success at unambiguously representing a category member. Icons are also useful in menus as a means for achieving incidental learning for later use in toolbars or other isolated conditions (Dix, 1995).

Baecker and colleagues (1991) asserted that the meaning of an icon should be obvious to experienced users and self-evident to novice users. They declare that many icons do not meet the former criterion, and most do not meet the latter. As a result, they evaluated the impact of animated icons in reducing this problem, and found that animated icons were useful and helpful. In every case in their study where a static icon was not understood, the animated icon was successful. Again, the generalizability of this result is dependent on the animated icon design.

The overall value of icons, however, is still debated. Some believe their greatest value is in visual entertainment, adding some interest to an otherwise dull piece of equipment, and that in terms of usability, they are at best irrelevant and at worst distracting (Jones & Marsden, 2006).

Menu Alternatives

Menus are a definitive improvement over command line interfaces or function key–based interfaces (e.g., pressing the FCN key simultaneously with another key to see the phonebook, a common early interaction with cell phones). However, can they be improved? Jones and Marsden (2006) proposed an alternative method for mobile phones—a structure inspired by B+Tree interfaces used in databases. This structure involves enabling users of mobile phones to begin entering the name of the function they wish to access and then to select that function from the best-match list. For example, if users wanted to access Call Divert, they would start spelling the function name using the keypad (in this case, pressing the 2 key for a *C* and pressing it again for an *A*). The system would then display a scrollable list of all possible choices, such as Call Divert, Call Identification, Call Barring, Banner, Backlight, and so on. The users could continue spelling to reduce the number of matches, or select the desired option from the list at any time. Users can also move through the entire list using the scroll keys without attempting to spell anything first. Thus, this design supports the novice user's desire to explore every function, as well as the expert user's need to rapidly choose a known function.

Using the Nokia 5110 mobile phone as a test case, they found that users in their study required an average of 9.54 keystrokes to complete their tasks in the

B+Tree-inspired structure, compared to 16.52 keystrokes on the Nokia 5110, and that the mean task time was reduced from 42.02 seconds to 33.42 seconds.

Menu Design Guidelines

Based on the research discussed above and other known principles for menu design, the following comprise a few best-practice guidelines for menu design:

1. Organize menu items into orthogonal categories that follow the target users' perceived organization. Unfortunately, being realistic, at times this can be impossible to fully achieve. Often, through careful consideration, repeated attempts, and with the help of user studies, near orthogonal categories can be created. It is at this stage that the user studies will further aid menu design because menu items that could appear under multiple categories can be assigned to the category to which most of the users in the study perceived it to belong. For example, should Ring Volume for a mobile phone appear under Ringtones or Sound Settings? Valid arguments could be made either way, but if the majority of users in a study placed it under Sound Settings, then that is where is should go. Of course, this problem only occurs when categories are not orthogonal, so when possible, identifying orthogonal categories is always the best solution for issues like this.

2. Create menu titles that are clearly worded and accurately convey their contents. Utilize user data in creating the menu titles.

3. Use simple action verbs to describe menu options.

4. For submenus, title the top of each menu (perhaps in the header) with the name of the menu item previously selected. This will aid users in knowing their location within the menu hierarchy as well as confirm their selection.

5. When using looped menus, ensure that the users know when the menu has looped and returned to the first item. One way to accomplish this is to place the first menu item only at the top of the screen, even if it is scrolled to by pressing the down arrow key. Thus, the first menu item will never appear at any position on the screen but the top position. This will result in a jump of the highlight from the bottom of the screen to the top when the user presses the down button from the last menu item to the first menu item. This jump will cause the user to take note that something different has happened and signify that the user is now looking at the first item in the list.

6. Make it very easy for users to navigate upward in the menu hierarchy. Doing so will enable users to recover from a mistake if they realize the item they are looking for is not in the category they selected.

7. Provide a way for the user to exit the menu without making a selection. Doing so will also enable the users to recover from a mistake if they realize the item they are looking for is not in the menu at all.

10.4.4   Reading and Comprehension on Small Screens

Research into reading and comprehension on small screens has found that reading will be only slightly slower on small screens and that the experience is not dramatically different from reading on a large screen. Duchnicky and Kolers (1983) studied the readability and comprehension of text when varying screen size. When varying screen height (1, 2, 4, and 20 lines), they found that when only 1 to 2 lines were displayed, performance was significantly poorer than when 4 or 20 lines of text were visible, and that there was no difference between 4 and 20 lines. Although the difference was statistically significant, the magnitude of the performance decrement was not large. They found that reading was only 9 percent slower than when display height was increased from 1 to 20 lines, and there was no difference in comprehension. Varying screen width, however, had a larger impact. The full-width and $\frac{2}{3}$-width displays read 25 percent faster than the $\frac{1}{3}$-width display. Elkerton and Williges (1984) also studied the impact of varying screen height and found a significant performance decrement with screens of one line of text. Based on these results, if displays will be used heavily for reading, preserve as much of the width of the display for reading as possible. This also implies placing navigational controls, menu bars, and so on, in the vertical plane.

Even if there is little performance degradation in reading and comprehension on small screens, experience has taught us that users prefer not to read on small screens. Marshall and Ruotolo (2002) provided two classrooms of students the same class materials in e-book format on Pocket PCs and other electronic and paper formats and examined how students used the Pocket PCs. They found that the students employed them primarily for reading secondary materials, excerpts, and shorter readings. They used them when portability was important, such as waiting in line, in the classroom, when traveling, when waiting for the bus, and so on. They found that on the Pocket PC, users rarely read linearly; instead they tended to skim to particular spots and then read short passages. Finally, the study found that search was a key function for finding familiar materials and for navigating within the document. This study gives some insight into how people will use small screens to read when such materials are in competition with other sources of the same material. This confirms the belief that the true strength of small-screen reading is in casual, opportunistic reading.

10.4.5   Use of Common GUI Elements

User interface designs that build on people's previous experiences are often quicker to learn through transfer of training and increased initial familiarity. It is often wise when designing a user interface to examine the background knowledge and experiences of the target user group and determine whether there are opportunities to benefit from this transfer of training. In many markets around the world, the target user group of small-screen GUIs often has experience with desktop computer GUIs. Thus, the common desktop GUI elements can often be utilized in small-screen design to enable quicker learning. However, some of the common elements in their current implementation for desktop GUIs require adaptation to the small screen. This section will present some of the valuable adaptations of common GUI elements that can be used in small-screen GUIs.

Windows

In their implementation on desktop GUIs, windows are resizable and movable. On small screens, the screen size is generally limited and the ability to move and resize is less valuable. In addition, these activities often require two hands, which is also not ideal for handheld small-screen devices. Thus, generally speaking, it is wise to avoid movable and resizable windows.

Tabs

Tabs are an element often found in desktop GUIs. They can be similarly valuable in small screens; however, tabs with two or three rows may require too much screen real estate and should be avoided.

Pull-Down and Pop-Up Menus

Pull-down and pop-up menus are established widgets in small-screen design. For touch screen interfaces, pop-up menus are often preferred over pull-down menus because the hand will not cover the menu. It is best to avoid scrolling menus if possible because they require a great deal of manual skill from the user, particularly for touch screens (Zwick et al., 2005).

Dialog Boxes

Dialog boxes are often used in small screens and are easily transferred from desktop GUIs. They can be used for persistent system notes ("You have a new text message"), for queries ("Please enter your PIN" or "Delete this message?"), for list selection (call Jane's mobile number, work number, or home number), and for many other purposes. Mobile phones introduce a new type of dialog box, typically called a transient. Transients are pop-up dialog boxes that are automatically dismissed after a specified period of time. They are used to convey system status changes such as "File saved" or "Messages deleted." Transients are valuable additions to the mobile phone user interface because they are able to provide important but noncritical information to the user without requiring specific user action to dismiss. However, they should only be used when it is not essential that the user sees the information they convey because it is possible that the user will miss the transient.

Lists

Lists are a primary interaction method for small-screen GUIs. Lists are often used for menus, single selection of an item from a list, multiple selection of items, and in forms.

Scroll Bars

Scroll bars can be effectively used in small-screen design to indicate to the user that scrolling is possible. However, there is a tendency to shrink their size to a point where they are difficult to notice. As a result, secondary indicators of the ability to scroll can be helpful (such as presenting a partial line of text at the bottom of the screen). Combining vertical scrolling with horizontal scrolling, however, is difficult for users, particularly on small screens, and should be avoided if possible. If avoiding is not possible, Zwick et al. (2005) recommend using a navigation border that runs all the way around a screen to enable the content to be moved freely in any direction. Scrolling on small screens will be discussed in more detail later in this chapter.

Zoom

Zoom is a technique that holds promise for small screens in some applications provided that the hardware can perform this function smoothly. It involves presenting an overview of a large space and enabling the user to "zoom" into an area of interest. Zooming is able to create a logical link between an overview and detailed content and is therefore beneficial in presenting large amounts of information in limited space (Zwick et al., 2005).

Dynamic Organization of Space

One heavily researched area involves techniques for displaying websites on small screens, most notably PDAs. No dominant design has emerged to date, as many of these techniques are still under investigation. A quick look into a few of these techniques, though, can provide unique and inventive ideas for presenting large amounts of information on small screens and provide a means for users to interact with it.

Some of the most researched techniques provide users with an overview visualization of the larger web page while allowing them to rapidly zoom in on tiles of relevant content. The overview visualization allows users to get an understanding of their context within the web page and zooming allows the user to view sections of the page at a readable size. Different techniques utilize different approaches for zooming. The collapse-to-zoom technique (Baudish et al., 2004) allows users to select areas of the web page that are irrelevant, such as banner advertisements or menus, and remove them, thereby enabling the relevant content to utilize more screen space. The collapse-to-zoom technique uses gestures for collapsing and expanding content.

Another technique, summary thumbnails (Lam & Baudish, 2005), is essentially thumbnail views that are enhanced with readable text fragments. The readable text within the thumbnail is intended to enable users to more quickly identify the area of interest without having to zoom in and out of the page. A zooming capability is still provided, allowing users access to the full, readable text. Other techniques include WebThumb (Wobbrock et al., 2002) and fisheye, zoom, or panning techniques (Gutwin & Fedak, 2004).

10.5   TECHNIQUES FOR TESTING THE INTERFACE

Usability testing of small-screen devices is an important step in the user-centered design process. The methods and processes for testing small-screen interface designs are similar to those for other products. Test preparation, protocol development, participant selection and recruiting, conducting of the test itself, and analysis of results are the same for usability testing of small screens as they are for usability testing of any other user interface. Since these items are covered in depth in a number of sources, a full discussion here is beyond the scope of this chapter (see instead Dumas & Redish, 1994, and Rubin, 1994). However, the adaptations and variations of equipment setup needed to evaluate small-screen devices are specialized. As a result, the current section will focus on this aspect of testing.

One of the most challenging aspects of usability testing of small-screen devices is enabling the moderator to see what is happening on the screen and video-record it. There are many possible solutions to this challenge, but three solutions that have proven successful in the past involve the use of an over-the-shoulder camera, a document camera, or a software utility that will replicate the screen of a device on a computer. There are several such utilities for some models of mobile phones, for example. This section will discuss each of these techniques in some detail.

10.5.1   Over-the-Shoulder Camera

With the over-the-shoulder camera option, typically the camera is mounted on a tripod behind the participant's shoulder and raised to a height where it can be focused on the screen. In a laboratory where small screens are often tested, this camera can be mounted on the wall or ceiling instead of on a tripod. With a camera capable of high zoom (e.g., in the range of 16×), a clear and close picture of the device's screen can be obtained.

If the moderator wishes to sit in the room with the participant, the video output from this camera can be split and run to the recording device as well as to a television or monitor. This monitor can be placed in the participant room to enable the moderator to see the participant's screen on the monitor without

FIGURE   Example setup of an over-the-shoulder camera.
                    The camera is mounted on a tripod in a conference room with an in-room
10.16   moderator.

having to crowd the participant by trying to view his/her screen directly (see Figure 10.16).

The additional challenge for handheld small-screen devices that an over-the-shoulder camera must solve is that the device can be moved out of the focus of the camera as the participant naturally moves to adjust posture and so on. One approach to mitigate this issue is to place a marker on the table in front of the participant (with instructions to try to hold the device over the marker) and also to orient the moderator's monitor so both the participant and the moderator can view it. Orienting the monitor in this way enables the participant to better understand the need to hold the device over the marker and eases any thoughts she may have as to where that camera is focused. To achieve a picture-in-picture view of the screen and the participant's face, a camera can be located across the room focused on the participant (Figure 10.16).

Lighting is another item to consider and control. The screens on the devices can reflect an overhead light or other light source. Test the lighting of the participant's location by sitting down at that location while watching the camera image on the monitor. Check to ensure that the image is glare-free and not washed out.

A disadvantage of the over-the-shoulder camera is that despite the marker and the monitor facing the participant, users will still move out of the designed areas at times and need to be reminded to keep the device in the focus of the camera. The advantage of this setup is that the user is able to hold the device naturally without anything attached to it or in front of it, and a view of the fingers pressing the keys can also be captured by the camera.

10.5.2   Document Camera

A document camera (Figure 10.17) can be used instead of the over-the-shoulder camera. There are multiple possibilities for this camera, ranging from a lightweight wireless camera that can be attached to the device itself, to placing the device in a fixed holder (e.g., a vise) and utilizing a stationary camera. In fact, it is even possible to design and develop a custom version of this equipment to suit specific needs (Schusteritsch et al., 2007). Any of these options, and other variations of these options, will give a very good view of the device's screen and, depending on the camera angle, the fingers as well. They can also be used in testing setups similar to that described previously with an in-room monitor for an in-room moderator to view, and another camera for capturing video of the participant's expressions.

The advantage of the document camera is that the camera will move with the participant, always staying in focus on the device and screen. The disadvantage is

FIGURE   Document camera.

10.17   This is a wireless mobile device camera by Noldus (2007). (Courtesy Noldus Information Technology.)

that it does not allow the device under test to be held as it was designed to be held (although the example in Figure 10.17 does attempt to approximate it), and it places the camera in the direct view of the participant, which can be annoying or get in the way of the participant's view. A further disadvantage for mobile phones is that the camera can get in the way of placing phone calls, if that is one of the tasks under test.

10.5.3   Screen Replication Utility

If the device supports it, a screen replication utility can provide a detailed and clear view of the small screen on a larger computer monitor. Once on a computer monitor, numerous screen-recording utilities can be used to record what is happening on the screen for later viewing. However, this option requires support by the small-screen device. There are several utilities that can serve this purpose for Windows Mobile Smart phones and PocketPCs, as well as Nokia Series 60 devices and select Sony Ericsson phones. The advantage to this option is that it provides an unobstructed view of the screen without having to remind a participant to put it in a certain position and without having to attach a camera to the device. However, the disadvantage is that it is dependent on support by the device under test, and it cannot show the participant's fingers.

10.5.4   Usability Laboratories

Many usability tests occur in a usability laboratory. Usability laboratories often utilize two rooms: a participant room and an observer room. A one-way mirror often separates the two rooms and enables the observers to see into the participant room, but prevents the participants from seeing into the observer room. Microphones can be mounted in various locations within the participant room to enable the observers to hear conversations, and cameras can be mounted in various locations to enable video recordings of the events.

10.5.5   Testing without a Laboratory

Usability testing does not require an elaborate laboratory, however. Many times, usability tests can be run just as effectively in conference rooms with minimal equipment. This guideline is just as true for usability testing of small-screen interfaces (see Figure 10.16). The techniques described above can be used in any conference room: The camera can be a camcorder (unless the document camera is used), and the microphone can be directly connected to the camcorder (or can be the camcorder's microphone itself). In fact, much of the equipment described above has options for the use of inexpensive equipment. The main point is that usability testing can be run on any equipment budget. It is just important that it is run.

10.6   DESIGN GUIDELINES

The fundamentals of software design remain constant across various implementations whether they are small-screen, web, or traditional PC applications. The focus of this section will be on design guidelines that are unique or important to small-screen design. All design principles must be adapted to the strengths and limitations of the implementation medium.

As with all interface design projects, user interface design for small displays should follow the user-centered design (UCD) process. This means involving users in each step of the requirements-gathering and design process, plus performing iterative design with usability testing. A full discussion of the UCD process is outside the scope of this chapter; see Lewis and Reiman (1993) for a detailed exploration of the subject.

As demonstrated in Section 10.4, the size of the visual display of small-screen devices is a primary restriction to interface design. However, other factors must also be considered, such as the imperfection of current input technologies and the often limited attention of the user in a mobile setting. With those restraints in mind, several overarching design principles emerge: Design for simplicity, design with the small screen in mind, provide useful feedback, maintain interaction design standards, and respect the user's physical and mental effort.

10.6.1   Design for Simplicity

As discussed in Section 10.3, small-screen devices tend toward two extremes: the "does-everything" mobile PC replacement and the "does-one-thing-well" information appliance. In both cases, the underlying technology is usually complex and the interface must strive toward the principal goal of simplicity. In his classic book, *The Design of Everyday Things*, Donald Norman defines the paradox of technology as follows: "Whenever the number of functions and required operations exceeds the number of controls, the design becomes arbitrary, unnatural and complicated" (1998:29). The hardware design trend toward more complicated and feature-filled devices makes the guideline of keeping things simple a very difficult task indeed.

Relate Visual Precedence to Task Importance

Understanding what the user wants the device to do should be the first step in the simplification process. Many interface design projects have gone wrong by making uninformed assumptions about user needs (Nielsen, 1993). The overarching need for simple design makes understanding both the primary tasks and their priority of use absolutely critical to good design. Once tasks and task priority are understood, the design of the interface can be organized accordingly. The most frequently used tasks and those that are used by the majority of users should be

More clicks ↓	Less visible ──────────────→	
	By many	**By few**
Frequent	**Frequent by many** Visible, few clicks	**Frequent by few** Suggested, few clicks
Occasional	**Occasional by many** Suggested, more clicks	**Occasional by few** Hidden, more clicks

FIGURE

10.18

Visibility tasks based on usage.
The tasks in the interface are based on the amount of usage and the number of users. *Source:* From Isaacs and Walendowski (2002).

the most visible in the interface (Isaacs & Walendowski, 2002). Figure 10.18 shows the suggested visibility of tasks based on their usage.

Reduce Functionality

Strongly consider removing functionality that falls into the "Occasional by few" category in the figure. In fact, the designer should agonize over the addition of each individual feature to the design no matter where it falls on the frequency of use scale (37signals, 2006). Bear in mind that each additional feature makes the overall product more complex and more difficult to use.

Keep Navigation Narrow and Shallow

Small-screen devices are generally used in settings where the user's attention is at a premium, such as making a call on a busy bus or making a meal in the microwave. In these cases, users do no want to choose from a dozen options; they want to complete their primary task as quickly as possible. Task efficiency is an extremely important design goal and often helps enforce the principle of simplicity. This implies that the information hierarchy of the interface should be kept both narrow (fewer choices at each level) and shallow (fewer levels of choice to the bottom of the hierarchy). As discussed in Section 10.4, if the number of options must exceed a narrow and shallow hierarchy, then it is more efficient to design a deeper structure than a wider one.

Avoid Extraneous Information on Each Screen

The principle of simplicity extends not only to the overall ease of use of the device but also to the complexity of each individual screen. Note the limited number of controls, interface widgets, and text used in the Now Playing view of the Apple iPod shown in Figure 10.19.

The Now Playing screen could show other potentially useful data items, such as detailed music metadata, next and previous track information, track ratings, repeat/shuffle status, and playlist name, among others. However, the designer

FIGURE

10.19

Now Playing view of Apple's iPod.
This view demonstrates how each screen of the interface can be made both simple and useful at the same time. *Source:* From Apple.com, 2007; courtesy Apple Inc.

has intentionally shown only the information critical to the user's primary task. Note also the simple iPod hardware controls: Each hardkey on the device serves a dedicated and important purpose within each context, and there are no extraneous controls.

Reduce or Remove Preferences

Advanced customization is a standard aspect of traditional application design that is not suited to the small screen. Many traditional applications allow the user to customize the user experience through complex (and often unused) controls such as preference dialogs. While keeping the user in control of the interface is important, providing many different ways to customize the way the application works incorporates too much complexity into a small-screen interface. Typically, small-screen applications are simpler than traditional applications; thus customization is less important even to advanced users. The exception to this rule is potentially annoying features such as warning messages that often pop up. In these cases, a simpler customization, such as allowing the user to permanently dismiss the dialog, can be useful because it simplifies the entire experience.

Use Progressive Disclosure

Perhaps the most powerful tool in interface simplification is progressive disclosure. Progressive disclosure involves breaking down complex tasks into separate understandable parts. Each step in the task is split into screens that lead the users toward their goal. Many interface guidelines have their basis in progressive disclosure; perhaps the best example is the hierarchical navigation found in nearly every modern interface. Instead of giving the user 50 different options to choose from, those choices are broken up into meaningful groups and subgroups. On the surface, progressive disclosure appears to break the guideline of efficiency; however, in practice, progressive disclosure actually improves the user's speed in completing tasks (Buyukkokten, Garcia-Molina, & Paepcke, 2001).

With a flat list of options, the user is forced to decide between myriad (often ambiguous) options. It takes time for the user to decide which option to choose and, if the chosen option is incorrect, the user must scan the long list and choose again. As discussed in Section 10.4, a properly designed progressive disclosure employs mutually exclusive (orthogonal) choices at each step to ensure that the user is always progressing forward toward her goal. On a small-screen device, progressive disclosure is even more important than in traditional applications because of the limited screen space available to display information.

10.6.2   Design with the Small Screen in Mind

Minimize User Input

As discussed in Section 10.2, user input is an even more difficult design challenge than visual output for small-screen devices. This impediment implies that the designer must be thoughtful whenever adding user input widgets to the interface. If possible, avoid user input all together. For example, if the device is giving feedback to the user that is noncritical, it may make sense to use a transient dialog box and have the dialog automatically dismiss itself after a short period of time.

In cases where user input is required, keep the required user effort to a minimum. If a particular screen requires the user to enter his city (say to determine time zone on a mobile phone), do not present a free-form text entry field. While free-form entry might be fairly easy for a user with a full-sized QWERTY keyboard, using the DTMF keypad found on a typical mobile phone to enter a long city name will require a great deal of effort and time. Widgets that use the cognitive principle of recognition (such as radio buttons, drop-downs, and checkboxes), rather than recall (such as free-form entry fields), are much more effective. In the previous example, using a drop-down menu that displays city names or a field that enables ZIP/postal code entry (and performs a lookup to determine city name) would be much more appropriate.

Minimize Vertical Scrolling and Avoid Horizontal Scrolling

Ideally screens should be designed so that there is no need for scrolling of any kind. In large-screen applications and on the Web, applications are often heavy with instructional or informational text. The small-screen designer does not have this luxury and instructional text should be less necessary since functionality is normally simpler on the small screen. A design that requires lengthy textual instructions is an indication that the interface is too complex and should be revised.

In some more complicated applications, it is unavoidable to present a scrolling screen. In these cases it's important to use vertical rather than horizontal scrolling. The proliferation of the Web has made vertical scrolling a well-understood interface paradigm for the majority of users. Western users read from top to bottom and from left to right, so vertical scrolling maps translate well to the real-world task of reading a book. Horizontal scrolling is far less understood by users and breaks both the metaphor of book reading and the perceptual process of visual scanning. In some cases, horizontal scrolling is unavoidable (e.g., when scrolling around a map or an image too large to fit on the screen). In these cases, use a scrolling border (as discussed in Section 10.4).

Use Hyperlinking Effectively

In many cases, it is difficult to provide enough information to the user inside a single small screen (e.g., some error messages require detailed textual advice). In these situations, hyperlinking can be used. The beauty of the hyperlink is its ability to hide details while still making them available to those who want more information. Hyperlinking is especially useful in cases where users may see the same text many times (e.g., a common error message). In these cases, the user may need detailed information the first time the message is displayed, but does not want to be forced to scroll through the same long text every time the error occurs. Of course, if an error is so common that the user is seeing it all the time, this is a good indication that the design is flawed. Always strive for a design that minimizes the number of possible user-initiated errors.

Provide Useful Error Messages

Simplified content can sometimes lead to confusion if taken to an extreme. Error messages need to be helpful even on the small screen. Ideally an error message contains three distinct parts: (1) a unique error identifier, (2) a description of the problem, and (3) possible solutions. Item 1 is important in the case where the user is unable to quickly figure out a solution. A unique identifier allows users to easily seek help about their specific problem from an external source (e.g., a website or customer service). Items 2 and 3 need to be written without technical jargon and in a language that the user can understand. Properly written error messages can make a significant difference in the usability of a complicated system.

Prioritize Information on Each Screen

Even minimal content can be confusing to a user if it is not presented in a meaningful way. As mentioned earlier, display the most important information first and largest on the screen (always based on your knowledge of the user's primary tasks). If no priority can be given to the information, order the data in a logical manner (e.g., alphabetically or chronologically) to reduce the user's memory load and increase task efficiency. Similarly, grouping like data elements together makes visually scanning items faster.

10.6.3   Provide Useful Feedback

Visual, auditory, and sensory feedback to the user is one of the most critical aspects of the interaction design of a small-screen device. Providing just the right amount of feedback without overwhelming the user can be a tricky task.

Identify Critical Feedback

Another guideline based on the limited attention that users typically provide their small-screen devices is that user feedback must be very strong. Users need to be made continually aware of what the system is doing while they interact with the device. Identify the information that is critical to the user's needs and present only that information to the user. For the relatively simple example of a microwave display, it is important that users know the heat setting of the microwave and the remaining time left in the heating process. Removing or obscuring either of those two critical items from the screen to display other information will confuse and frustrate users.

Employ Alternative Feedback Modalities Intelligently

The use of alternative forms of feedback can be very powerful on small-screen devices (often in contrast to traditional computer applications). Using sound and tactile feedback is a standard technique on many small-screen devices because the user's attention is rarely focused exclusively on the device. Imagine how useless a mobile phone would be if it only presented visual feedback of an incoming call! However, the decision to use alternative feedback should not be made lightly. Users do not want their attention diverted unless an important event has occurred. The guideline here is to use alternative feedback intelligently and sparingly.

Ensure Quick System Response Time

Related to task efficiency is the amount of time that it takes the interface itself to respond to the user. There are very few tasks that are technically complex enough that the user should have to wait for the interface to respond. The *BlackBerry Style Guide* states, "user interface response time should be, at worst, 200 ms (1/5 of a second) or better. If an operation is expected to take a long time (e.g., a text search or a database query), then it should be executed in the background" (Research in Motion, 2003). If for some reason a long task cannot be performed in the background and the user is forced to wait, ensure that detailed feedback is presented. Ideally such feedback includes an estimate of the amount of time the process will take to finish. The progress bar widget is an ideal feedback control for such situations since it gives the user an indication of how long the process will take and provides feedback that the process is still moving forward. For any process that forces the user to wait, even with strong feedback, ensure that you provide a way to cancel the task.

10.6.4   Maintain Interaction Design Standards

Across the wide range of small-screen devices, many different user interface styles are used. However, standards do exist and in some cases they are very well developed and documented. Where standards do not exist, using concepts that users already understand can dramatically improve the usability of a complex interaction.

Use Existing Standards

When designing for the small screen, as with any other type of interface design, the design should begin with existing standards. Many of the more advanced small-screen devices (such as the Windows Mobile platform) have detailed interface guideline documents that are created to ensure consistency across all platform applications. Less complex devices also have standards, but they are often implicit and rarely written down. For example, users have an expectation that pressing a button on a watch will cause some action to occur in the watch display. This may seem obvious, but many watches break this basic standard by including (often unlabeled) buttons that only work in certain contexts.

It is important to explore the presentation, widget, and interaction standards of the device before design begins. Avoiding these fundamental building blocks of the interface can cause significant rework when user testing uncovers problems with the nonstandard design. Similarly, existing device visual-design standards should only be broken in the most extreme cases, such as when the existing framework simply will not support the task you are trying to achieve.

Use Real-World Metaphors

Existing real-world metaphors that translate into the world of software are some of the most powerful user interface paradigms. Take for example the play, pause, track forward, and back buttons found on almost all digital music players (Figure 10.19). These controls have been carried forward since stand-alone tape players were first released. Making the decision to break a well-understood metaphor will take the interface into unknown waters; the designer will have no idea whether the new interface will work effectively until user testing is performed. The same goes for control widgets. Making a decision to create or modify an existing, well-understood widget is extremely dangerous to the usability of the device. In his famous book *Tog on Interface*, Apple Macintosh GUI designer Bruce Tognazzini (1992) chronicles his experience trying to create a hybrid checkbox/radio button widget. After many frustrating iterations of design and testing, he completely abandoned the seemingly simple control because users could never figure out how the new widget worked.

10.6.5   Respect Both Mental and Physical Effort

Due to the limited amount of attention users give to their small-screen devices, it is important to respect the amount of both mental and physical effort the user must employ to interact with the interface. Instead of forcing the user to do the work, a good design will make the user's tasks almost effortless. For example, several cameras provide panorama modes that aid the user in the creation of panoramic pictures by showing the side of the previous picture on the screen to help the user line up the next picture in the series (Isaacs & Walendowski, 2002).

Use Wizards to Simplify Complex Interactions

The software wizard is a potent navigation paradigm that embodies the guideline of progressive disclosure and reduces the user's memory load by splitting up complex tasks into multiple steps. A well-designed wizard includes the ability to move forward and back through the various steps and also keeps the users informed about where they are in the wizard process. The key with small-screen wizard design is to eliminate as many input fields as possible through simplification by putting the onus on the system software to do as much of the work as possible. For example, if the interface is capturing an address, ask for the ZIP/postal code first, then prepopulate the city and state fields with the appropriate data to reduce user input.

Use Multitasking Operating Systems Effectively

Many modern small-screen operating systems natively support multitasking. While multitasking is a terrific boon to software design, care must be taken in its interface implementation. Due to fact that users have limited attention for their devices, it is not a good idea to force a user to multitask to complete a task. When using a small-screen device, users typically have a single task in mind and they want to complete that task as quickly as possible. Multitasking is a powerful tool, but it should be used primarily by the system and not by the user. For example, system tasks that cannot be resolved in a timely manner should be performed in the background so that they do not interfere with the user's continued use of the device.

Design for Efficiency

The Nokia Series 60 Contacts application shown in Figure 10.20 is a strong example of efficient and simple small-screen design. The user is able to scroll vertically through the list of alphabetized contacts. However, if the user has a large number of contacts, the list gets too long to scroll through. In this case the designers have included an alphabetical filter that works with the keypad. Pressing the letter M will filter the entire list to show only those names that have an M beginning the first or last name. This technique is surprisingly effective at improving task-completion time and is intuitive to learn even by novice users.

FIGURE

10.20

Nokia Series 60 Contacts application.
The field with the magnifying glass allows the user to quickly find a contact using filtering. *Source:* From allaboutsymbian.com, 2007.

10.7   CASE STUDY

A case study of a small-screen interface can be found at *www.beyondthegui.com*.

10.8   FUTURE TRENDS

As discussed in Section 10.3 of this chapter, there are several trends that will continue into the future of small-screen devices. The most prevalent one affects all technological fields: increased capability of both software and hardware. When making purchase decisions, users often buy based on the feature list of the device (Sarker & Wells, 2003). This trend encourages hardware manufacturers to integrate additional functionality into their small-screen devices. High-end smart phones and hybrid devices are constantly incorporating functionality previously found only in dedicated devices, such as location-based services, high-resolution cameras, multimedia messaging, and improved text input. Technologies formerly found solely in traditional personal computers will also continue to be integrated into small-screen devices: multimedia editing, increased wireless data bandwidth, larger device memory, and open software architectures.

The opposite trend will also continue in the future: simple dedicated information appliances that perform a limited number of functionalities but do so in a flexible, feature-rich, and usable manner. While those on the cutting edge will continue to purchase leading-edge devices, the average consumer wants less sophisticated and easier-to-use technology. The maturing world of multimedia involves the blending of several complex technologies, making the consumer's learning curve steep. Media center systems will continue to present straightforward, uncomplicated interfaces hiding the intricate media playback, wireless networking, media sharing, and increased software control found under the surface. The number of information appliances targeted toward all areas of small-screen technology will continue to blossom over time.

The trend to larger, higher-resolution displays will also continue. Interfaces that use vector graphics to scale to any resolution will become more common. Flexible OLED displays will mean that big screens will become portable and use less battery power than LCDs while displaying richer data and media. Figure 10.21 shows two concept devices using rollable displays. In addition, eyewear enabling users to privately view a big-screen-like display has already become available. In 2005, Orange SA launched video eyewear with a Samsung phone enabling users to view TV, movies, photos, and broadband Internet content with a big-screen viewing effect (DigitalCamera@101review, 2005). The small-screen device will continue to catch up to the PC in terms of CPU speed and functionality, and mobile devices will run an increasingly complex variety of software.

In the medium-term future, displaying information on a traditional screen will be eclipsed by new display technologies. The problem with current larger-but-portable screen technology is that no matter how small individual components can be made, the screen will always limit the reduction in size of the form factor. People want small devices that are portable. Making the display external to the device removes this restriction. Three interesting technologies are projection displays, heads-up displays, and open-air holographic-like displays. Projection technology means that the screen itself can be as large as the user requires, but the device itself is still very small and thus highly portable. Heads-up displays use an external device to display important information to the user. The open-air holographic-like display projects an image in the air just above it. One example of this is the heliodisplay developed by IO2 Technology that is capable of projecting an interactive two-dimensional image (Genuth, 2006). All of these display technologies will see further refinement in coming years.

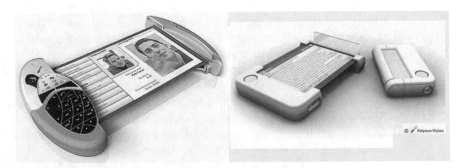

FIGURE

10.21

Concept devices using rollable displays.

The concept at right shows device open and closed. *Source:* From De Rossi (2005); courtesy Polymer Vision.

As demonstrated in Section 10.2, the paramount challenge in small-screen device design is not output but user input. The primary methods of input on a personal computer (the mouse and keyboard) do not translate well into the realm of mobile devices. In the next 10 years, three areas of research will dominate small-device input: spatial interfaces, eye-tracking interfaces, and natural-language comprehension.

A spatial interface is an input technology that is aware of device movement. As the device (or a dedicated input device) is moved, the interface responds appropriately. Spatial interfaces are easy to use and well understood even by novice users. Typically, a spatial interface works in three dimensions, allowing the input device to be moved anywhere in space. The sudden-motion sensor built in to Apple laptops (originally designed to prevent damage to the hard disk if dropped) has already been co-opted as a spatial input. Enterprising engineers have already built software for the motion sensor to scroll windows and play games (Singh, 2006). Dedicated spatial-interface devices like the Wii Remote included with the Nintendo Wii provide a completely new way to interact with games by including a gyroscope to track hand movement. Motion-sensing technology is a practical input methodology for small devices because the hardware required can be made very small and the interaction is natural for users.

Another technology with considerable potential requires even less user effort than a spatial interface. Eye tracking (sometimes called a gaze interface) monitors the movement of the eye across a visual display. The tracked movement of the eye is translated into input within the interface. Eye tracking has been traditionally used in perception and advertising studies but shows promise for visual search tasks such as exploring images and for software navigation. The downside of eye-tracking interfaces is that they require some training to use effectively. However, eye-tracking interfaces can be made small enough for small-screen devices and are also appealing from an accessibility standpoint.

A technology with great potential, both from an ease-of-use and an accessibility perspective, is natural-language recognition and comprehension. As discussed in Section 10.2, some mobile phones already incorporate rudimentary language recognition with limited success. As linguists better understand how to teach computers semantic and grammatical understanding, language interfaces will improve to the point that they are a powerful input modality (Yates, Etzioni, & Weld, 2003). Already technologies such as IBM's Via Voice for the PC are approaching functional everyday use, and it is only a matter of time before the technology is found in smaller devices.

Even further in the future, the quintessential human–computer interface will communicate directly with the brain. While virtual reality pumped directly into the brain (like the interface depicted in the movie *The Matrix*) may be many years away, elementary brain input interfaces are already being studied in labs such as the University of British Columbia's Brain–Computer Interface lab (Birch, 2007). Clearly a direct-to-brain interface involves almost no effort to control input or output data, and is accessible to anyone with or without physical disabilities. The brain interface truly represents the panacea of computing interfaces.

REFERENCES

37signals. (2006). Getting Real; available at *http://gettingreal.37signals.com/toc.php*.

Alibaba.com. (2007). Car Parking Sensor System with LED Display; available at *http://cyberfair.trustpass.alibaba.com/product/10885882/Car_Parking_Sensor_System_With_LED_Display.html*.

Allaboutsymbian.com. (2007). Nokia e61 Contacts Application; available at *http://allaboutsymbian.com/images/devices/e61/nokaie61_review2_07.JPG*.

American National Standards Institute (ANSI)/Human Factors and Ergonomics Society (HFES) 100-1988. (1988). *American National Standard for Human Factors Engineering of Visual Display Terminal Workstations*. Santa Monica: Human Factors and Ergonomics Society.

Apple. (2007). Sound and Hearing; available at *http://www.apple.com/sound/*.

AppleInsider.com. (2006). iPod: How big can it get? *AppleInsider.com*, May—*http://www.appleinsider.com/article.php?id=1770*.

Av.it.pt. (2007). Tektronix TDS 3052B; available at *http://www.av.it.pt/wcas/images/PIC0014.JPG*.

Baecker, R., Small, I., & Mander, R. (1991). Bringing icons to life. *Proceedings of the SIGCHI Conference on Human Factors in Computing Systems: Reaching through Technology*. New York: ACM Press, 1–6.

Baudish, P., Xie, X., Wang, C., & Ma, W. (2004). Collapse-to-zoom: Viewing Web pages on small screen devices by interactively removing irrelevant content. *Proceedings of Seventh Annual ACM Symposium on User Interface Software and Technology*. New York: ACM Press, 91–94.

Bernard, M. L. (2002a). Examining a metric for predicting the accessibility of information within hypertext structures. PhD diss. Wichita State University.

Bernard, M. L. (2002b). Examining the effects of hypertext shape on user performance. *Usability News* 4.2—*http://psychology.wichita.edu/surl/usabilitynews/42/hypertext.htm*.

Blackwell, H. R. (1946). Contrast thresholds of the human eye. *Journal of the Optical Society of America* 36:624.

Birch, G. (2000). The Brain–Computer Interface Project; available at *http://www.ece.ubc.ca/~garyb/BCI.htm*.

Bohan, M. (2000). Entering text into hand-held devices: Comparing two soft keyboards. *Usability News* 2.1; available at *http://psychology.wichita.edu/surl/usabilitynews/2W/softkeyboards.htm*.

Board of Standards Review (BSR)/HFES 100. (2006). *Human Factors Engineering of Computer Workstations: Second Canvass Review, 2006*. Santa Monica, CA: Human Factors and Ergonomics Society.

Buyukkokten, O., Garcia-Molina, H., & Paepcke, A. (2001). *Seeing the Whole in Parts: Text Summarization for Web Browsing on Handheld Devices*. Digital Library Project (InfoLab), Stanford University. Available at *http://www10.org/cdrom/papers/594/*.

Cambridgeincolour.com. (2007). Basics of Digital Camera Pixels; available at *http://www.cambridgeincolour.com/tutorials/digital-camera-pixel.htm*.

Chapanis, A. (1949). How we see: A summary of basic principles. In *Human Factors in Undersea Warfare*. Washington, DC: National Research Council, 12.

Cooke, N. J. (1994). Varieties of knowledge elicitation techniques. *International Journal of Human–Computer Studies* 41:801–49.

De Rossi, L. C. (2005). Portable rollable displays. Foldable screens that fit in your pocket; available at *http://www.masternewmedia.org/video_display/portable_screens/rollable_portable_screens_prototypes_first_looks_20050712.htm.*

DigitalCamera@101reviews. (2005). Big screen viewing effect for mobile phone videos; available at *http://digitalcamera.101reviews.com/news/big-screen-viewing-effect-for-mobile-phone-videos.*

Dix, D. A. (1995). Acceleration and toolbars: Learning from the menu. *Proceedings of People and Computers IX (Adjunct): British Computer Society Conference on Human Computer Interaction.* London: British Computer Society, 138–43.

Duchnicky, R. L., & Kolers, P. A. (1983). Readability of text scrolled on visual display terminals as a function of window size. *Human Factors* 25(6):683–92.

Dumas, J. S., & Redish, J. C. (1994). *A Practical Guide to Usability Testing.* Norwood, NJ: Ablex.

Elkerton, J., & Williges, R. (1984). Information retrieval strategies in a file search environment. *Human Factors* 26(2):171–84.

Enablemart.com. (2007). Hearing; available at *http://www.enablemart.com/productGroupDetail.aspx?store=10&dept=15&group=29.*

Engadget.com. (2005). Morse Code Trumps SMS in Head-to-Head Speed Texting Combat. May 26—*http://www.engadget.com/2005/05/06/morse-code-trumps-sms-in-head-to-head-speed-texting-combat/.*

Engadget.com. (2004). Watch this Wednesday: The Samsung GPRS Class 10 Watch Phone. November 17—*http://www.engadget.com/2004/11/17/watch-this-wednesday-the-samsung-gprs-class-10-watch-phone/.*

Genuth, I. (2006). Heliodisplay—Floating Free-Space Interactive Display; available at *http://www.tfot.info/content/view/101/.*

Geven, A., Sefelin, R., & Tscheligi, M. (2006). Depth and breadth away from the desktop—the optimal information hierarchy for mobile use. In *Proceedings of Eighth Conference on Human–Computer Interaction with Mobile Devices and Services.* New York: ACM Press, 157–64.

Gohring, N. (2006). Mobile phone sales topped 800 million in 2005, says IDC. *InfoWorld*—*http://www.infoworld.com/article/06/01/27/74861_HNmobilephonesales_1.html.*

Grether, W. F., & Baker, C. A. (1972). Visual presentation of information. In Van Cott, H. P., & Kinkade, R. G., eds., *Human Engineering Guide to Equipment Design.* New York: John Wiley & Sons, 41–121.

Gutwin, C., & Fedak, C. (2004). Interacting with big interfaces on small screens: A comparison of fisheye, zoom, and panning techniques. *Proceedings of 2004 Conference on Graphics Interfaces.* Waterloo, Ontario, Canada, 145–52.

Gwmicro.com. (2007). GW Micro—Small Talk Ultra; available at *http://www.gwmicro.com/Small-Talk_Ultra/.*

Hayhoe, D. (1990). Sorting-based menu categories. *International Journal of Man–Machine Studies* 33:677–705.

Isaacs, E., & Walendowski, A. (2002). *Design from Both Sides of the Screen.* Indianapolis: New Riders.

Jones, M., & Marsden, G. (2006). *Mobile Interaction Design.* London: John Wiley & Sons.

Lam, H., & Baudisch, P. (2005). Summary thumbnails: Readable overview for small screen web browsers. *Proceedings of SIGCHI Conference on Human Factors in Computing Systems*. New York: ACM Press, 681–90.

Lee, E. S., & Raymond, D. R. (1993). Menu-driven systems. In Kent, A., & Williams, J., eds., *The Encyclopedia of Microcomputers* 11:101–27.

Lewis, C., & Reiman, J. (1993). Task-Centered User Interface Design; available at *http://hcibib.org/tcuid/index.html*.

Logitech.com. (2007). Harmony® 670 Advanced Universal Remote; available at *http://www.logitech.com/index.cfm/products/detailsharmony/US/EN,CRID = 2084, CONTENTID = 12701*.

MacGregor, J. N. (1992). A comparison of the effects of icons and descriptors in videotext menu retrieval. *International Journal of Man–Machine Studies* 37:767–77.

Marics, M. A., & Engelbeck, G. (1997). Designing voice menu applications for telephones. In Helander, M., Landauer, T. K., & Prabhu, P., eds., *Handbook of Human–Computer Interaction*. Amsterdam: Elsevier Science, 1085–102.

Marshall, C. C., & Ruotolo, C. (2002). Reading-in-the-small: A study of reading on small form factor devices. In *Proceedings of Second ACM/IEEE-CS Joint Conference on Digital Libraries*. New York: ACM Press, 56–64.

MiniFonts.com. (2006). Home page. *http://www.minifonts.com/*.

Muto, W. H., & Maddox, M. E. (2007, in press). Visual displays. In Weinger, M. B., ed., *Handbook on Human Factors in Medical Device Design*. Hillsdale, NJ: Erlbaum.

Nielsen, J. (1993). *Usability Engineering*. San Francisco: Morgan Kaufmann.

Noldus. (2007). Mobile device camera; see *http://www.noldus.com/site/doc200402054*.

Norman, D. (1999). *The Invisible Computer*. Cambridge, MA: MIT Press.

Norman, D. (1988). *The Design of Everyday Things*. New York: Basic Books.

Norman, K. L., & Chin, J. P. (1988). The effect of tree structure on search performance in a hierarchical menu selection system. *Behaviour and Information Technology* 7:51–65.

Nystedt, D. (2006). Mobile subscribers to reach 2.6B this year. *InfoWorld—http://www.infoworld.com/article/06/11/10/HNmobilesubscribers_1.html*.

Onetouch.ca. (2007). OneTouch Ultra; available at *http://www.onetouch.ca/english/product_detail.asp?gr = 0&cat = 1&pid = 47*.

Paap, K. R., & Cooke, N. J. (1997). Design of menus. In Helander, M., Landauer, T. K., & Prabhu, P., eds., *Handbook of Human–Computer Interaction*. Amsterdam: Elsevier Science, 533–72.

Research in Motion. (2003). Blackberry Wireless Handheld User Interface Style Guide; available at *http://www.blackberry.com/developers/na/java/doc/bbjde/BlackBerry_Wireless_Handheld_User_Interface_Style_Guide.pdf*.

Roeder-johnson.com. (2007). Projection Keyboard PDA; available at *http://www.roeder-johnson.com/RJDocs/CAProjectionKeyboardPDA.jpg*.

Roske-Hofstrand, R. J., & Paap, K. R. (1986). Cognitive networks as a guide to menu organization: An application in the automated cockpit. *Ergonomics* 29(11):1301–12.

Rubin, J. (1994). *Handbook of Usability Testing: How to Plan, Design, and Conduct Effective Tests*. New York: John Wiley & Sons.

Rugg, G., & McGeorge, P. (1997). The sorting techniques: A tutorial paper on card sorts, picture sorts, and item sorts. *Expert Systems* 14(2):80–93.

S60.com. (2007). Nokia 95; available at *http://www.s60.com/life/s60phones/displayDevice-Overview.do?deviceId=1065.*

Sanders, M. S., & McCormick, E. J. (1993). *Human Factors in Engineering and Design.* New York: McGraw-Hill.

Sarker, S., & Wells, J. D. (2003). Understanding mobile handheld device use and adoption. *Communications of the ACM* 46(12):35–40.

Schiffman, H. R. (1990). *Sensation and Perception: An Integrated Approach.* New York: John Wiley & Sons.

Schusteritsch, R., Wei, C., & LaRosa, M. (2007). Towards the perfect infrastructure for usability testing on mobile devices. *Proceedings of Conference on Human Factors in Computing Systems. CHI '07 Extended Abstracts on Human Factors in Computing Systems.* New York: ACM Press, 1839–44.

Singh, A. (2006). The Apple Motion Sensor as a Human Interface Device; available at *http://www.kernelthread.com/software/ams2hid/ams2hid.html.*

Tognazzini, B. (1992). *Tog on Interface.* Reading, MA: Addison-Wesley.

U.S. Department of Defense. (1999). *Department of Defense Design Criteria Standard: Human Engineering.* MIL-STD-1472F. Washington, DC: U.S. DOD.

Wikimedia.org. (2007). Motorola RAZR Keypad; available at *http://upload.wikimedia.org/wikipedia/en/thumb/1/10/RazrKeypad.jpg/180px-RazrKeypad.jpg.*

Wikipedia.org. (2007). Vacuum fluorescent display; available at *http://en.wikipedia.org/wiki/Vacuum_fluorescent_display.*

Williams, M., & Cowley, S. (2006). Update: PC market achieved double-digit growth in 2005. *InfoWorld*; available at *http://www.infoworld.com/article/06/01/19/74317_HNgrowthin2005_1.html.*

Wobbrock, J. O., Forlizzi, J., Hudson, S. E., & Myers, B. A. (2002). WebThumb: Interaction techniques for small-screen browsers. *Proceedings of Fifteenth Annual ACM Symposium on User Interface Software and Technology.* New York: ACM Press, 205–08.

Wright, S. L., & Samei, E. (2004). Liquid-crystal displays for medical imaging: A discussion of monochrome versus color. Medical Imaging 2004: Visualization, Image-Guided Procedures, and Display. *Proceedings of SPIE* 5367:444–55.

Yates, A., Etzioni, R., & Weld, D. (2003). A reliable natural language interface to household appliances. In *Proceedings of Eighth International Conference on Intelligent User Interfaces.* New York: ACM Press, 189–96.

Zwick, C., Burkhard, S., & Kühl, K. (2005). *Designing for Small Screens.* Switzerland: AVA.

Multimode Interfaces: Two or More Interfaces to Accomplish the Same Task

Aaron W. Bangor, James T. Miller

11.1 NATURE OF THE INTERFACE

Today, technology is providing consumer and business users with an increasing array of devices and software applications that can be used to organize, entertain, and inform. Because these technology solutions have become more ubiquitous, flexible, and capable, it is reasonable to ask if the user interface (UI) that allows the user to take advantage of the technology has kept pace. A couple of scenarios may help illustrate this.

In the first scenario, let us suppose that you are riding the bus home from work and, just as Murphy's Law might have predicted, you get stuck in traffic and will not be home until much later than you expected. You will, in fact, be so late that you are going to miss your favorite TV show. You did not set it up to record because you thought you would be home in plenty of time. No one is home to record it for you, so you now wish you had a way to record the show, such as calling your digital video recorder (DVR) from your cell phone to set up the recording.

Another scenario is that your best friend just had a big argument with her roommate and you want to call her to give her some support. She has moved to another place and emailed you her new phone number. You know you changed the phone number when you got a message from her, but you cannot remember where you changed it—in your e-mail address book, your cell phone speed dial, your voice dial list on your home phone, and/or your PDA. And you definitely do not want to call the old number and speak with the former roommate.

The first scenario points to users' need to be able to manage their services using a number of different devices, depending on the situation. The second scenario illustrates the frustrations that many users can experience when these needs are met, but without forethought about how the multiple devices that provide the same functionality might work together.

That is the topic of this chapter—how to design applications that provide users with more than one UI modality, so that they can interact with the same application and do the same tasks with whatever device and interface are most convenient at the time. For example, in the first scenario the visual on-screen interface for the home television is sufficient for the most common user situations. However, another visual interface accessed by a computer, a wireless phone, or PDA would be useful in other situations. It might even be better to provide an auditory interface via a phone so that the user would not be forced to enter the relevant information but could just speak it.

In the second scenario, the multiple modalities are already present, but the design of each device was done without considering that the device—and the information used by the device—might also apply to several other devices used to accomplish the same tasks. Ideally, the information common to all of these devices should be available to all of them so that a change in the information for one device would result in a change for all.

The nature of the individual UIs that are a part of multimode UIs (MMUIs) is, for the most part, not different from the other UI modalities discussed in this book. And much has been researched and written about single-user, single-mode UIs compared to MMUIs (Kray, Wasinger, & Kortuem, 2004). However, the focus of this chapter is on the unique nature of designing UIs for the same application where the user is expected to use multiple modalities to interact with the same service and accomplish the same tasks. Research requirements, criteria to consider when electing to create multiple modalities, and design and testing methods are included in this chapter.

As a start, Seffah, Forbrig, and Javahery (2004) provide the following elements that define an MMUI:

◆ Allows a single or a group of users to interact with server-side services and information using different interaction/UI styles.
◆ Allows an individual or a group to achieve a sequence of interrelated tasks using multiple devices.
◆ Presents features and information that behave the same across platforms, even though each platform/device has its specific look and feel.
◆ Feels like a variation of a single interface for different devices with the same capabilities.

Note that these elements are technology independent and focus on a user—or group of users—accomplishing tasks, although there may be variations on how this is implemented, depending on the exact device and UI technology. That is not to say that an MMUI is implemented without considering the specific UI technologies available (Paterno, 2004).

Typically, an MMUI is implemented using visual and/or auditory UIs. Other chapters of this book cover these in more detail, including speech (Chapter 6), touch-tone (Chapter 7), and small-screen visual displays (Chapter 10). Chapter 12 also has useful information on designing for more than one modality, although it covers integrating more than one modality into the same UI, rather than providing multiple, yet separate, modalities. There are numerous sources available for how to design visual user interfaces. Please see the end of this chapter for general references regarding graphical UIs (GUIs) (Carroll, 2003; ISO, 1998; Thatcher et al., 2006; Weinschenk, Jamar, & Yeo, 1997).

11.2 TECHNOLOGY OF THE INTERFACE

Any UI technology can be a candidate for an MMUI. Generally some combination of graphical, web, touch-tone, speech recognition, and small-screen visual UIs is considered when designing MMUIs. It is possible that other modalities, such as tactile, olfactory, and gustatory, could be used, but these generally have not been considered because they lack the powerful input/output capabilities that the visual and auditory technologies have.

Other chapters of this book discuss the technologies used in nonvisual UI technologies and many other works can provide the same for visual UIs. However, there are still some important technology considerations for MMUIs.

First, it should be understood that with MMUIs there is a greatly increased chance of simultaneous interaction between application and user or multiple users on the same account—either intentionally or unintentionally. While these situations occur with single-modality UIs, greater attention needs to be given to them in the case of MMUIs because of the multiple access methods.

Major concerns in these scenarios follow:

✦ Ensure accurate presentation of information to the user because of updates made in other UIs.
✦ Prevent user confusion about displayed information or application behavior because of changes made through other UIs that changed the state of the information or application.
✦ Establish the precedence given to changes made to the account or account data from the different UIs.
✦ Prevent data corruption or lost work when accessing the same data from more than one UI.
✦ Allow the user to quickly switch between UIs to accomplish a task or a series of tasks, if it is appropriate to the task(s).

As an example of the first consideration, think of a situation in which changes to an address book on one device are propagated to other devices that have address books.

If all devices are not quickly updated as soon as a change to an address has been made, and the user tries to use the address on another device, then the subsequent selection of that address may not be accurate. Additional confusion may occur, depending on whether the user recognizes that the update has not yet occurred. Consequently, it is important that information be updated in a timely manner so a user can trust that changes made using one device will be available on all devices.

Confusion about data or application behavior because of changes made in other UIs can happen in many situations. For instance, in a voice mail application that can be reached over the Web or the phone, one member of a household might listen to a message streamed over the Web and then delete it. At about the same time another member of the household is checking messages from her cell phone and hears that there is a new message. However, when she chooses to listen to the message, there will be no message to play because it has been deleted.

Typically, the precedence for changes to an account or account data from different UIs is whichever is the most recent. So, in the previous example, usually the household member who gave the command to delete or save the message first would establish the precedence for this specific set of information. However, other criteria for precedence could be established based on user needs and contexts of use for the application. For example, if there are master and secondary accounts, changes made by the master user would override changes by the secondary account users.

Preventing data corruption or lost work when accessing the same data from more than one UI can be critical. There are few situations more disheartening and frustrating for the typical user who has just spent time and effort accomplishing a task, only to see it lost. Worse yet is when a user thinks he had successfully finished a task, than finding out later that he did not. For example, a case worker in the field calls into an automated system to make updates to a patient's file immediately after an in-home visit. She makes changes and hangs up, believing that the file has been updated. Meanwhile, her procrastinating coworker is at their office computer and just getting around to completing updates from the previous week's visit to the same patient. The coworker's changes are made later and overwrite the changes of the first case worker. In this case, a lockout system to prevent accessing the same record at the same time could prevent this scenario, or an application that could track a history of changes and the individuals who made them could help.

Although most of these scenarios have been about preventing negative consequences of a user or users accessing the application through multiple UIs, MMUIs can be beneficial, if taken advantage of, for the appropriate tasks. A user may want the ability to repeatedly switch back and forth between the UI modalities to more quickly or easily accomplish a task. For example, although the primary method for setting up a voice mail application is a setup wizard on a website, it may be much more convenient for the user to record greetings using a phone interface at the appropriate time. In this case, accomplishing the task (setting up the voice mail interface) uses the appropriate interface modality to facilitate the components of the task.

Another major technological consideration for MMUI applications is the underlying software and systems that enable the UI. Although not directly related to UI design, the supporting infrastructure can have a major impact on the quality of user interaction with the application. This is a practical consideration for any UI design, but it is even more important for MMUIs. This is because there is a far greater possibility for a disjointed user experience if an MMUI is created by stitching together several independently developed UIs to create the multimodal application. If the requirement to develop the necessary infrastructure to support MMUIs is adhered to, then it is likely that the resulting MMUIs will be characterized by descriptions like "seamless," "fully integrated," and "converged." If not, then these terms will only be buzzwords for applications that will ultimately disappoint users. Consequently, good MMUI design is inextricably linked to understanding the ability of the underlying systems to aggregate, store, and provide data to multiple UIs in a consistent, timely, and device-independent manner.

To help understand these complex relationships among users, contexts of use, multiple UIs, and underlying infrastructure, Figure 11.1 depicts an example of how an application that uses multiple UIs might be implemented.

The uniqueness of designing an MMUI is factored in from the beginning with consideration of the user (although sometimes there are multiple users) and their different contexts of use and tasks. Because of this, there may be a need to interact with multiple devices and, consequently, multiple UIs, although some contexts of use may limit such interaction. These devices interact with various software applications to provide service functionality. Depending on the implementation, the same application may support only one device/UI, but it could support more than one. If there are multiple applications, their ability to communicate with each other is a factor in the overall level of integration that the user experiences. Finally, applications network with back-end systems (e.g., databases) to carry out user requests. Again, some implementations may use multiple back-end systems, and their ability to synchronize data also will have an impact on the user.

11.3 CURRENT IMPLEMENTATIONS OF THE INTERFACE

The prevalence of MMUIs is much less than single-mode UIs. Seffah, Forbrig, and Javahery (2004) suggest that this may be due to the "difficulty of integrating the overwhelming number of technological, psychological, and sociological factors" needed to create a good application. But while not as prevalent as single-modality UIs, the use of more than one interface modality to complete the same task or set of tasks is more common than one might think. For example, checking on the showtimes for a movie can easily be done over the phone (using speech or even

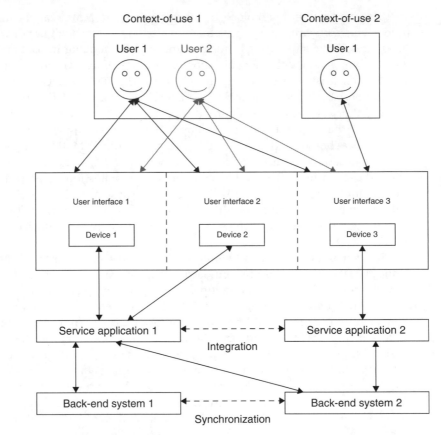

FIGURE

11.1

Example of an MMUI.

Interrelationship among users, contexts of use, multiple UIs, and the underlying infrastructure for an example MMUI.

touch-tone), on the Web, and using a small-screen device like a cell phone or PDA. In fact, information retrieval is currently one of the most common uses of MMUIs.

11.3.1 Information Portals

Web portals, such as Yahoo!, are hubs on the Internet that collect and organize some of the vast array of information sources and streams available to web users (Figure 11.2). Over the phone, voice portals perform much the same function, offering users the ability to hear information about news, stocks, weather, sports scores, movie listings, phone number lookup, and information on other topics that a user would want to know at any time or place. The latter motivation is

FIGURE

11.2

Traditional information portal for the Web.

The portal organizes disparate sources of information for the user. (Courtesy Yahoo! Inc.)

one of the reasons that voice portals have traditionally been associated with wireless phone services. However, the lower information-processing bandwidth over the phone generally lends itself to far fewer topics in voice portals, as compared to Web portals.

11.3.2   Automated Self-Help

Although some companies, such as Amazon.com, have established their business with little or no direct human interaction, many traditional companies with major call centers have been substituting automated self-help for the traditional care provided by customer service representatives in order to reduce company costs. The relatively high cost of providing live-agent call center support for a wide array of products and services, especially in a 24/7 environment, has resulted in the use of automated systems to provide support for tasks related to billing, payments, technical support, account inquiries, and sales.

Many of the early automated self-help applications were touch-tone interactive voice response (IVR) systems (see Chapter 7 for more details on designing this kind of application). The reason for this, simply, was that customers were already calling and the simplest way to engage them in an automated interaction was to handle their task over the phone. Nowadays, many of these tasks are also being integrated into the company's website. Customers can now call or visit a website to pay a bill, order products and services, get technical support, and do other routine tasks.

11.3.3   Address Books

Users have address books in myriad locations in the electronic world. They have them in a GUI environment within their e-mail programs and instant messengers. Some are in cell phones for voice dialing and text messaging. Others are in online accounts on the Web for shopping. The diverse tasks that benefit from having an address book have driven the need to have a single, synchronized address book that can be accessed from almost any end point: computer, mobile device, telephone, or TV.

11.3.4   Message Mailboxes

Handling messages, such as e-mail, short message service (SMS) text, instant messages (IMs), voice mails, and faxes, has become a part of everyday life. Messages allow people to keep in contact even when they are not available at that moment. This need for ubiquity, combined with very different origins for each type of message, has led to the use of MMUIs to better handle and, more importantly, expand their usefulness by giving users more flexibility in how and when they message other people.

11.3.5   Remote Access to Services

In the business setting many network managers have to leave their desks to troubleshoot, but need to take their access to services with them. Others in business need mobility in order to do their jobs, but need access to the same resources that they have in their office. And, increasingly, home automation is providing more capabilities for the home that can be remotely controlled and/or monitored. In all of these cases, the user has a normal environment in which a robust set of applications or services is available to them, but also need to move away from a wired device. This has resulted in a corresponding need to remotely access these services. The usual medium of interaction, typically a GUI or full-screen website, is not portable in these cases and so other interfaces, such as phone-based IVRs or small-screen devices, become a good alternative.

Figure 11.3(a) shows a common task—checking the weather radar—as a user might see it when using a PC with a broadband connection. Figure 11.3(b) shows what might be seen when using a mobile wireless device for the same task.

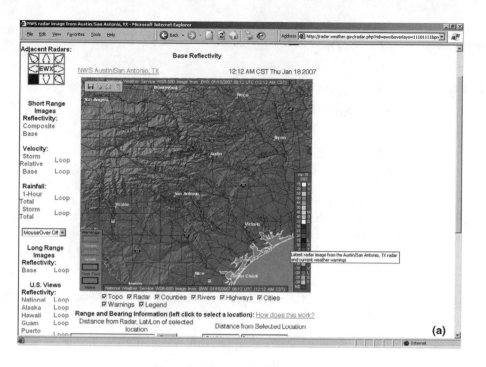

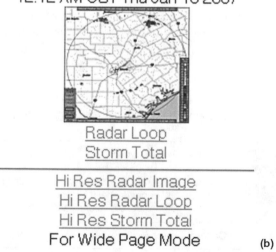

FIGURE 11.3   Checking the weather radar.

(a) Using a PC with a broadband connection. The weather radar from the National Weather Service on the website has numerous options and a large display.

(b) Using a mobile wireless device. The weather radar from the National Weather Service for mobile devices has few options and a small display. (Courtesy NOAA/NWS.)

11.3.6   Applications of the Interface to Universal Design and Accessibility

It is worth taking some time to discuss the impact of MMUIs on people with disabilities. Having a disability, whether inherent (e.g., vision or hearing difficulties, dyslexia, etc.) or due to one's circumstances (e.g., inability to see secondary tasks while driving, to operate small controls while wearing gloves, etc.), can present barriers to the use of computer technology. However, computer technology can be one of the great enablers for people of all abilities. When the design has considered the full range of user capabilities and limitations, this is called *universal design*. And the degree to which someone can use the design is part of accessibility.

A major tenet of universal design and accessibility is flexibility of user input/output (ANSI/HFES, 2005). The distinguishing characteristic of MMUIs is that, by their very nature, they offer multiple input/output modalities to the user.

One good example of MMUIs helping to achieve a more universal design is the use of touch-tone commands in conjunction with a speech-enabled UI (Figure 11.4). Whether offered concurrently as a backup (although this is more of a combined interface, as discussed in Chapter 12) or as an optional setting, offering these two UI technologies can greatly improve the performance for a wide range of users.

The use of speech recognition helps those with a cognitive disability who would have trouble remembering the mapping of touch-tone options played in a menu, while another user with limited use of the hands (either because they have

	System	Okay, which are you calling about: your bill, an order, or technical support?
	Caller	*<says nothing>*
	System	I'm sorry, I didn't hear. If you're calling about your bill, say "my bill" or press one. If it's about an order, say "order" or press two. Otherwise, say "technical support" or press three.
	Caller	<presses the 3 key>

FIGURE

11.4

MMUI example.
This is a brief caller exchange with an automated system that demonstrates the use of both speech and touch-tone input options.

a disability or are wearing gloves or holding something) can use speech to provide the input. In contrast, offering touch-tone input as an option helps a user with a speech impairment or a difficult accent, and provides other users who are in a noisy environment or have a need for privacy, the ability to respond to the application. Offering both modes of interaction enables as many users as possible to interact with the application in the manner that works best for them.

While MMUIs should not be thought of as a means to circumvent making each UI accessible, the inclusion of multiple modalities for interaction greatly increases the flexibility for users to interact in ways that match their particular capabilities and limitations.

11.4 HUMAN FACTORS DESIGN OF THE INTERFACE

Beginning to design an application that uses more than one UI modality is based on the assumption that more than one is needed. Often this is based on a belief that because it *can* be done it *should* be done. As with the choice of an interface modality and technology for any application, the expected users and their needs, tasks, and contexts of use first need to be identified and evaluated.

11.4.1 Research Requirements before Choosing the Interface Modality

The user-centered design process begins with users, not technology. This is the core belief of user-centered design, but it is mentioned here because when dealing with the myriad UI technologies and the complex technological infrastructure needed for MMUIs, it is even more likely that this point will be lost.

The first step in the design of MMUIs is the same as designing any other UI: Define the user population. This process defines the characteristics of the potential users, including their capabilities and limitations and the environments in which the application is expected to be used. This process includes considering people with disabilities. Once the target users have been defined, then user research can begin.

While doing the usual needs and contextual analysis, be aware that some user needs may be specific to different environments in which the application will be used—for example, checking e-mail from the office while at one's desk and checking e-mail on a crowded subway to and from the office. Some needs may be based on the goals that users might want to accomplish using the proposed application.

Once the needs are well defined and understood, application features can be added, removed, or refined. Matching these features with user needs creates a set of common goals for the user and the application. Next, the usual task analysis should be performed to determine how users accomplish these goals and to look for opportunities to improve existing methods. The task analysis should also be used to determine which goals can or cannot be accomplished using specific interface modalities. For example, setting up a group of address book contacts having a common property, such as your daughter's soccer team, may be relatively easy to do with a visual interface, keyboard, and mouse, but very difficult or impossible to do using a telephone keypad. However, it may be possible to edit a team member's e-mail address using speech recognition.

There is some research into the area of development considerations particular to MMUIs. For example, Kray, Wasinger, and Kortuem (2004) identify a taxonomy of management (e.g., conflict resolution among users and/or tasks), technical (e.g., synchronization), and social issues (e.g., privacy), but note that "they are often hard to classify as due to overlapping categories." Seffah, Forbrig, and Javahery (2004) discuss the use of various models for development, including task model–based development, UI pattern–driven development, and device-independent development languages. They found that pattern models do help with understanding the full complexity of tasks and the interrelationships particular to MMUIs. To assist design, Paterno (2004) describes an authoring environment for MMUIs. Some of its key components are grouping, relation, ordering, and hierarchy.

While these techniques are not peculiar to MMUIs, the overriding theme is that the design and development of applications using MMUIs are complex and require a planned, methodical approach. To prevent becoming overwhelmed when designing and developing an MMUI, the technique used is less important than whether it is thorough.

Critically evaluating these points will provide a basis for an informed decision about what modality and associated technology can best enable the users to perform their tasks in the expected situations.

11.4.2 Choosing One or More User Interface Modalities

Once background research and analysis have been conducted, then it is appropriate to select of one or more UI modalities. The questions in Table 11.1 help guide the decision of whether multiple UIs are needed, although they do not make up an exhaustive list.

If the conclusion is that multiple modalities are warranted, then choosing which modalities to use is the next step.

TABLE 11.1   Questions to Aid in Choosing an MMUI

Evaluation Question	Explanation	Example
Can more than one modality be cost-justified?	Designing and developing more than one user interface is costly. Can the added expense be balanced with a greater customer base, greater use of the service, increased customer retention, and so forth? (See Bias and Mayhew [2005].)	Some customers want to pay their bill over the Web, but others call to pay their bill and could use an IVR. Offering both will automate more bill payments overall, saving money and increasing revenue collection.
Can all of the tasks be performed in only one modality?	Based on the task analysis, can all of the tasks be adequately performed using only one modality? If so, there may not be a compelling reason to use multiple modalities.	All voice mail tasks can be done over the phone using touch-tones.
Would more than one interface modality directly fulfill user needs?	Sometimes one user interface modality cannot support all of the needs and tasks of the user population.	Sending a long e-mail in festive colors to an entire list of party guests with numerous details and a map may call for a website or GUI. Alternatively, sending a quick e-mail to a friend letting them know you are stuck in traffic and will be 15 minutes late may call for something mobile and simple, like a cell phone or PDA UI.
Do the contexts of use and/or environments necessitate different modalities?	When users perform the same task (or set of tasks) in different situations they may not have the same needs, so the same modality may not work, depending on where they are and what they are doing.	A tool to help customers resolve problems with their Internet service offers both web-based help for those who can still access the Internet and over-the-phone help for those whose Internet connections are totally down.
Will the new application's functionality be composed of myriad tasks currently performed by different modalities?	Are the user needs and tasks currently handled by different applications using different modalities that users will not easily give up?	An application that provides movie listings uses a web user interface and a telephone user interface because some users have always used one of these and do not want to be forced to change.

(Continued)

TABLE 11.1   *Cont'd*

Evaluation Question	Explanation	Example
Are the users so diverse so as to warrant multiple modalities?	Some applications are narrowly focused with what they do and so their users tend to be almost as narrowly focused. However, other applications have broad appeal and their users are diverse.	An address book is used by some people only in conjunction with sending e-mails to a close circle of family and friends. Others use their address book to keep track of all business acquaintances, including multiple phone numbers, e-mail addresses, street addresses, personal details, and so forth.
Do the capabilities and limitations of the users preclude using any modality?	Will users with disabilities, either inherent or based on their situation, have trouble using one or more of the modalities? If so, are the other modalities accessible to them?	Someone with hand tremors or someone who is driving can interact using speech rather than pressing keys on a phone to check messages.

11.4.3   Selecting Appropriate User Interface Modalities

Sometimes the choice of UI modality may be clear-cut, and sometimes further analysis is necessary to ensure that the right tool is being used for the job. This section provides some of the relative strengths and weaknesses of the UI modalities (Table 11.2) that are typically used together to form MMUIs. Please see the other chapters of the book and the references at the end of this chapter for a more exhaustive list of the strengths and weaknesses of the various UI modalities.

There are other strengths and weaknesses associated with each modality, but Table 11.2 should provide a good starting point for the trade-offs involved when deciding which are appropriate for a particular application.

11.4.4   Task Allocation to Multiple Modalities

Once the appropriate UI modalities have been selected for the application, the next step is to assign which tasks will be performed with which modality. Just as human factors have traditionally performed function allocation (see Wright, Dearden, & Fields, 2000; Kantowitz & Sorkin, 1987) for the human operator and machines, task allocation among the selected UI modalities needs to be

TABLE 11.2   Strengths and Weaknesses of UI Modalities

User Interface	Strengths	Weaknesses
GUI/Web	Uses the high-bandwidth visual channel with a large screen. Usually not time dependent. Can provide multimedia interaction.	Hardware associated with these UIs usually is not easily portable. Usually requires vision and manual dexterity for the user to interact with it, unless designed with accessibility in mind.
Small-screen devices	Easily portable. Typically enabled for audio input and output.	Limited screen real estate to work with. Input methods are usually more difficult. Small size makes it difficult to see and manipulate.
Touch-tone IVR	Recognition of user input is nearly perfect. Not susceptible to background noise (Paterno, 2004). Lends itself to simplified menu choices and linear tasks (Lee & Lai, 2005).	Input options are constrained and have to be mapped to arbitrary numbers. Entering characters other than numerals is difficult. Contexts of use or disability can prevent the use of hands for input.
Speech-enabled IVR	A natural means of communication for most users (Cohen et al., 2005). Better at performing nonlinear tasks (Lee & Lai, 2005). "Hands-free" and good for when the visual channel is overloaded (Paterno, 2004).	Recognition of user input is not 100%. Susceptible to noisy environments (Cohen, Giangola, & Balogh, 2005). Not all users can produce recognizable speech.

performed. This allocation should be based on the needs and tasks of the users, the reasons behind selecting multiple modalities, and the strengths and weaknesses of those modalities to fulfill the needs of the users.

In most cases, all tasks will be assigned to each of the UI modalities. The most important benefit of this practice is that users have full choice of which modality they can use for any task that the application can perform. This can be very important for universal design objectives. It also eliminates the need for users to remember whether a specific function is supported within the current UI and avoids the accompanying confusion that arises if a user is in the middle of a task and then realizes that he needs to stop and switch modalities. An illustration of this might be a bill payment application that allows payment with a credit card, debit card, checking account, or gift card. If both a touch-tone IVR and a website permit all forms of payment, the user is free to choose to call or go online to pay a bill. However, if the IVR cannot accept gift cards, then the user may call, enter his account number, the amount he would like to pay, and only then find out that the gift card he wanted to use will not work and he must hang up and visit the Web.

Although there are many advantages to implementing all functions in all of the UIs, this strategy is often costly. In addition, there are tasks that may not be appropriate for every UI modality. Furthermore, including all features from GUIs and websites (i.e., high information-processing bandwidth modalities [Wickens, 1992]) into small-screen devices or IVRs that use audio (i.e., lower information-processing modes) may result in user overload in the latter. Figure 11.3(a) and (b) (on page 367) illustrate this point. While the weather map for the full website has controls to add or remove numerous features, the weather map for mobile devices picked only those that are needed for the essential task and removed the customization controls.

Whatever the choice, it must be made during the early stages of analysis and design and then tested for validation as a workable solution.

11.4.5   Perform Detailed Design, Test, and Iterate Design

The last steps in the design process for MMUIs are not much different than in any other user-centered design process. The inputs from the previous human factors activities provide the basis for detailed UI design, the design must be tested with representative users, and the design iterated based on user input. While the steps may be the same, considerable effort must be devoted to these steps.

One special consideration during the design process for MMUIs is that, to the extent possible, all UIs should be designed at the same time. Doing this helps to ensure that there is consistency among the UIs from the very beginning. Each design "should take into account the specific capabilities and constraints of the device while maintaining cross-platform consistency and universal usability" (Seffah, Forbrig, & Javahery, 2004). In addition, the learning process that occurs while designing task flows for one UI can be transferred to the other UIs. Concurrent design also helps to ensure that the user needs supported by one UI do not marginalize the needs supported by the other UIs. If this cannot be done, then the predominant UI should be designed first. The predominant UI can be determined based on which UI will support the highest-frequency tasks and any other tasks that are critical to the success of the application (e.g., setup).

While performing concurrent design, Seffah, Forbrig, and Javahery (2004) proposed the following usability factors specific to MMUIs:

+ Various interaction styles among UIs
+ Information scalability/adaptability to various devices and UIs
+ Task structure adaptability
+ Consistent application behavior

✦ User awareness of trade-offs due to technical capabilities and constraints
✦ Application availability
✦ Overall "inter-usability"

The last point is one Seffah, Forbrig, and Javahery (2004) defined as "horizontal usability." Either name is suitable, but the key point is that it is a dimension specific to MMUIs. It is important to understand that while the usability of the UIs may be sufficient separately, to truly know the usability of an MMUI, all UIs must be assessed together to determine their integrated user friendliness.

Another special consideration for MMUIs is that all the UIs should be jointly user-tested. For these studies, exposing all participants to all of the designs provides the most flexibility, allowing the same participant to make direct comparisons among the various UIs being tested. This also allows the designer to see which UI, if any, is initially picked by the users or to see if users will spontaneously abandon one of the UIs for another when completing tasks.

Beyond the usual benefits of user testing (Hix & Hartson, 1993), concurrently testing the UIs helps in several ways. It can be used to validate the task allocation that was performed, either by evaluating whether the same task in all UIs can be accomplished or whether the choice to allocate a task to only one UI is appropriate or too limiting. Depending on the reason for needing more than one UI, joint testing provides realistic scenarios for participants. In many cases, user preferences for accomplishing tasks with particular UIs can be very helpful when considering application defaults and further understanding user behavior.

Finally, just as concurrently designing the UIs offers one the ability to learn from the experience with one modality and apply lessons to the other modalities, the same is true for the data collected during user testing. Having participants perform the same task in all of the modalities allows one to observe inconsistencies in the design, as well as the strengths and weaknesses of the UI modalities themselves.

11.4.6   A Special Case: Adding a Modality to an Existing Application

Sometimes the design process for a multimode application does not start from the beginning. There are times when an application has already been developed that uses one modality, but where a decision has been made to add another mode of interaction. As mentioned before, the decision may be based on a misguided belief that because another UI can be done, it is a good idea to do it. Other times the reasons for adding another UI are based on the types of data mentioned in the previous section: users' needs, their contexts of use, and/or the environments in which they are currently using the application.

One of the opportunities in this situation that may not be as available when designing from scratch is that the users, their contexts of use, and tasks are already available for study. At some point after deployment of an application, studies of the application may uncover significant gaps in its usefulness to the users or its usability. Analyzing how to address these problems may show that iterative improvements to the original UI are sufficient. Other times, however, the original modality may not be the best choice for the best solutions and the selection of another modality is appropriate. But as with designing all the modalities initially, the decision to use another modality—and which one to use—should be based on data about the users, what they need, and how they perform their tasks.

Task allocation among modalities for an MMUI designed from the beginning is somewhat different than when adding another modality to an existing application. The crux of the difference is that one of the UIs has already been designed and thus has already been allocated its tasks. The principal design problem for the new UI is determining what other tasks should be included in the new UI. There is no definite answer, but time and cost considerations usually mean that only the tasks that would make the necessary improvements to the original application are included in the new modality. In addition, tasks allocated to the original UI could be added or removed, but usually not at this point. Alternatively, a complete task allocation can be performed, treating the inclusion of a new UI as a chance to perform a complete redesign. This is usually more time consuming and costly, but sometimes a thorough redesign is needed to achieve the desired improvements.

The other major difference between developing an MMUI from scratch and adding another modality after application deployment is that joint design is not possible. In this case, the new UI should be designed in light of the existing modality, understanding that the original UI probably will not fundamentally change but also recognizing that many users are accustomed to the existing interface and that any changes to that interface are likely to have negative consequences.

Joint testing of the original and new UIs should take place in the same way as for a totally new application, but with one caveat: Both existing users and new users with no knowledge of the application should be participants in these studies. Perspectives of both user groups are very important because while it is important that current customers are not alienated by the new design, it is also crucial that new users are attracted to the improved application. When testing current users, pay attention to any negative transfer to the new UI from their experiences with the existing one, especially regarding consistency. With users who have no knowledge of the application—new or old—concentrate on whether the addition of the new UI meets the requirements that were established for adding the new modality.

This early testing should be designed to decide whether the task allocation is correct, especially if the minimalist approach was taken due to time and cost. Moreover, it will also point out whether the original UI needs to be significantly changed to accommodate the new modality. Finally, as is the case with designing from scratch, this testing, redesign, and retesting should continue until a satisfactory design has been achieved.

11.4.7   Summary of Design Steps for Multimode User Interfaces

Steps for analyzing and designing multiple UI modalities are

1. Define the user population, including people with disabilities.
2. Determine capabilities and limitations of users.
3. Collect user needs, including specific contexts of use and environments.
4. Match user needs with system functionality.
5. Perform task analysis.
6. Evaluate the need for more than one UI.
7. Select the specific interface modality(s) based on strengths and weaknesses of each.
8. Assign tasks/functionality to all modalities or only some of them.
9. Concurrently perform the detailed design of the UIs.
10. Test with representative users, collecting data about usability/accessibility for tasks and modalities.
11. Iterate design and retest as needed.

11.5   TECHNIQUES FOR TESTING THE INTERFACE

Because this chapter does not discuss a unique interface but rather a combination of interfaces, the testing techniques described here will focus on testing an application that uses multiple UIs and how to test whether the correct UIs are being used rather than how to test a particular modality. See other chapters in this book for a discussion of non-GUI/Web testing techniques, or the references at the end of this chapter for discussions of testing GUI/Web UIs.

When evaluating MMUIs, many of the standard usability evaluation techniques are used. Early testing will generally be done with prototypes of the application and fewer participants; the focus is on finding ways to make the UI designs better. Later testing will generally be with a fully developed application and have more participants; the goal here will be to iron out details and begin to predict overall user performance. In both cases, comparative techniques are especially important. As with any application, the most important testing is that done early and iteratively, with representative users.

11.5.1   Early Testing

The primary goals of early testing are to validate the choice of modalities and any task allocation that has been done. The first round of testing ideally should be conducted while the detailed UI design process is under way. At this stage, testing with throwaway (or "exploratory") prototypes is best because they can be built and changed quickly, even during testing. For visual interfaces, these prototypes are usually "wireframes" or simple clickable prototypes created in HTML or rapid prototyping languages. For speech recognition systems, a Wizard of Oz technique can be very useful for creating quick mock-ups of early designs (Cohen, Giangola, & Balogh, 2005). (See also Chapter 6.) In this same way, even touch-tone IVRs can be prototyped (i.e., by having the caller say the key they want to press).

The essential theme for this testing is that the earlier representative users have a chance to provide meaningful, valid input into the development process. The closer these early interfaces resemble the target UI design, the greater the likelihood that the product will be designed with fewer iterations. This is especially important for MMUIs because with multiple UIs, any one UI will likely get fewer resources to make changes than if it were a stand-alone application. In addition, MMUIs typically have a more complex support infrastructure, whose own design cannot change easily.

Because fine details are not important at this stage, a small sample size of representative users is appropriate. The test should expose all of the participants to each of the design alternatives so direct comparisons can be made. The tasks for participants should be selected to focus on the ones most relevant to the ongoing detailed design, which include the following:

+ Most frequently performed
+ Most critical to the overall success of the application
+ Most fully explore the choice of a particular modality
+ Specifically allocated to only one of the UIs
+ Most likely to use more than one modality at a time or consecutively

Testing the most frequent and critical tasks is common in usability testing and will not be discussed here (Hix & Hartson, 1993). When testing MMUIs, the goal should be to choose a sample of tasks that allows the experimenter to observe the behavior with each of the UI modalities. Not only does this provide performance data for the tasks, but it also provides the opportunity to get a better sense of what does and does not work for that modality in the application, as well as a chance to ask participants which modality they preferred and why.

Tasks that have been allocated to only one modality, especially if those are tasks that can only be performed using that modality, should be thoroughly tested. These tasks are included because there is no alternative modality in the application for them and therefore they absolutely must work in the chosen modality. If not, one or more of the other modalities needs to be considered for the task.

Tasks that require using more than one modality at a time or consecutively may or may not be relevant to a particular application. Sometimes there are tasks that may require a user to interact with more than one modality at the same time (e.g., registering a small-screen device with the application while logged in to a GUI/Web UI) or consecutively (e.g., once setup using a telephone UI reaches a logical end point, the user may be required to use the Web to finish the installation). If this is the case, then these tasks should be tested early with the best prototypes available (even if the study participants are asked to "make believe" about some things) because these tasks can be very complex and the transition between modalities may not be clear. It is important to determine whether the interactions in both modalities are consistent with users' expectations and that the UIs in both modalities are consistent and meet the requirements for good usability. If this is not the case, then one or both interfaces may need to be redesigned, and this should be done as early in the design process as possible.

Based on the results of early testing, it is important to apply the results to the ongoing design process. Adding, changing, or removing functionality may be required, as well as revising the UI designs. At this point, even starting over and trying a new approach may be appropriate.

11.5.2   Later Testing

After early testing has helped shape the designs of the UIs, a workable design should be ready to be built. Testing on a semi-functioning system—either an evolutionary prototype or an early alpha version of the application—should take place as soon as possible. When this system is ready, the next major round of testing should take place.

There are two major goals for this round of testing: validating the UI designs that were based on early user testing and testing tasks that could not be easily or accurately tested with lower-fidelity prototypes. Again, standard usability techniques are used, although more participants are probably warranted here as the need grows to become more confident with the results.

The same type of tasks that were used in early testing are still applicable, especially the high-frequency and critical tasks. Tasks for exploring a modality and those assigned to a single modality are less critical because that part of the design needs to be mostly settled at this point but still should be validated.

Some of the new tasks important in this round of testing include

+ Tasks that are heavily dependent on back-end systems or other dependencies that made them difficult or unreliable to evaluate in early testing
+ Tasks that can be performed on more than one UI

The former is for tasks that could not adequately be captured in early testing with lower-fidelity prototypes. In many cases with MMUIs, the user interaction is heavily dependent on integrated systems and/or dynamic data that cannot always be sufficiently replicated in early testing. This is the time to test users with those tasks.

This is also where in the testing process to determine the relative performance of the same task with multiple UIs. This can be important when assessing the relative strengths and weaknesses of the various UIs of the application, for setting application defaults, and for projecting usage patterns.

A test during which all participants experience all of the experimental options is still appropriate at this point. This time the direct comparisons are usually made between performance on the same task with different modalities, although if alternative designs are available at this stage, a comparative evaluation is also useful. However, if there are numerous tasks with several UIs, a test in which various groups of participants perform diverse tasks with different modalities may be the only practical experimental design at this point. With multiple UI modalities to be factored in, the test size can quickly grow out of hand when all participants do all of the tasks.

Up to this point the user testing could be done just about anywhere because the focus was on validating the fundamentals of the UI designs. But often the needs and tasks of the users that drove the need for multiple UI modalities are unusual and/or varied. In these cases a more naturalistic setting may be necessary. The context of use and the environments in which the users are expected to use the application need to be considered as part of the testing methodology. For example, if a UI for a small-screen device is part of the application because of a strong need for mobility/ubiquity, then some of the testing should be conducted in an environment that replicates the expected uses that originally required the use of multiple UIs.

11.5.3   Inclusion of People with Disabilities in User Testing

During all phases of design and testing the inclusion of the needs and input of people with disabilities is important. Overlooking this part of design and testing is often accompanied by the excuse of trying to get most of the UI right for nondisabled users, with the promise of making the necessary adjustments to accommodate the disabled user later. This belief is even less true with MMUIs because the multiple modes often are a boon to people with disabilities. Moreover, MMUIs tend to be used more often in diverse contexts of use and environments, creating situations that are functionally equivalent to a disability (e.g., not being able to hear because of a noisy environment).

People with disabilities not only provide feedback during testing about accessibility features, but frequently are more susceptible to marginal designs and can uncover usability problems that would not necessarily be uncovered with a small sample size. As such, it is valuable to consider these needs during design, but it is also important to understand that the improvements to the design resulting from user testing with this group improve the overall user-centered process.

11.6 DESIGN GUIDELINES

Multimode UIs are extremely varied. Because they are composed of more than one UI, they are generally more complex and yet have all of the same issues as their component UIs. The sum can be greater than its individual parts, but only if done right.

Although there is no recipe for how to design a good MMUI, this section lists and describes selected guidelines on ways to improve the chance for creating a useful, usable, and accessible one.

11.6.1 Only Choose MMUIs If There Is a Compelling, Real, and Beneficial Need

Having multiple UI modalities for the same application makes almost everything about the design and development more difficult, complex, and costly. And if the UI designer is not careful, this burden can be transferred to the users. The goal of design is to let the users benefit from the choice of multiple UI modalities, but not let them suffer because of it. To do that, first be judicious about the decision to use multiple UI modalities and about which are chosen. If the data support the need to have an MMUI (i.e., based on the user population and analysis of their needs and contexts of use) is strong enough, then the UI design should follow that course.

Example: A diverse user population, including individuals with vision and hearing loss, need to be able to find out the showtimes for numerous movies at multiple theaters, whether they are at home, work, or in the car. Based on these varied users and contexts of use, web and speech recognition IVR UIs are selected.

11.6.2 Choose Modalities Whose Strengths and Weaknesses Match the Needs and Tasks of the Intended User Population

The entire purpose of having more than one UI modality for an application is to take advantage of the capabilities that multiple modalities provide. Consequently, users' needs and the tasks they perform that cannot adequately be handled by one modality must be assigned to a modality whose capabilities are a good fit.

Example: For a movie showtimes application, a speech recognition IVR is chosen over touch-tone because there are numerous movie names, so saying the name of the movie is much easier than selecting from multiple hierarchical menus.

11.6.3   Concurrently Design User Interfaces

As discussed in Section 11.4, the analysis that a human factors specialist conducts before the design of the MMUI is irrespective of having more than one UI modality and any particular technology. Once the need for more than one modality is established, the tendency to silo their design and development must be resisted. The multiple UIs may have different tasks, design documents, development teams and timelines, and different technologies, but the human factors specialist must champion the fact that it is still a user performing tasks with a single application. To help prevent false assumptions and design errors and to bolster consistency, the high-level and detailed design of the UIs should be executed at the same time.

> *Example:* For the movie showtimes application, design documents for a web UI (e.g., wireframes) and speech recognition IVR UI (e.g., call flows) are created at the same time.

11.6.4   Conduct Iterative User Testing Early and Often

This is a valuable guideline for any UI design, but it is especially important when designing an application with multiple UIs. Not only is it necessary to make more than one UI usable and accessible, but the interaction between the user and the UIs goes from being a simple back-and-forth style to a multifaceted, multimode interaction whose interplay is very difficult to predict. This amount of design work necessitates multiple iterations of user testing to ensure optimal performance on any of the interfaces, but it is also critical to do it early because there is even more infrastructure to an MMUI application that cannot easily be changed if development has progressed too far. On the positive side, though, early testing with users likely will provide unique and valuable perspectives about how to make the UIs work better together because the participants will be less focused on the UI technology and more on what they want to do.

> *Example:* The user testing of a movie showtimes application begins with a paper prototype of a web UI and a Wizard of Oz prototype for the speech recognition IVR. Results of these tests are used to refine the designs, either by improving and then retesting the low-fidelity prototypes or by creating higher-fidelity prototypes based on the findings.

11.6.5   Jointly Test Multiple User Interfaces as Early as Possible

Just as it is important to concurrently design the UIs, it is important to jointly test them. And as with any user-centered design process, generally the earlier and more often testing is done, the better. Joint testing helps to provide realistic scenarios for performing tasks. It also helps to provide data about user preferences for accomplishing certain tasks with the various UIs. Lastly, the lessons learned from testing one UI can be transferred to the other UIs within the product to validate whether they are appropriate in all cases.

Example: A messaging application that combines voice mails, faxes, and e-mails is tested using a prototype of a web UI and a touch-tone IVR UI. Participants perform the same tasks with both UIs, such as setting up a new mailbox, checking for a voice mail from a coworker, deleting an e-mail, changing the voice mail greeting, and setting up new message notification. The relative performance and user preference for each are collected for comparison.

11.6.6 Be Consistent among the Interfaces to the Greatest Extent Possible Given Limits of Modalities

Consistency is vital for any UI, but for an MMUI it is critical. "Emphasize robustness and consistency even more than in traditional interface design" (Kray, Wasinger, & Kortuem, 2004). It is important that the user not be required to learn—in fact or in perception—how to use different UIs for the same application. The overriding philosophy in designing tasks for all of the UIs is to make them as similar as possible. Although it can be tempting to optimize tasks for each UI modality, in most cases the value to the user from cross-learning among all of the application's UIs, the predictability of consistency, and the quicker formation of an accurate mental model usually outweighs the gain from adapting tasks to each modality.

The caveat, however, is that the different UI modalities are, in fact, different (after all, more than one was needed), and the various capabilities and constraints of each modality and UI technology have to be considered (Seffah, Forbrig, & Javahery, 2004). Consequently, the implementation of functionality cannot be exactly the same for each modality. Occasionally, it will be clearly better to design a task specific to a single UI knowing that the same design cannot be applied to another UI. Nevertheless, the goal is to achieve a Pareto-optimal design among the UI modalities, which is much easier to achieve if concurrent design of the UIs is done.

Example: An automated billing application has the same task flow for the web UI and for a speech recognition IVR by first presenting the bill amount and due date, and then prompting for the payment information in the same order.

11.6.7   Use Consistent Labels among Modalities

One of the most important dimensions of consistency among the different UI modalities is to use the same names and descriptions for the same items, functions, data, and even the modalities themselves. By doing this, users will not only be able to cross-learn the modalities, but will also have a level of comfort with the application because, even though their tasks may look and/or sound different among the UIs, the same terms are used to refer to the various components of the application.

Example: The term "cell phone" is used for the web, IVR, and small-screen UIs as well as the user documentation. The terms "mobile phone" or "wireless phone" are not used anywhere within the application.

11.6.8   Adhere to Conventions of Modalities Used While Maintaining Consistency with Other Modalities in the Application

When designing more than one UI modality for an application, there can be such a thing as too much consistency among the UIs. As discussed earlier, consistency helps users to learn and transfer knowledge from one UI to another. However, each UI modality has its own strengths, weaknesses, idiosyncrasies, and sometimes expectations of the user about how it works (Seffah, Forbrig, & Javahery, 2004). Sometimes being consistent with the rest of the application will produce a suboptimal design for one of the UIs. In those cases, it may be more effective to create a UI for one modality that is not the same as the other modalities.

Example: When checking voice mail messages over the phone, the new messages are first played together, followed by all saved messages. When viewing a list of voice mail messages on the web, the new and saved messages are shown mixed together, in reverse chronological order.

11.6.9   Do Not Make Users Reenter Information across Modalities

If the user enters information, including user data, preferences, and so on, then that information should be available when the application is accessed via any of the modalities. Not only does this save the users time and reinforce the fact that they are using one application (rather than several applications cobbled together), but it also prevents having different versions of the same data within the application.

Example: The user enters his favorite sports teams into a web portal so that he can see a customized list of scores, which the voice portal also uses to create a list of scores.

11.6.10   Make Sharing and Updating of User Data among User Interface Modalities Transparent to Users

One of the major benefits of an MMUI is that it supplants multiple applications each with its own UI. But to achieve the benefits of a single application, the data cannot be segmented or handled on a per-modality basis. If the application needs to use multiple databases or other data stores, then the application needs to unobtrusively perform the necessary transactions so as to appear as if it were a single application to the user.

Example: When a user changes the phone number for one of her contacts in her cell phone's address book, the phone automatically sends a message to a centralized database for updating, which in turn schedules it for synchronization with her PDA.

11.6.11   Provide Global Preferences That Apply to All Modalities

Some user preferences should apply to the entire application, regardless of which UI is being used. This practice saves the user time and reduces complexity because any changes need to be made in only one place and there are fewer overall settings to track.

Example: There is a user setting to employ the same language (e.g., English) in all UIs.

11.6.12   Judiciously Provide Preferences That Apply to One or Only Some Modalities

Although in most cases it is beneficial to have global preferences, due to the nature of the various modalities there is sometimes a need to have special preferences for a single modality. In addition, preferences with different settings may be useful, depending on which modality is being used.

Example: By default when viewing stock quotes on a web portal, the opening price, daily high, daily low, 52-week high, 52-week low, closing price, change in price, and percentage change are all shown for each stock. Alternatively, by default when listening to stock quotes over the phone via a speech recognition IVR, only the closing price and change in price are given, with an option to hear more details. The user is also given the preferences about which pieces of information to display on the web portal and the IVR.

11.6.13   Use a Multimode User Interface to Enhance Universal Design and Accessibility, Not as a Substitute for Making Each User Interface Modality Accessible

Providing more than one modality for user interaction can greatly benefit people with disabilities, as well as all users who need or want to use the application in restricted contexts of use. But just because one type of disability may not often use one of the modalities, it does not mean that reasonable effort should not be made to make all of the UI modalities accessible. Doing so will mean that people with disabilities are not limited to one modality, but may choose the modality that best suits their needs at the time that they use the application.

> *Example:* In a messaging application, web and speech recognition IVR UIs are used to fulfill the needs of users and give them more flexibility over how they interact with their messages. However, the *W3C Web Content Accessibility Guidelines* are followed for the website to make it accessible to blind users with screen readers, rather than forcing them to use only the speech recognition IVR.

11.7   CASE STUDY

A case study of an MMUI can be found at *www.beyondthegui.com*.

11.8   FUTURE TRENDS

In the future it is likely that the use of MMUIs for the same application will grow. The growth probably will not be phenomenal because of the complexity and cost involved in developing them. However, their increased use probably will be surprising because of the innovative ways they will be used to fulfill the needs of users.

In fact, the growth will be spurred by the growing need of users for both mobility and flexibility. As the need for information and services while on the go increases, new methods of supplying the information in the future will be required, including wireless broadband data and the maturation of the technology used in wearable computing. This, in turn, will be driven by a larger range of tasks that people will be able to perform and also by the greater diversity of people and contexts of use.

One of the key drivers of flexibility will be the increased focus on universal design and accessibility. To some extent this will be driven by civil rights legislation for people with disabilities, such as the Americans with Disabilities Act, Section 255 of the Telecommunications Act, Section 508 of the Rehabilitation Act, and similar legislation around the world. But from the other side, consumer demand will grow for more inclusive designs. This demand is mostly going to come from the aging population, as Baby Boomers age and longer life expectancies increase the relative number of older consumers.

This fundamental change for almost all applications for the user population will make accessibility issues part of the mainstream design process. As these needs, both regulatory and market driven, come to the forefront, UI design will need to provide consumers with products that meet those needs. As has been discussed earlier in this chapter, MMUIs are not a substitute for making UIs accessible. However, when selected properly, they provide greater flexibility and choice over how users interact with an application, which can be a boon for people with disabilities and others with temporary impairments or reduced abilities because of their environment.

From a technical standpoint, the growth in demand for applications with more than one UI modality needs to be—and is being—supported by a corresponding improvement in technology to enable MMUIs. The most important of these advancements is the increasing use of standards. The widespread adoption of standards helps in two important ways. First, making data device independent results in easier manipulation by any UI technology. For example, eXtensible Markup Language (XML) can be rendered on the Web using hypertext markup language (HTML) or with an IVR using VoiceXML (Paterno, 2004). It also enables back-end system interoperability, which helps to achieve much of the "seamlessness" that is critical for the ease of use of MMUIs.

The other major technological change that is occurring to increase the development of MMUIs is in computing technology itself. As computing power rapidly grows, with an opposite drop in price and size, the demand of consumers for access to information and services at home or work (where a richer interaction can be designed) and on the go (where mobility restricts interaction techniques) will increase. Context-dependent and "migratory" UIs that exploit different modes of input will be a major driver of the use of MMUIs (Paterno, 2004).

Even with this increase in computing ability and other technological developments, it is still likely that MMUIs will continue to be almost exclusively combinations of visual and auditory interfaces. Although the future no doubt holds improvements in our understanding and innovations with the other senses, this is not likely to be enough in the near term to be sufficient for a stand-alone component of an MMUI.

One final thought about the future of MMUIs is in order. As technology becomes more ubiquitous and the tools to build these interfaces become more widely available, there will likely be an increase in competition. When the technology and features are no longer the defining reasons for a purchase and price points converge, consumers will demand—and start making purchasing decisions based on—the ease of use of the product or service. The well-trained and data-driven human factors professional can be the bridge from grand ideas of what an application can be to making it simple and enjoyable for people to use.

REFERENCES

American National Standards Institute & Human Factors and Ergonomics Society. (2005). *Human Factors Engineering of Software User Interfaces*. Santa Monica, CA: Human Factors and Ergonomics Society.

Bias, R. G., & Mayhew, D. J. (2005). *Cost-Justifying Usability: An Update for the Internet Age*, 2nd ed. San Francisco: Morgan Kaufmann.

Brooke, J. (1996). SUS: A "quick and dirty" usability scale. In Jordan, P. W., Thomas, B., Weerdmeester, B. A., & McClelland, A. L., eds., *Usability Evaluation in Industry*. London: Taylor and Francis.

Carroll, J. M., ed. (2003). *HCI Models, Theories, and Frameworks: Toward a Multidisciplinary Science*. San Francisco: Morgan Kaufmann.

Cohen, M. H., Giangola, J. P., & Balogh, J. (2005). *Voice User Interface Design*. Boston: Addison-Wesley.

Hix, D., & Hartson, H. R. (1993). *Developing User Interfaces: Ensuring Usability Through Product & Process*. New York: John Wiley & Sons.

International Standards Organization. (2006). *Ergonomics of Human–System Interaction—Part 110: Dialogue Principles*. Geneva: ISO.

International Standards Organization. (1998). *Ergonomic Requirements for Office Work with Visual Display Terminals (VDTs). Part 11: Guidance on Usability*. Geneva: ISO.

Kantowitz, B. H., & Sorkin, R. D. (1987). Allocation of functions. In Salvendy, G., ed., *Handbook of Human Factors*. New York: Wiley.

Kray, C., Wasinger, R., & Kortuem, G. (2004). Concepts and issues in interfaces for multiple users and multiple devices. *Proceedings of Workshop on Multi-User and Ubiquitous User Interfaces (MU3I) at IUI 2004*, Funchal, Madeira, Portugal, 7–11.

Laurel, B. (1990). *The Art of Human–Computer Interface Design*. Reading, MA: Addison-Wesley.

Lee, K. M., & Lai, J. (2005). Speech versus touch: A comparative study of the use of speech and DTMF keypad for navigation. *International Journal of Human–Computer Interaction* 19(3):343–60.

Paterno, F. (2004). Multimodality and multi-device interfaces. *W3C Workshop on Multimodal Interaction*. Sophia Antipolis, France: World Wide Web Consortium.

Raskin, J. (2000). *The Humane Interface: New Directions for Designing Interactive Systems*. New York: ACM Press.

Seffah, A., Forbrig, P., & Javahery, H. (2004). Multi-devices "Multiple" user interfaces: Development models and research opportunities. *Journal of Systems and Software* 73:287–300.

Seffah, A., & Javahery, H. (2004). *Multiple User Interfaces: Cross-Platform Applications and Context-Aware Interfaces*. Hoboken, NJ: Wiley.

Thatcher, J., Burks, M. R., Heilmann, C., Henry, S. L., Kirkpatrick, A., Lauke, P. H., Lawson, B., Regan, B., Rutter, R., Urban, M., & Waddell, C. D. (2006). *Web Accessibility: Web Standards and Regulatory Compliance*. Berkeley, CA: Friends of Ed.

Weinschenk, S., Jamar, P, & Yeo, S. C. (1997). *GUI Design Essentials*. New York: John Wiley & Sons.

Wickens, C. D. (1992). *Engineering Psychology and Human Performance*, 2nd ed. New York: HarperCollins.

World Wide Web Consortium. (2006). Web Content Accessibility Guidelines—*http://www.w3.org/WAI/intro/wcag.php*.

Wright, P., Dearden, A., & Fields, R. (2000). Function allocation: A perspective from studies of work practice. *International Journal of Human–Computer Studies* 52(2):335–55.

<table>
<tr><td>12</td><td rowspan="2"></td><td rowspan="2"># Multimodal Interfaces: Combining Interfaces to Accomplish a Single Task</td></tr>
<tr><td>CHAPTER</td></tr>
</table>

12 Multimodal Interfaces: Combining Interfaces to Accomplish a Single Task

CHAPTER

Paulo Barthelmess, Sharon Oviatt

12.1 NATURE OF THE INTERFACE

An essential distinguishing capability of multimodal systems is the use these systems make of two or more natural input modalities such as speech, handwriting, gestures, facial expressions, and other body movements (Oviatt, 2007). Using such systems, users may for example deal with a crisis management situation using speech and pen input over a map. An emergency response route can for example be established by sketching a line on a map along the desired route, using a digital pen or a Tablet PC stylus, while speaking "Create emergency route here" (Oviatt et al., 2003). In this example, a multimodal system analyzes the speech and sketching captured via sensors (a microphone and an instrumented writing surface), and interprets the user's intention by combining (or fusing) the complementary information provided via these two modalities. The result in this case is an interpretation that takes location information from the sketched line and the attributes of the object being created (that this is an "emergency route") from the speech.

Multimodal interfaces may take many forms and aspects (Figure 12.1). Systems may be hosted by small portable devices, such as on a PDA or cell phone that is taken to the field, or on a tablet computer used within offices or in cafeterias. Groups of people may interact multimodally via large interactive boards or sheets of digital paper. The common aspect in all these cases is the interface support for interaction via natural modes of communication involving combinations, for example, of speech, pen input, gestures, gaze, or other modalities.

12.1.1 Advantages of Multimodal Systems

The intrinsic advantage of multimodal systems is that they allow users to convey their intentions in a more expressive way, better matched to the way they naturally

(a)   (b)

FIGURE

12.1

Multimodal interfaces span a variety of use contexts.
(a) Mobile interface. (b) Collaborative interface.

communicate. A well-designed multimodal system gives users the freedom to choose the modality that they feel best matches the requirements of the task at hand. Users have been shown to take advantage of multimodal system capabilities without requiring extensive training. Given a choice, users preferred speech input for describing objects and events and for issuing commands for actions (Cohen & Oviatt, 1995; Oviatt & Cohen, 1991). Their preference for pen input increased when conveying digits, symbols, graphic content, and especially when conveying the location and form of spatially oriented information (Oviatt, 1997; Oviatt & Olsen, 1994; Suhm, 1998).

As a result of the choice of input they provide to users, multimodal systems make computing more accessible, lowering input requirement barriers so that a broader range of users can be accommodated (e.g., by allowing a user with a sensory deficit to use the modality she is most comfortable with). As a consequence, users become able to control a broader range of challenging, complex applications that they might not otherwise be able to command via conventional means. These factors lead to the strong user preference for interacting multimodally that has been documented in the literature. When given the option to interact via speech or via pen input in a map-based domain, 95 to 100 percent of the users chose to interact multimodally (Oviatt, 1997). While manipulating graphic objects on a CRT screen, 71 percent of the users preferred to combine speech and manual gestures (Hauptmann, 1989).

Error Reduction via Mutual Disambiguation

A rich set of modalities not only provides for enhanced freedom of expression, allowing users to choose the modalities that best fit their particular situation, but may also lead to better recognition when compared for instance to speech-only interfaces (Oviatt, 1999a). *Mutual disambiguation* is used to refer to the positive effect that recognition in one mode may have in enhancing recognition in another mode. Analysis of an input in which a user may say "Place three emergency hospitals here <point> here <point> and here <point>" provides a multimodal system with multiple cues indicating, for instance, the number of objects to be placed, as the spoken word "three" can be matched to the number of pointing gestures that are provided via another modality (Oviatt, 1996a). Similarly, misrecognized commands in each modality may in some cases (e.g., tutorial lectures) be corrected by exploiting redundancy.

Empirical results demonstrate that a well-integrated multimodal system can yield significant levels of mutual disambiguation (Cohen et al., 1989; Oviatt, 1999a, 2000). Studies found mutual disambiguation and error suppression ranging between 19 and 41 percent (Oviatt, 1999a, 2000, 2002).

Similar robustness is found as well in systems that integrate speech and lip movements. Improved speech recognition results can occur during multimodal processing, both for human listeners (McLeod & Summerfield, 1987) and systems (Adjoudani & Benoit, 1995; Tomlinson, Russell, & Brooke, 1996). A similar opportunity for disambiguation exists in multiparty settings—for instance, when communicating partners handwrite terms on a shared space, such as a whiteboard or a shared document, while also speaking these same terms. The redundant delivery of terms via handwriting and speech is something that happens commonly in lectures or similar presentation scenarios (Anderson, Anderson et al., 2004; Anderson, Hoyer et al., 2004; Kaiser et al., 2007). By taking advantage of redundancy when it occurs, a system may be able to recover the intended terms even in the presence of misrecognitions affecting both modalities (Kaiser, 2006).

Performance Advantages

Empirical work has also determined that multimodal systems can offer performance advantages (e.g., during visual-spatial tasks) when compared to speech-only processing or conventional user interfaces. Multimodal pen and speech have been shown to result in 10 percent faster completion time, 36 percent fewer task-critical content errors, 50 percent fewer spontaneous disfluencies, and shorter and simpler linguistic constructions (Oviatt, 1997; Oviatt & Kuhn, 1998). When comparing the performance of multimodal systems to a conventional user interface in a military domain, experiments showed a four- to nine-fold speed improvement over a graphical user interface for a complex military application (Cohen, McGee, & Clow, 2000).

Multimodal systems are preferable in visual-spatial tasks primarily because of the facilities that multimodal language provides to specify complex spatial information such as positions, routes, and regions using pen mode, while providing additional descriptive spoken information about topic and actions (Oviatt, 1999b).

12.2   TECHNOLOGY OF THE INTERFACE

Multimodal systems take advantage of recognition-based component technologies (e.g., speech, drawing, and gesture recognizers). Advances in recognition technologies make it increasingly possible to build more capable multimodal systems. The expressive power of rich modalities such as speech and handwriting is nonetheless frequently associated with ambiguities and imprecision in the messages (Bourguet, 2006). These ambiguities are reflected by the multiple potential interpretations produced by recognizers for each input. Recognition technology has been making steady progress, but is still considerably limited when compared to human-level natural-language interpretation. Multimodal technology has therefore developed techniques to reduce the uncertainty, attempting to leverage multimodality to produce more robust interpretations.

In the following, a historic perspective of the field is presented (Section 12.2.1). Sections 12.2.2 and 12.2.3 then present technical concepts and mechanisms and information flow, respectively.

12.2.1   History and Evolution

Bolt's "Put that there" is one of the first demonstrations of multimodal user interface concepts (1980). This system allowed users to create and control geometric shapes of multiple sizes and colors, displayed over a large-format display embedded in an instrumented media room (Negroponte, 1978). The novelty introduced was the possibility of specifying shape characteristics via speech, while establishing location via either speech or deictic (pointing) gestures. The position of users' arms was tracked using a device attached to a bracelet, displaying an x on the screen to mark the position of a perceived point location when a user spoke an utterance. To create new objects, users could for instance say "Create a blue square here" while pointing at the intended location of the new object. Such multimodal commands were interpreted by resolving the use of "here" by replacing it with the coordinates indicated by the pointing gesture. By speaking "Put that there," users could select an object by pointing, and then indicate with an additional gesture the desired new location (Figure 12.2). This demonstration system was made possible by the emergence of recognizers, particularly for speech, that for the first time supported multiple-word sentence input and a vocabulary of a few hundred words.

Early Years

Initial multimodal systems made use of keyboard and mouse as part of a traditional graphical interface, which was augmented by speech. Spoken input provided an alternative to typed textual input, while the mouse could be used to provide

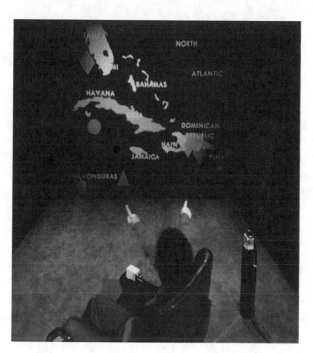

FIGURE

12.2

Bolt's multimodal demonstrator.
A user moves an object by pointing to it and then the desired new location while saying, "Put that there." *Source:* From Bolt (1980). (Courtesy ACM.)

pointing information. Examples of early systems include XTRA, a multimodal system for completing tax forms (Allegayer et al., 1989), Voice Paint and Notebook (Nigay & Coutaz, 1993), Georal (Siroux et al., 1995), MATIS (Nigay & Coutaz, 1995), CUBRICON (Neal & Shapiro, 1991), and ShopTalk (Cohen, 1992). Other systems of this period include Virtual World (Codella et al., 1992), Finger-Pointer (Fukumoto, Suenaga, & Mase, 1994), VisualMan (Wang, 1995), Jeannie (Vo & Wood, 1996), QuickTour, Jeannie II, and Quarterback (Vo, 1998), and a multimodal window manager (Bellik & Teil, 1993).

The sophistication of the speech recognition components varied, with some systems offering elaborate support for spoken commands. CUBRICON and Shop-Talk allowed users to ask sophisticated questions about maps and factory production flow, respectively. The interfaces of these systems activated extensive domain knowledge so that users could point to a location on a map (in CUBRICON) and ask "Is this an air base?" or point to a specific machine in a diagram (in ShopTalk) and command the system to "Show all the times when this machine was down."

Speech and Pen Systems

Systems started to move away from augmenting graphical interfaces by incorporating input with a full spoken-language system. During this time, equipment became available to handle pen input via a stylus, providing a natural second source of rich information that complements speech (Oviatt et al., 2000). Pen input made it possible for users to input diagrams and handwritten symbolic information, in addition to pointing and selection, replacing similar mouse functionality. The availability of more powerful recognition components led in turn to the development of new techniques for handling parallel, multimodal input.

Pen-based input over a tablet-like surface started to be used to provide additional commands via sketched gestures (e.g., in TAPAGE [Faure & Julia, 1994]). Pen input then evolved to provide a second semantically rich modality, complementing and reinforcing the spoken input. Handwriting recognition (e.g., in Cheyer and Julia, 1995) and more sophisticated gesture recognition (in MVIEWS [Cheyer & Luc, 1998] and Quickset [Cohen et al., 1997]) provided users with significant expressive power via input that did not require the keyboard or the mouse to be used, providing instead a flexible, truly multimodal alternative input via speech and pen.

Empirical work in the early 1990s (Oviatt et al., 1992) laid the foundation for more advanced speech and pen systems such as Quickset (Figure 12.3). Quickset provides a generic multimodal framework that is still currently used (e.g., by Charter [Kaiser et al., 2004], a system for collaborative construction of project schedules). One of the initial Quickset applications was a map-based military system (which evolved into a commercial application, NISMap [Cohen & McGee, 2004]) (see Section 12.3.3). Using this system, military commanders can enter battle plan fragments using multimodal language (Figure 12.3). More specifically,

FIGURE

12.3

Quickset/NISMap interface on a Tablet PC.
The keyboard and mouse are replaced by pen and speech input. Sketches and handwriting on a map are fused with spoken input.

users of the system can draw standard military symbols comprising sketched and handwritten elements, complemented by speech. For example, a commander can establish the position of a fortified line by sketching the line on a map and speaking "Fortified line."

MVIEWS (Cheyer & Luc, 1998) demonstrated multimodal functionality related to video analysis activities for military intelligence purposes. MVIEWS integrated audio annotations of video streams, allowing users to mark and describe specific targets of interest. By circling an object on a video frame using a stylus, the system could be commanded via speech to track the object in subsequent frames. Users could also request to be notified whenever an object selected via a stylus moved, or whenever activity within a region was detected (e.g., by speaking "If more than three objects enter this region, alert me.").

Other systems of the same generation include IBM's Human-centric Word Processor (Oviatt et al., 2000), Boeing's Virtual Reality (VR) Aircraft Maintenance Training (Duncan et al., 1999), Field Medic Information System (Oviatt et al., 2000), and the Portable Voice Assistant (Bers, Miller, & Makhoul, 1998).

The Human-centric Word Processor provides multimodal capabilities for dictation correction. Users can use a pen to mark a segment of text, issuing spoken commands such as "Underline from here to here <draw line>," or "Move this sentence <point to sentence> here <point to new location> (Oviatt et al., 2000). The Field Medic Information System is a mobile system for use in the field by medics attending to emergencies. Two components are provided—one allowing spoken input to be used to fill out a patient's record form, and a second multimodal unit—the FMA—that lets medics enter information using speech or alternatively handwriting to perform annotations of a human body diagram (e.g., to indicate position and nature of injuries). The FMA can also receive data from physiological sensors via a wireless network. Finally, the patient report can be relayed to a hospital via a cellular, radio, or satellite link.

More recently, tangible multimodal interfaces have been explored in Rasa (Cohen & McGee, 2004; McGee & Cohen, 2001). Rasa is a tangible augmented-reality system that enhances existing paper-based command and control capabilities in a military command post (McGee & Cohen, 2001). Rasa interprets and integrates inputs provided via speech, pen, and touch. The system reacts by projecting information over a paper map mounted on a board, producing spoken prompts, and inserting and updating elements in a database and military simulators.

Rasa does not require users to change their work practices—it operates by transparently monitoring users' actions as they follow their usual routines. Officers in command posts operate by updating information on a shared map by creating, placing, and moving sticky notes (Post-its) annotated with military symbols. A sticky note may represent a squad of an opposing force, represented via standardized sketched notation (Figure 12.4). As users speak among themselves while performing updates, sketching on the map and on the sticky notes placed on the map, Rasa observes and interprets these actions. The map is affixed to a

Users collaborating with Rasa.
                      Sketches on sticky notes represent military units, placed on specific locations on
a map. *Source:* From McGee and Cohen (2001). (Courtesy ACM.)

touch-sensitive surface (a SmartBoard), so that when sticky notes are placed on it, the system is able to sense the location.

The attributes of introduced entities are integrated by fusing the location information with the type information provided via the sticky note sketch. A sketch representing an enemy company may be integrated with a spoken utterance such as "Advanced guard," identifying type, location, and identity of a unit. Rasa confirms the interpretation via a spoken prompt of the kind "Confirm: Enemy reconnaissance company called 'advanced guard' has been sighted at nine-six nine-four." Users can cancel this interpretation or confirm it either explicitly or implicitly, by continuing to place or move another unit.

Speech and Lip Movements

A considerable amount of work has been invested as well in examining the fusion of speech and lip movements (Benoit & Le Goff, 1998; Bernstein & Benoit, 1996; Cohen & Massaro, 1993; Massaro & Stork, 1998; McGrath & Summerfield, 1985; McGurk & MacDonald, 1976; McLeod & Summerfield, 1987; Robert-Ribes et al., 1998; Sumby & Pollack, 1954; Summerfield, 1992; Vatikiotis-Bateson et al., 1996). These two modalities are more tightly coupled than speech and pen input or gestures. Systems that process these two highly correlated modalities strive to achieve more robust recognition of speech phonemes based on the evidence provided by *visemes* during physical lip movement, particularly in challenging situations such as noisy environments (Dupont & Luettin, 2000; Meier, Hürst, & Duchnowski, 1996; Potamianos et al., 2004; Rogozan & Deléglise, 1998; Sumby & Pollack, 1954; Vergo, 1998).

Comprehensive reviews of work in this area can be found elsewhere (Benoit et al., 2000; Potamianos et al., 2004).

Multimodal Presentation

The planning and presentation of multimodal output, sometimes called *multi-modal fission*, employs audio and graphical elements to deliver system responses across multiple modalities in a synergistic way. Some of the systems and frameworks exploring multimodal presentation and fission are MAGIC (Dalal et al., 1996), which employs speech and a graphical display to provide information on a patient's condition; WIP (Wahlster et al., 1993); and PPP (Andre, Muller, & Rist, 1996). SmartKom (Wahlster, 2006) is an encompassing multimodal architecture that includes facilities for multimodal fission as well.

Multimodal presentation sometimes makes use of animated characters that may speak, gesture, and display facial expressions. The study of the correlation of speech and lip movements discussed in the previous section is also relevant in informing the construction of realistic animated characters able to synchronize their lip movements with speech (Cohen & Massaro, 1993). Further work on animated characters has explored ways to generate output in which characters are able to gesticulate and use facial expressions in a natural way, providing for rich multimodal output (Nijholt, 2006; Oviatt & Adams, 2000). For overviews of the area, see, for example, André (2003) and Stock and Zancanaro (2005).

Vision-Based Modalities and Passive Interfaces

Facial expressions, gaze, head nodding, and gesturing have been progressively incorporated into multimodal systems (Flanagan & Huang, 2003; Koons, Sparrell, & Thorisson, 1993; Lucente, Zwart, & George, 1998; Morency et al., 2005; Morimoto et al., 1999; Pavlovic, Berry, & Huang, 1997; Turk & Robertson, 2000; Wang & Demirdjian, 2005; Zhai, Morimoto, & Ihde, 1999). Poppe (2007) presents a recent overview of the area.

Further incorporation of modalities is expected as vision-based tracking and recognition techniques mature. Of particular interest are the possibilities that vision-based techniques introduce a means for systems to unobtrusively collect information by passively monitoring user behavior without requiring explicit user engagement in issuing commands (Oviatt, 2006b). Systems can then take advantage of the interpretation of natural user behavior to automatically infer user state, to help disambiguate user intentions (e.g., by using evidence from natural gaze to refine pointing hypotheses), and so forth.

Eventually, it is expected that systems may take advantage of the information provided by multiple sensors to automatically provide assistance without requiring explicit user activation of interface functions. The ability to passively observe and analyze user behavior and proactively react (e.g., Danninger et al., 2005; Oliver & Horvitz, 2004) are of special interest for groups of users that may be collaborating, as elaborated in the next subsection.

Collaborative Multimodal Systems

Recently, considerable attention has been given to systems that interpret the communication taking place among humans in multiparty collaborative scenarios, such as meetings. Such communication is naturally multimodal—people employ speech, gestures, and facial expressions, take notes, and sketch ideas in the course of group discussions.

A new breed of research systems such as Rasa (McGee & Cohen, 2001), Neem (Barthelmess & Ellis, 2005; Ellis & Barthelmess, 2003), and others (Falcon et al., 2005; Pianesi et al., 2006; Rienks, Nijholt, & Barthelmess, 2006; Zancanaro, Lepri, & Pianesi, 2006) have been exploiting multimodal collaborative scenarios, aiming at providing assistance in ways that leverage group communication and attempt to avoid adversely impacting performance.

Processing unconstrained communication among human actors introduces a variety of technical challenges. Conversational speech over an open microphone is considerably harder to recognize than more constrained speech directed to a computer (Oviatt, Cohen, & Wang, 1994). The interpretation of other modalities is similarly more complex. Shared context that may not be directly accessible to a system is relied on very heavily by communicating partners (Barthelmess, McGee, & Cohen, 2006; McGee, Pavel, & Cohen, 2001).

Devising ways to extract high-value items from within the complex group communication streams constitutes a primary challenge for collaborative multimodal systems. Whereas in a single-user multimodal interface, a high degree of control over the language employed can be exerted, either directly or indirectly, systems dealing with group communication need to be able to extract the information they require from natural group discourse, a much harder proposition.

Collaborative systems are furthermore characterized by their vulnerability to changes in work practices, which often result from the introduction of technology (Grudin, 1988). As a consequence, interruptions by a system looking for explicit confirmation of potentially erroneous interpretations may prove too disruptive. This in turn requires the development of new approaches to system support that are robust to misrecognitions and do not interfere with the natural flow of group interactions (Kaiser & Barthelmess, 2006).

Automatic extraction of meeting information to generate rich transcripts has been one of the focus areas in multimodal meeting analysis research. These transcripts may include video, audio, and notes processed by analysis components that produce transcripts (e.g., from speech) and provide some degree of semantic analysis of the interaction. This analysis may detect who spoke when (which is sometimes called "speaker diarization") (Van Leeuwen & Huijbregts, 2006), what topics were discussed (Purver et al., 2006), the structure of the argumentation (Verbree, Rienks, & Heylen, 2006b), roles played by the participants (Banerjee & Rudnicky, 2004), action items that were established (Purver, Ehlen, & Niekrasz, 2006), structure of the dialog (Verbree, Rienks, & Heylen, 2006a), and high-level turns of a meeting

(McCowan et al., 2005). This information is then made available for inspection via meeting browsers (Ehlen, Purver, & Niekrasz, 2007; Nijholt et al., 2006). Some of this work still relies primarily on the analysis of speech signals, although some additional work focuses on incorporation of additional modalities.

Multimodal collaborative systems have been implemented to monitor users' actions as they plan a schedule (Barthelmess et al., 2005; Kaiser et al., 2004). Other systems observe people discussing photos and automatically extract from multimodal language key terms (tags) useful for organization and retrieval (Barthelmess et al., 2006). Pianesi and colleagues (2006) and Zancanaro and coworkers (2006) have proposed systems that analyze collaborative behavior as a group discusses how to survive in a disaster scenario. These systems passively monitor group interactions, analyze speech, handwriting, and body movements, and provide results after the observed meetings have concluded.

The Charter system (Kaiser et al., 2004, 2005), derived from the Quickset system (Cohen et al., 1997), observes user actions while building a Gantt chart. This diagram, a standard representation of schedules, is based on placing lines, representing tasks, and milestones, represented by diamond shapes, within a temporal grid. Users build a chart by sketching it on an instrumented whiteboard or a Tablet PC (Figure 12.5). Sketch recognition identifies the graphical symbols. A multimodal approach to label recovery (Kaiser, 2006) matches handwritten labels to the redundant speech in the temporal vicinity of the handwriting, as users say, for instance, "The prototype needs to be ready by the third week <write prototype next to milestone symbol>." The mutual disambiguation provided by this redundant speech and handwriting accounts for up to 28 percent improvement in label recognition in test cases (Kaiser et al., 2007). Once the meeting is over, the system automatically produces a MS-Project rendition of the sketched chart.

The Charter system has been more recently extended to support remote collaboration by letting users who are not colocated visualize sketches made at a remote site and contribute elements to the shared Gantt chart, by using a stylus on a Tablet PC or sheets of digital paper (Barthelmess et al., 2005). The support for remote collaboration also incorporates a stereo vision–based gesture recognizer—MIT's VTracker (Demirdjian, Ko, & Darrell, 2003). This recognizer identifies natural gestures made toward the whiteboard on which the Gantt chart is sketched, and automatically propagates the gesture as a halo so that remote participants become aware of pointing performed at remote sites.

The multimodal photo annotator described by Barthelmess, Kaiser, and colleagues (2006) exploits social situations in which groups of people get together to discuss photos to automatically extract tags from the multimodal communication taking place among group members. These tags are representative terms that are used to identify the semantic content of the photos. Tags can be used to perform clustering or to facilitate future retrieval. By leveraging the redundancy between handwritten and spoken labels, the system is able to automatically extract some of these meaningful retrieval terms (Kaiser, 2006; Kaiser et al., 2007).

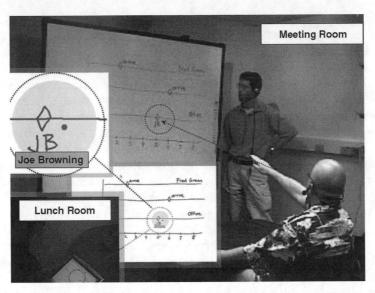

FIGURE

12.5

A group of colocated and remote participants (*inset*) using Charter to define a project schedule.

Sketches over a shared space are propagated to remote users, who can add their contribution. Charter monitors this interaction and produces an MS-Project chart at the end of the meeting.

Automatic tag extraction addresses the reluctance many users have of spending time performing manual annotation of digital photos.

Multimodal group communication has also been used to analyze group behavior in meetings (McCowan et al., 2005; Pianesi et al., 2006; Rienks & Heylen, 2005; Wainer & Braga, 2001; Zancanaro et al., 2006). Pianesi and Zancanaro and respective colleagues have been exploring a multimodal approach to the automatic detection and classification of social/relational functional roles assumed by people during meetings. Detected roles do not correspond to the hierarchical roles participants may have in an organization, but rather reflect attitudes during a meeting. Two main categories are identified: task and social-emotional. Whereas the former has to do with facilitation and coordination of task performance (e.g., the definition of goals and procedures), the latter is concerned with the relationship among group members (e.g., moderation of group discussions or cooperative attitude toward other participants).

Using information about speech activity (who is talking to whom at each moment) and localized repetitive motions (fidgeting) obtained from a visual channel, the behavior of participants is classified within five task roles and five social–emotional roles (Zancanaro et al., 2006). The ultimate product of this system is a semiautomated "coach" that presents to participants episodes of the meeting during which dysfunctional behavior is detected (e.g., dominant or aggressive behavior), with the goal of improving meeting participation behavior over time by allowing participants to reflect upon their own actions.

Other work looks into the role of speech amplitude, lexical content, and gaze to automatically detect who the intended addressee is in interactions involving groups of people and computational assistants (Jovanovic, op den Akker, & Nijholt, 2006; Katzenmaier, Stiefelhagen, & Schultz, 2004; Lunsford, Oviatt, & Arthur, 2006; Lunsford, Oviatt, & Coulston, 2005; van Turnhout et al., 2005). Speech amplitude is found to be a strong indicator of who participants intend to address in situations in which a computational assistant is available (Lunsford et al., 2005, 2006). Directives intended to be handled by the computer are delivered with amplitude significantly higher than speech directed to human peers, as indicated by studies of users engaged in an educational-problem–solving task. Leveraging such a technique, a system is able to automatically determine whether specific spoken utterances should be interpreted as commands requiring a response from the system, separating these from the remaining conversation intended to be responded to by human peers. This open-microphone engagement problem is one of the more challenging but fundamental issues remaining to be solved by new multimodal collaborative systems.

12.2.2 Concepts and Mechanisms

A primary technical concern when designing a multimodal system is the definition of the mechanism used to combine—or *fuse*—input related to multiple modalities so that a coherent combined interpretation can be achieved. Systems such as Bolt's (1980) "Put that there" and other early systems mainly processed speech, and used gestures just to resolve x, y coordinates of pointing events. Systems that handle modalities such as speech and pen, each of which is able to provide semantically rich information, or speech and lip movements, which are tightly correlated, require considerably more elaborate fusion mechanisms. These mechanisms include representation formalisms, fusion algorithms, and entirely new software architectures.

Multimodal fusion emerges from the need to deal with multiple modalities not only as independent input alternatives, but as also able to contribute parts or elements of expressions that only make sense when interpreted synergistically (Nigay & Coutaz, 1993). When a user traces a line using a pen while speaking "Evacuation route," a multimodal system is required to somehow compose the attributes of the spatial component given via the pen with the meaning assigned to this line via speech.

A well-designed multimodal system offers, to the extent possible, the capability for commands to be expressible through a single modality. Users should be able to specify meanings using only the pen or using just speech. It must be noted, though, that specifying spatial information via speech is not preferred by users in general (Oviatt, 1997; Oviatt & Olsen, 1994; Suhm, 1998). It may be required in situations in which users are constrained to using speech (e.g., while driving).

Early and Late Fusion

In order to be able to integrate interpretations generated by multiple disparate unimodal recognizers, these interpretations need to be represented in a common semantic representation formalism. A multimodal integration element then analyzes these uniform semantic representations and decides how to interpret them. In particular, it decides whether to fuse multiple individual elements, for example a spoken utterance and some pointing gestures or handwriting, or whether such input should be considered to represent individual commands that should not be composed into a combined interpretation. In the former case, we say that a *multimodal* interpretation was applied, and in the latter that the interpretation was *unimodal*.

Two strategies that have been explored to deal with the combination of modalities are late, or *feature fusion*, and early, or *semantic fusion*.

Early-fusion techniques are usually employed when modalities are tightly coupled, as is the case for instance for speech and lip movements. Early fusion is achieved by training a single recognizer over a combination of features provided by each of the modalities. Such features may be composed from some representation of phonemes from a speech modality and *visemes* characterizing lip movements provided by a visual modality. The advantage of early fusion is that parallel input from each modality is used concurrently to weight interpretation hypotheses as the recognition takes place. This results in many cases in enhanced recognition accuracy, even when one of the signals is compromised (e.g., speech in noisy environments) (Dupont & Luettin, 2000; Meier et al., 1996; Potamianos et al., 2004; Rogozan & Deléglise, 1998; Sumby & Pollack, 1954; Vergo, 1998).

Late- or semantic-fusion techniques employ multiple independent recognizers (at least one per modality). A sequential integration process is applied over the recognition hypotheses generated by these multiple recognizers to achieve a combined multimodal interpretation. Late fusion is usually employed when two or more semantically rich modalities such as speech and pen are incorporated.

This approach works well in such cases because each recognizer can be trained independently over large amounts of unimodal data, which is usually more readily available, rather than relying on typically much smaller multimodal corpora (sets of data). Given the differences in information content and time scale characteristics of noncoupled modalities, the amount of training data required to account for the wide variety of combinations users might employ would be prohibitively large. A system would for instance need to be trained by a large number of possible variations in which pen and speech could be used to express every single command accepted by a system. Given individual differences in style and use conditions, the number of alternatives is potentially very large, particularly for systems supporting expressive modalities such as pen and speech.

By employing independent unimodal recognizers, systems that adopt a late-fusion approach are able to use off-the-shelf recognizers. That makes it possible for late-fusion systems to be more easily scaled and adapted as individual recognizers are replaced by newer, enhanced ones. In the rest of this discussion, we concentrate primarily on late-fusion mechanisms.

12.2.3 Information Flow

In general terms, in a multimodal system multiple recognizers process input generated by sensors such as microphones, instrumented writing surfaces and pens, and vision-based trackers (Figure 12.6). These recognizers produce interpretation hypotheses about their respective input.

Temporal Constraints

Temporal constraints play an important role is deciding whether multiple unimodal elements are to be integrated or not (Bellik, 1997; Johnston & Bangalore, 2005; Oviatt, DeAngeli, & Kuhn, 1997). In most systems, a *fixed temporal threshold* determines, for instance, whether a spoken utterance is to be fused with pen gestures, or whether multiple pen gestures should be fused together. In general, if input from different modalities has been produced within two to four seconds of each other, integration is attempted. In practical terms, that means that systems will wait this much time for users to provide additional input that might be potentially used in combination with a preceding input.

While this approach has been successfully used in systems such as Quickset (Cohen et al., 1997) and MATCH (Johnston & Bangalore, 2005), its use introduces delays in processing user commands that equal the chosen threshold. More recently, learning-based models have been developed that adapt to user-specific thresholds (Huang & Oviatt, 2005). By employing user-specific models that take into account the empirical evidence of multimodal production, system delays are shown to be reduced 40 to 50 percent (Gupta & Anastasakos, 2004; Huang, Oviatt, & Lunsford, 2006).

Response Planning

In general, a multimodal integration element produces a potentially large number of interpretations, some of which may be multimodal and others may be unimodal. The most likely interpretation hypothesis is then chosen as the current input interpretation, and the other less likely ones are removed from consideration. This choice is sometimes influenced by a *dialog manager*, which may exploit additional contextual information to select the most appropriate interpretation (Johnston & Bangalore, 2005; Wahlster, 2006).

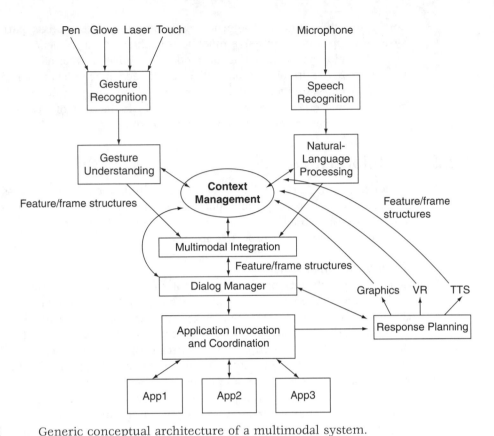

FIGURE

12.6

Generic conceptual architecture of a multimodal system.
Multiple recognizers are used for gesture and speech input. *Source:* From Oviatt
et al. (2000).

Once an interpretation is chosen, a system response may be generated. Response planning depends very heavily on the specifics of each system and its domain and intended functionality. A response may consist of multimodal/multimedia display during which users are presented with graphical and audio/spoken output on a computer or within a virtual-reality environment, sometimes embodied as an animated character (Cassell et al., 2000; Nijholt, 2006). Other responses may include interacting with another application, such as updating a military planning information system or driving simulation (Cohen & McGee, 2004; Johnston et al., 1997), updating an MS-Project schedule (Kaiser et al., 2004), or supporting remote collaboration (Barthelmess et al., 2005).

A considerable amount of work on planning and presentation has been pursued by systems and frameworks such as MAGIC (Dalal et al., 1996), WIP (Wahlster et al., 1993), PPP (Andre, Muller, & Rist, 1996), and SmartKom (Wahlster, 2006).

12.3 CURRENT IMPLEMENTATIONS OF THE INTERFACE

Even though few commercial multimodal applications are available, multimodality has been explored in a variety of different areas and formats, including information kiosks, mobile applications, and ultramobile applications based on digital-paper technology. Multimodal interfaces have also been used to promote accessibility. This section examines some representative examples of interfaces in each of these areas and form factors.

12.3.1 Information Kiosks

Kiosks providing a variety of services, such as tourist and museum information, banking, airport checking, and automated check-out in retail stores, are becoming more prevalent. The majority of these devices use touch and keypad as data entry mechanisms, producing responses via graphical displays and sometimes speech and audio. Given the requirements for robustness and space constraints, keyboards and mice are not usually available (Johnston & Bangalore, 2004). Multimodal interfaces provide additional means of interaction that do not require keyboard or mice as well, and may therefore provide an ideal interface option for kiosks. In this section, some representative examples are presented.

MATCHKiosk

MATCHKiosk (Johnston & Bangalore, 2004) is a multimodal interactive city guide for New York City and Washington, DC, providing restaurant and subway/metro information. The kiosk implementation is based on the mobile MATCH multimodal system (Johnston et al., 2002).

This interactive guide allows users to interact via speech, pen, and touch. Responses are also multimodal, presenting synchronized synthetic speech, a life-like virtual agent, and dynamically generated graphics (Figure 12.7). The system helps users find restaurants based on location, price, and type of food served. A user may ask for information using speech, as in "Find me moderately priced Italian restaurants in Alexandria." The same query can be expressed multimodally, for instance by speaking "Moderate Italian restaurants in this area <circle area on the map>," or by using just the pen, for instance by circling an area in the map and handwriting "Cheap Italian." It also provides subway directions between locations, for instance, when a user asks, "How do I get from here <point to map location> to here? <point to map location>," or just circles a region with the pen and handwrites "Route."

System output combines synthetic speech synchronized with the animated character's actions and coordinated with graphical presentations. The latter may

408

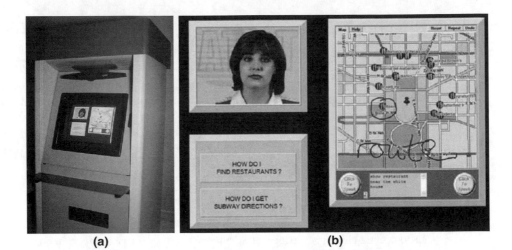

(a)   (b)

FIGURE
12.7

MATCHKiosk.
(a) Hardware. (b) User interface. *Source:* From Johnston and Bangalore (2004).
(Courtesy Association of Computational Linguistics.)

consist of automatic panning and zooming of the map portion of the display, such
as showing route segments.

A variety of mechanisms are implemented to let users fine-tune their queries
and correct misrecognitions. An integrated help mechanism is also available, and
is automatically activated when repeated input failures are detected.

SmartKom-Public

SmartKom is a flexible platform that includes facilities for handling multimodal input and output generation (Wahlster, 2006). SmartKom-Public is a kiosk
instantiation (Figure 12.8) of the architecture (Horndasch, Rapp, & Rottger, 2006;
Reithinger et al., 2003; Reithinger & Herzog, 2006). This kiosk, mounted inside a
telephone booth, provides gesture and facial recognition via cameras, microphone,
graphical display, and audio output.

The kiosk provides information about movies in the city of Heidelberg,
Germany, as well as communication facilities, such as document transmission.
This discussion will focus on the movie information functionality. A user may initiate a dialog with the system by speaking, "What's playing at the cinemas tonight?"
The system responds by showing a listing of movies, as well as a map displaying
the locations of the theaters (Figure 12.9 on page 410). An animated character—
Smartakus—provides spoken information such as "These are the movies playing
tonight. The cinemas are marked on the map." Users can then ask for specific information multimodally—"Give me information about this one <point to the display>
for example"—causing the system to display the text showing details

SmartKom-Public.

This platform is a kiosk instantiation mounted in a telephone booth. From *http://www. smartkom.org/eng/project_en_frames.pl?public_en.html*. (Courtesy German Research Center for Artificial Intelligence GmbHo.)

of the selected movie. The system also provides ticket reservations, supported by dialogs in which users may choose seats. Finally, the system is able to provide information about how to get to a specific location, such as a particular movie theater.

12.3.2   Mobile Applications

Multimodal interfaces provide a natural way to overcome the intrinsic limitations of displays (usually small) and input mechanisms of mobile devices. The ability to use speech or pen is also important in accommodating the variety of mobile use contexts. Speech can be used, for instance, while walking or driving; a pen interface

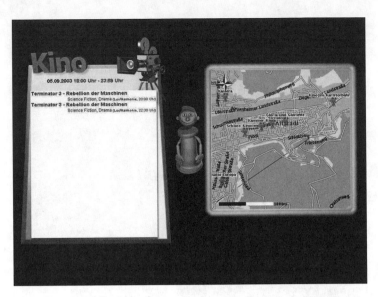

FIGURE

12.9

SmartKom-Public interface.
The interface presents a listing (in German) of the locations showing
Terminator 3. Source: From Reithinger and Herzog (2006); courtesy of
Springer.

may provide the privacy required in public settings in which speech is not appropriate. The challenge in developing mobile multimodal interfaces is in building
systems that are compatible with the low-powered processing units found in most
mobile devices. The following two interfaces illustrate the characteristics and
approaches used by mobile multimodal systems.

Microsoft MiPad

Microsoft developed a multimodal mobile application—MiPad (Multimodal Interactive Notepad, *http://research.microsoft.com/srg/mipad.aspx*) (Deng et al.,
2002; Huang et al., 2001)—that demonstrates pen and speech input on a portable
digital assistant (PDA). Pen input is used to select icons and to activate voice
recognition via a "tap-to-talk" software button (Figure 12.10).

The system interprets spoken commands, and is able to initiate e-mail, set up
appointments, and manage contact lists. Using this system, users are able to say,
for example, "Send mail to Nicky," causing the system to open up an e-mail dialog
with the recipient already filled out (K. Wang, 2004). The user can then dictate the
content of the message. Users can also tap with the stylus on a form field, such as
the "To" field of an e-mail. This provides the system with contextual information
that will help the system select interpretations, as in, for example, preferring
"Helena Bayer" to "Hello there" as an interpretation when an e-mail Recipient field

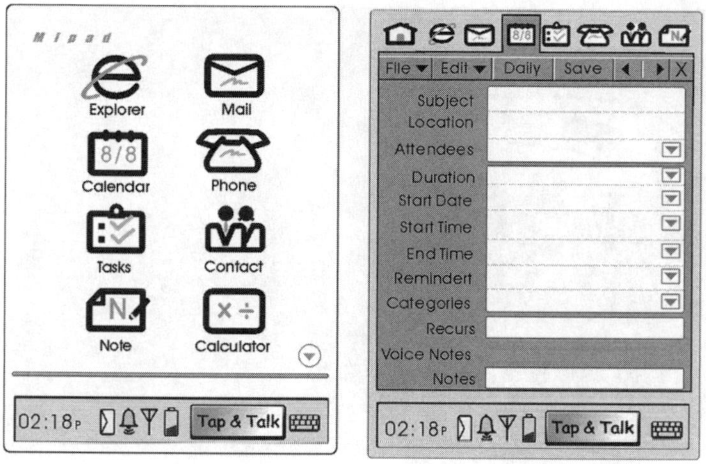

FIGURE

12.10

Microsoft's MiPad.
The interface, displaying icons and a form that can be activated multimodally.
From *http://research.microsoft.com/srg/mipad.aspx*. (Courtesy Microsoft.)

is selected. Finally, correction can be performed multimodally by selecting a misrecognized word by tapping with the stylus and then speaking it again.

Kirusa's Multimodal Interface

Kirusa offers a platform for the development of multimodal mobile applications intended for wireless phone providers. Its functionality is illustrated by a sports sample application (*http://www.kirusa.com/demo3.htm*). This application provides an interface structured as menus and forms that can be activated and filled out multimodally. Menu items can be selected by pen tap or by speech. Form fields can be selected via taps and filled via speech; fields can also be selected by speaking the name of the field.

Besides the structured graphical information (menus and forms) the interface presents synthesized speech, hyperlinks, and videos. Multimodal input is demonstrated by video control commands, such as zooming by issuing the command "Zoom in here <tap on location>" while a video is being displayed (Figure 12.11).

12.3.3   Ultramobile Digital-Paper–Based Interfaces

Despite the widespread introduction of technology, a considerable number of users still prefer to operate on paper-based documents. Rather than diminishing, the worldwide consumption of paper is actually rising (Sellen & Harper, 2003). Paper is lightweight, high definition, highly portable, and robust—paper still "works" even after torn and punctured. Paper does not require power and is not

FIGURE   Kirusa's sports demo.

12.11   The demo shows the selection of a location while the user commands the interface to "zoom in" via speech. From *http://www.kirusa.com/demo3.htm*. (Courtesy Kirusa, Inc.)

subject to system "crashes," which makes its use ideal in safety-critical scenarios (Cohen & McGee, 2004). These aspects make this medium highly compatible with mobile field use in harsh conditions.

NISMap is a commercial multimodal application produced by Adapx (*http://www.adapx.com*) that captures standard military plan fragments and uploads these plans to standard military systems such as CPOF. Officers operate the system by speaking and sketching on a paper map (Figure 12.12). The system interprets military symbols by fusing interpretations of a sketch recognizer and a speech recognizer. A user may for instance add a barbed-wire fence multimodally by drawing a line and speaking "Barbed wire." As an alternative, this command can be issued using sketch only, by decorating the line with an "alpha" symbol. The multimodal integration technology used by NISMap is based on Quickset (Cohen et al., 1997).

NISMap uses digital paper and pen technology developed by Anoto (2007). Anoto-enabled digital paper is plain paper that has been printed with a special

FIGURE

12.12

NISMap application using a digital pen and paper. *Source:* From Cohen and McGee (2004); courtesy ACM.

pattern, like a watermark. The pattern consists of small dots with a nominal spacing of 0.3 mm (0.01 inch). These dots are slightly displaced from a grid structure to form the proprietary Anoto pattern. A user can write on this paper using a pen with Anoto functionality, which consists of an ink cartridge, a camera in the pen's tip, and a Bluetooth wireless transceiver sending data to a paired device. When the user writes on the paper, the camera photographs movements across the grid pattern and can determine where on the paper the pen has traveled. In addition to the Anoto grid, which looks like a light gray shading, the paper itself can have anything printed on it using inks that do not contain carbon.

12.3.4   Applications of the Interface to Accessibility

Multimodal interfaces have the potential to accommodate a broader range of users than traditional interfaces. The flexible selection of modalities and the control over how these modalities are used make it possible for a wider range of users to benefit from this kind of interface (Fell et al., 1994). The focus on natural means of communication makes multimodal interfaces accessible to users of different ages, skill levels, native-language status, cognitive styles, sensory impairments, and other temporary illnesses or permanent handicaps. Speech can for instance be preferred by users with visual or motor impairments, while hearing impaired or heavily accented users may prefer touch, gesture, or pen input (Oviatt, 1999a).

Some multimodal systems have addressed issues related to disability directly. Bellik and Burger (1994, 1995) have developed multimodal interfaces for the blind. This interface supports nonvisual text manipulation and access to hyperlinks.

The interface combines speech input and output with a Braille terminal and keyboard. Using this interface, users can point to places in the Braille terminal and command the system via speech to underline words or to select text that can then be cut, copied, and pasted (Figure 12.13).

Multimodal locomotion assistance devices have also been considered (Bellik & Farcy, 2002). Assistance is provided via haptic presentation of readings coming from a laser telemeter that detects distances to objects. Hina et al. (Hina, Ramdane-Cherif, & Tadj, 2005) present a multimodal architecture that incorporates a user model which can be configured to account for specific disabilities. This model then drives the system interpretations to accommodate individual differences.

Facetop Tablet (Miller, Culp, & Stotts, 2006; Stotts et al., 2005) is an interface for the Deaf. This interface allows deaf students to add handwritten and sketched notes, which are superimposed by a translucent image of a signing interpreter (Figure 12.14). Deaf students are able to continue to monitor ongoing interpretation while still being able to take notes. Conventional systems would require students to move their eyes away from the interpreter, which causes segments of the interpretation to be lost. Other projects, such as the Visicast and the eSign Projects (*http://www.visicast.sys.uea.ac.uk/*) address sign language production via an animated character driven by a sign description notation. The objective of these projects is to provide deaf users access to government services—for instance, in a post office (Bangham et al., 2000) or on websites. Verlinden and colleagues (2005) employed similar animation techniques to provide for a multimodal

FIGURE

12.13

Braille terminal.
The terminal is in use with a multimodal interface for the blind. *Source:* From Bellik and Burger (1995); courtesy RESNA.

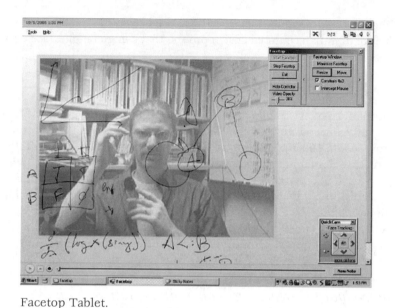

FIGURE

12.14

Facetop Tablet.
Handwritten notes are superimposed with the image of a signing interpreter.
Source: From Stotts et al. (2005).

educational interface, in which graphical and textual information is presented in synchrony with signing produced by an animated character.

12.4 HUMAN FACTORS DESIGN OF THE INTERFACE

Multimodal interface design depends on finding balance between the expressive power and naturalness of the language made available to users on the one hand and the requirements of processability on the other hand. The recognition technology that multimodal systems depend on is characterized by ambiguities and uncertainty, which derive directly from the expressiveness of communication that is offered. Current recognition technology is considerably limited when compared to human-level natural language processing. Particularly for naïve users, the fact that a system has some natural language capabilities may lead to expectations that are too high for a system to fulfill.

The challenge in designing multimodal interfaces consists therefore of finding ways to transparently guide user input in a way that agrees with the limited capabilities of current technology. Empirical evidence accumulated over the past decades has provided a wide variety of important insights into how users employ multimodal language to perform tasks. These findings are overviewed in Section 12.4.1, in which linguistic and cognitive factors are discussed. Section 12.4.2 then presents the principles of a human-centered design process that can guide interface development.

12.4.1   Linguistic and Cognitive Factors

Human language production, on which multimodal systems depend, involves a highly automatized set of skills, many of which are not under full conscious control (Oviatt, 1996b). Multimodal language presents considerably different linguistic structure than unimodal language (Oviatt, 1997; Oviatt & Kuhn, 1998). Each modality is used in a markedly different fashion, as users self-select their unique capabilities to simplify their overall commands (Oviatt et al., 1997). Individual differences directly affect the form in which multimodal constructs are delivered (Oviatt et al., 2003; Oviatt, Lunsford, & Coulston, 2005; Xiao, Girand, & Oviatt, 2002), influencing timing and preference for unimodal or multimodal delivery (Epps, Oviatt, & Chen, 2004; Oviatt, 1997; Oviatt et al., 1997; Oviatt, Lunsford, & Coulston, 2005).

Individual Differences in Temporal Integration Patterns

Accumulated empirical evidence demonstrates that users employ two primary styles of multimodal integration. There are those who employ a predominantly *sequential* integration pattern, while others employ a predominantly *simultaneous* pattern (Oviatt et al., 2003; Oviatt, Lunsford, & Coulston, 2005; Xiao et al., 2002). The main distinguishing characteristic of these two groups is that simultaneous integrators overlap multimodal elements at least partially, while sequential integrators do not. In a speech–pen interaction, for example, simultaneous integrators will begin speaking while still writing; sequential integrators, on the other hand, will finish writing before speaking (Figure 12.15).

These patterns have been shown to be remarkably robust. Users can in most cases be characterized as either simultaneous or sequential integrators by observing their very first multimodal command. Dominant integration patterns are also very stable, with 88 to 97 percent of multimodal commands being consistently delivered according to the predominant integration style over time (Oviatt, Lunsford, & Coulston, 2005). These styles occur across age groups, including children, adults, and the elderly (Oviatt et al., 2003), within a variety of different domains and interface styles, including map-based, real estate, crisis management, and educational applications (Oviatt et al., 1997; Xiao et al., 2002; Xiao et al., 2003). Integration patterns are strikingly resistant to change despite explicit instruction and selective reinforcement encouraging users to switch to a style that is not their dominant one (Oviatt, Coulston, & Lunsford, 2005; Oviatt et al., 2003). On the contrary, there is evidence that users further entrench in their dominant patterns during more difficult tasks or when dealing with errors. In these more demanding situations, sequential integrators will further increase their intermodal lag, and simultaneous integrators will more tightly overlap.

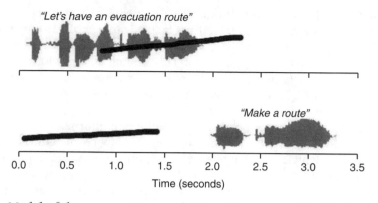

"Let's have an evacuation route"

"Make a route"

FIGURE

12.15

Model of the average temporal integration pattern for simultaneous and sequential integrators' typical constructions.
Top: Simultaneous integrator. *Bottom:* Sequential integrator. *Source:* From Oviatt, Lunsford, and Coulston (2005); courtesy ACM.

Integration patterns are also correlated to other linguistic and performance parameters. Sequential integrators present more precise articulation with fewer disfluencies, adopting a more direct command-style language with smaller and less varied vocabulary. Sequential integrators also make only half as many errors as simultaneous integrators (Oviatt, Lunsford, & Coulston, 2005).

These strong individual differences in patterns may be leveraged by a system to better determine which elements should be fused and the appropriate time to perform interpretation (Section 12.2.3). A system can reduce the amount of time that it waits for additional input before interpreting a command when dealing with simultaneous integrators, given prior knowledge that these users primarily overlap their inputs. More recently, integration patterns have been exploited within a machine learning framework that can provide a system with robust prediction of the next input type (multimodal or unimodal) and adaptive temporal thresholds based on user characteristics (Huang et al., 2006).

Redundancy and Complementarity

Multimodal input offers opportunities for users to deliver information that is either *redundant* across modalities, such as when a user speaks the same word she is handwriting, or *complementary*, such as when drawing a circle on a map and speaking, "Add hospital." Empirical evidence demonstrates that the dominant theme in users' natural organization of multimodal input to a system is complementarity of content, not redundancy (McGurk & MacDonald, 1976; Oviatt, 2006b; Oviatt et al., 1997; Oviatt & Olsen, 1994; Wickens, Sandry, & Vidulich, 1983). Users naturally make use of the different characteristics of each modality to deliver information to an interface in a concise way. When visual-spatial content is involved, for example, users take advantage of pen input to indicate location while using the strong descriptive capabilities of speech to specify temporal and other nonspatial information. This finding agrees with the broader observation by linguists that during interpersonal communication spontaneous speech and manual gesturing involve complementary rather than duplicate information between modes (McNeill, 1992).

Speech and pen input expresses redundant information less than 1 percent of the time even when users are engaged in error correction, a situation in which they are highly motivated to clarify and reinforce information. Instead of relying on redundancy, users employ a contrastive approach, switching away from the modality in which the error was encountered, such as correcting spoken input using the pen and vice versa (Oviatt & Olsen, 1994; Oviatt & VanGent, 1996).

Preliminary results indicate furthermore that the degree of redundancy is affected by the level of cognitive load users encounter while performing a task (Ruiz, Taib, & Chen, 2006), with a significant reduction in redundancy as tasks become challenging. The relationship of multimodal language and cognitive load is further discussed in following sections.

While complementarity is a major integration theme in human–computer interaction, there is a growing body of evidence pointing to the importance of redundancy in multiparty settings, such as lectures (Anderson, Anderson et al., 2004; Anderson, Hoyer et al., 2004) or other public presentations during which one or more participants write on a shared space while speaking (Kaiser et al., 2007). In these cases, a high degree of redundancy between handwritten and spoken words has been found. Redundancy appears to play a role of focusing the attention of a group and helping the presenter highlight points of importance in her message. Techniques that leverage this redundancy are able to robustly recover these terms by exploiting mutual disambiguation (Kaiser, 2006).

Linguistic Structure of Multimodal Language

The structure of the language used during multimodal interaction with a computer has also been found to present peculiarities. Users tend to shift to a locative-subject-verb-object word order strikingly different from the canonical English subject-verb-objective-locative word order observed in spoken and formal textual language. In fact, the same users performing the same tasks have been observed to place 95 percent of the locatives in sentence-initial positions during multimodal interaction and in sentence-final positions when using speech only (Oviatt et al., 1997).

The propositional content that is transmitted is also adapted according to modality. Speech and pen input consistently contribute different and complementary semantic information—with the subject, verb, and object of a sentence typically spoken and locative information written (Oviatt et al., 1997).

Provided that a rich set of complementary modalities is available to the users, multimodal language also tends to be simplified linguistically, briefer, syntactically simpler, and less disfluent (Oviatt, 1997), containing less linguistic indirection and fewer co-referring expressions (Oviatt & Kuhn, 1998). These factors contribute to enhanced levels of recognition in multimodal systems when compared to unimodal (e.g., speech-only) interfaces.

Pen input was also found to come before speech most of the time (Oviatt et al., 1997), which agrees with the finding that in spontaneous gesturing and signed languages, gestures precede spoken lexical analogs (Kendon, 1981; Naughton, 1996). During speech and three-dimensional–gesture interactions, pointing has been shown to be synchronized with either the nominal or deictic spoken expressions (e.g., "this," "that," "here"). The timing of these gestures can be furthermore predicted in a time window of −200 to +400 milliseconds around the beginning of the nominal or deictic expression (Bourguet, 2006; Bourguet & Ando, 1998).

Choice of Multimodal and Unimodal Input

Only a fraction of user commands are delivered multimodally, with the rest making use of a single modality (Oviatt, 1999b). The number of multimodal commands depends highly on the task at hand, and on the domain, varying from 20 to 86 percent (Epps et al., 2004; Oviatt, 1997; Oviatt et al., 1997; Oviatt, Lunsford, & Coulston, 2005), with higher rates in spatial domains. Figure 12.16 presents a graph showing the percentage occurrence of multimodal commands across various tasks and domains.

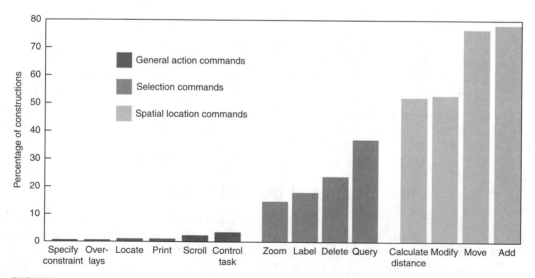

FIGURE

12.16

Percentage of commands that users expressed multimodally as a function of type of task command.

Negligible levels can be observed during general commands, with increasing levels for selection commands and the highest level for location commands (Oviatt, 1999b); courtesy ACM.

Influence of Cognitive Load

An aspect of growing interest is the relationship between multimodality and cognitive load. One goal of a well-designed system, multimodal or otherwise, should be to provide users with means to perform their tasks in a way that does not introduce extraneous complexity that might interfere with performance (Oviatt, 2006a). Wickens and colleagues' cognitive resource theory (1983) and Baddeley's theory of working memory (1992) provide interesting theoretical frameworks for examining this question and its relations to multimodal systems. Baddeley (1992) maintains that short-term or working memory consists of multiple independent processors associated with different modes. This includes a visual-spatial "sketch pad" that maintains visual materials such as pictures and diagrams in one area of working memory, and a separate phonological loop that stores auditory-verbal information. Although these two processors are believed to be coordinated by a central executive, in terms of lower-level modality processing they are viewed as functioning largely independently, which is what enables the effective size of working memory to expand when people use multiple modalities during tasks (Baddeley, 2003).

Empirical evidence shows that users self-manage load by distributing information across multiple modalities (Calvert, Spence, & Stein, 2004; Mousavi, Low, & Sweller, 1995; Oviatt, Coulston, & Lunsford, 2004; Tang et al., 2005). As task complexity increases, so does the rate at which users choose to employ multimodal rather than unimodal commands (Oviatt et al., 2004; Ruiz et al., 2006). In an experiment with a crisis management domain involving tasks of four distinct difficulty levels, the ratio of users' multimodal interaction increased from 59.2 percent during low-difficulty tasks to 65.5 percent at moderate difficulty, 68.2 percent at high, and 75.0 percent at very high difficulty—an overall relative increase of 27 percent. Analysis of users' task-critical errors and response latencies across task difficulty levels increased systematically and significantly as well, corroborating the manipulation of cognitive-processing load (Oviatt et al., 2004). In terms of design, these findings point to the advantage of providing multiple modalities, particularly in interfaces supporting tasks with higher cognitive demands.

12.4.2   A Human-Centered Design Process

Given the high variability and individual differences that characterize multimodal user interaction, successful development requires a user-centered approach. Rather than relying on instruction, training, and practice to make users adapt to a system, a user-centered approach advocates designing systems based on a deeper understanding of user behavior in practice. Empirical and ethnographic work should provide information on which models of user interaction can be based. Once users' natural behavior is better understood, including their ability to attend, learn, and perform, interfaces can be designed that will be easier to learn, more intuitive, and freer of performance errors (Oviatt, 1996b, 2006).

While the empirical work described in the previous section has laid the foundation for the development of effective multimodal interfaces, there is still a need for careful investigation of the conditions surrounding new applications. In the following sections, steps are described that can be followed when investigating issues related to a new interface design. The process described here is further illustrated by the online case study (see Section 12.7). Additional related considerations are also presented as design guidelines (Section 12.6).

Understanding User Performance in Practice

A first step when developing a new multimodal interface is to understand the domain under investigation. Multimodal interfaces depend very heavily on language as the means through which the interface is operated. It is important therefore to analyze the domain to identify the objects of discourse and potential standardized language elements that can be leveraged by an interface.

When analyzing a battle management task, for example, McGee (2003) used ethnographic techniques to identify the language used by officers while working with tangible materials (sticky notes over maps). When performed in a laboratory, this initial analysis may take the form of semistructured pilots, during which subjects perform a task with simulated system support. That is, to facilitate evolution, a simulated ("Wizard of Oz") system may be used (see Section 12.5.2 for additional details).

Identification of Features and Sources of Variability

Once a domain is better understood, the next step consists of identifying the specifics of the language used in the domain. In specialized fields, it is not uncommon for standardized language to be employed. The military employs well-understood procedures and standardized language to perform battle planning (McGee & Cohen, 2001; McGee et al., 2001). Similarly, health professionals (Cohen & McGee, 2004) and engineers have developed methods and language specific to their fields.

Even when dealing with domains for which there is no standardized language, such as photo tagging, the analysis of interactions can reveal useful robust features that can be leveraged by a multimodal interface. In a collaborative photo annotation task in which users discussed travel and family pictures, analysis revealed that the terms that are handwritten are also redundantly spoken. As a user points to a person in a photo and handwrites his name, she will also in most cases speak the person's name repeatedly while explaining who that is. Further analysis revealed that redundantly delivered terms were frequent during the discussion and that they were good retrieval terms (Barthelmess, Kaiser et al., 2006; Kaiser et al., 2007).

Many times languages and features identified during this phase prove too complex to be processable via current technology. An important task is therefore to recognize and model the sources of variability and error (Oviatt, 1995).

In Section 12.7 , this is examined in the context of identifying causes of speech disfluencies, which introduce hard-to-process features into the spoken language.

Transparent Guidance of User Input

A strategy for dealing with hard-to-process features (e.g., disfluencies) present in users' input is to design interfaces that transparently guide users' input to reduce errors (Oviatt, 1995). The essence of this approach is to "get users to say [and do] what computers can understand" (Zoltan-Ford, 1991).

The main goal of the design at this stage is to identify means to guide users' input toward simpler and more predictable language. How this is achieved may depend on the specifics of thc domain. Determining domain-specific factors requires detailed analysis of natural linguistic features manifested by users. Based on this analysis, experimentation with alternative interface designs reveals methods to reduce complexity in ways that are not objectionable to users. This is illustrated in the online case study (see Section 12.7).

Techniques that have led to transparent guidance of user input include choosing interface modalities that match the representational systems users require while working on a task (Oviatt, Arthur, & Cohen, 2006; Oviatt & Kuhn, 1998), structuring the interface to reduce planning load (Oviatt, 1995, 1996b, 1997), and exploiting users' linguistic convergence with a system's output (Larson, 2003; Oviatt et al., 2005). In terms of design, the language used in a system's presentation should be compatible with the language it can process. These aspects are further discussed in Section 12.6.

Development of Formal Models

Linguistic and behavioral regularities detected during the analysis of a domain can be represented via a variety of formal models that directly influence and drive the interpretation process of a system. Given these models, a system is able with higher likelihood to distinguish the actual users' intentions from a variety of alternative interpretations.

Formal models include primarily multimodal grammars used to drive interpretation. These grammars provide rules that guide the input interpretation, which in turn drives interface responses or activation of other applications.

More recently, there has been a growing interest in the application of machine learning techniques to model aspects of multimodal interactions. One example is the user-adapted model developed by Huang and Oviatt (2006) to predict the patterns of users' next multimodal interaction. Other machine learning models have been used to interpret user actions during multimodal multiparty interactions (e.g., McCowan et al., 2005; Pianesi et al., 2006; Zancanaro et al., 2006). The advantage of these systems is that they are able to adapt to individual users' characteristics and are therefore more capable of processing input with fewer errors.

12.5   TECHNIQUES FOR TESTING THE INTERFACE

Iterative testing throughout the development cycle is of fundamental importance when designing a multimodal interface. Empirical evaluation and user modeling are the proactive driving forces during system development.

In this section, the generic functionality required for successfully testing a multimodal interface is described. An initial step in testing an interface is the collection of user data for analysis. Given the need to examine a variety of modalities when designing a multimodal interface, an appropriate data collection infrastructure is required (Section 12.5.1). A strategy that has proven very fruitful for prototyping new multimodal interfaces is to exploit high-fidelity simulation techniques, which permit comparing trade-offs associated with alternative designs (Section 12.5.2). Analysis of multimodal data also requires synchronizing multiple data streams and development of annotation tools (as described in Section 12.5.3).

12.5.1   Data Collection Infrastructure

The focus of data collection is on the multimodal language and interaction patterns employed by users of a system. A data collection facility must therefore be capable of collecting high-quality recordings of multiple streams of synchronized data, such as speech, pen input, and video information conveying body motions, gestures, and facial expressions.

Besides producing recordings for further analysis, the collection infrastructure also has to provide facilities for observation during the collection in order to support simulation studies. Since views of each modality are necessary during some simulations, capabilities for real-time data streaming, integration, and display are required. Building a capable infrastructure to prototype new multimodal systems requires considerable effort (Arthur et al., 2006).

Of primary importance is that data be naturalistic and therefore representative of task performance that is expected once a system is deployed. Thus, any data collection devices or instrumentation need to be unobtrusive. This is challenging, given the requirement for rich collection of a variety of data streams that each can require a collection device (e.g., cameras, microphones, pens, physiological sensors). This can be particularly problematic when the aim is to collect realistic mobile usage information (Figure 12.17).

12.5.2   High-Fidelity Simulations

A technique that has been very valuable in designing multimodal interfaces is the construction of high-fidelity simulations, or Wizard-of-Oz experiments (Oviatt et al., 1992; Salber & Coutaz, 1993a, 1993b). These experiments consist of having subjects

424

FIGURE

12.17

Mobile data collection.
The subject (on the left) carries a variety of devices, including a processing unit
in a backpack. *Source:* From Oulasvirta et al. (2005); courtesy ACM.

interact with a system for which functionality is at least partially provided by a
human operator (i.e., "wizard") who controls the interface from a remote location.

The main advantage of prototyping via simulations is that they are faster and
cheaper than developing a fully functional system and they provide advanced
information about interface design trade-offs. Hard-to-implement functionality
can be delegated to the wizard, which may shorten the development cycle considerably via rapid prototyping of features. Simulations make it possible for interface
options to be tested before committing to actual system construction, and may
even support experiments that would not otherwise be possible given the state
of the art (one can, for instance, examine the repercussions of much enhanced
recognition levels on an interface).

One challenge when setting up a Wizard-of-Oz experiment is making it believable (Oviatt, 1992; Arthur et al., 2006). In order to be effective, working prototypes
that users interact with must be credible; for example, making some errors adds to
credibility. The wizard needs to be able to react quickly and accurately to user
actions, which requires training and practice. Techniques for further facilitating
this type of experiment include simplifying the wizard's interface through semi-
automation (e.g., by automatically initiating display actions) (Oviatt et al., 1992).
An automatic random error modules can be used to introduce system misrecognitions, which can be set at any level and contribute to the simulation's credibility.

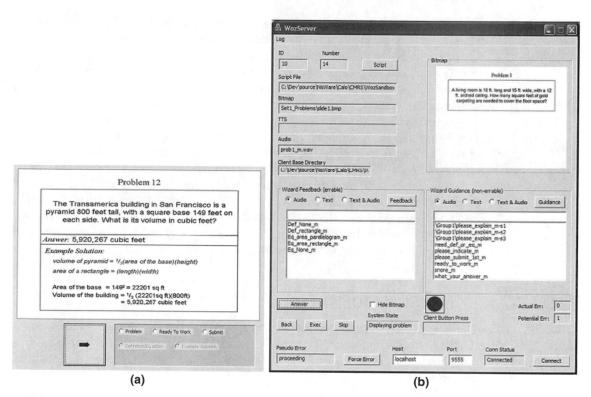

(a)   **(b)**

FIGURE

12.18

Simulated multimodal interface.

The interface as (a) seen by users and (b) used by the wizard to control the flow of interaction in a believable way. This interface was used during a collaborative geometry problem–solving interaction. *Source:* From Arthur et al. (2006); courtesy ACM.

Figure 12.18 shows an interface viewed by geometry students and the corresponding wizard interface.

To address the complexity of multimodal interactions, multiple wizards have been used in the past in some studies, especially in cases involving collaboration (Arthur et al., 2006; Oviatt et al., 2003; Salber & Coutaz, 1993a). Each wizard then concentrates on providing simulation feedback for particular aspects of the interaction, which assists in making the overall task of driving an interface manageable.

12.5.3 Support for Data Analysis

The assessment of the effectiveness of an interface design or the comparison of alternative designs requires detailed examination of the data that have been collected. Analysis tools therefore need to provide means for playback and navigation of multiple data streams. High-fidelity synchronization is also required. Audio, video, and other input (e.g., pen) should be aligned well enough that differences are not be perceptible to a human analyst.

The analysis process usually incorporates *mark-ups* (or annotations) of selected parts of an interaction. These annotations might include speech transcripts and semantic annotation of gestures, gaze, or prosodic characteristics of the speech. The specifics of what is annotated depend on the purpose of the research and interface being designed. Annotated data can be examined in terms of characteristics of their language production, performance, and error characteristics under varying circumstances.

A variety of different playback and annotation tools are available, such as Nomos (Gruenstein, Niekrasz, & Purver, 2005), Anvil (Martin & Kipp, 2002), and AmiGram (Lauer et al., 2005). Arthur et al. (2006) describe a tool for annotation and playback of multiple high-definition video and audio streams appropriate for the analysis of multiparty interactions.

12.6   DESIGN GUIDELINES

The guidelines presented in this section are grouped in two primary classes: those that have to do with issues related to the uncertainty of recognition (Section 12.6.1), and those concerned with the circumstances guiding the selection of modalities (Section 12.6.2).

12.6.1   Dealing with Uncertainty

One essential aspect that has to be considered when building systems that rely on ambiguous interpretation is how uncertain interpretations are dealt with (Bourguet, 2006; Mankoff, Hudson, & Abowd, 2000). Shielding users from errors and providing graceful ways for handling unavoidable misinterpretations are essential usability concerns in this class of interfaces.

As discussed (Section 12.4.1), human language production involves a highly automatized set of skills not under users' full conscious control (Oviatt, 1996b). The most effective strategy for error avoidance is to design interfaces that leverage users' engrained cognitive and linguistic behavior in order to transparently guide input that avoids errors. In fact, training and practice in an effort to change engrained behavior patterns often prove useless (Oviatt et al., 2003, 2005). While this strategy can be the most effective, it is also the most demanding in terms of user modeling and implementation effort. In order to determine the root cause of errors and design an effective strategy for avoiding and resolving them, a cycle of experiments is required, as illustrated by the online case study (see Section 12.7). In this section, the effective principles of multimodal interaction are distilled as a set of guidelines.

Support the Range of Representational Systems Required by the Task

The structural complexity and linguistic variability of input generated by users are important sources of processing difficulties. A primary technique to elicit simpler, easier-to-process language is related to the choice of modalities that an interface supports. Users will naturally choose the modalities that are most appropriate for conveying content. For example, users typically select pen input to provide location and spatially oriented information, as well as digits, symbols, and graphic content (Oviatt, 1997; Oviatt & Olsen, 1994; Suhm, 1998). In contrast, they will use speech for describing objects and events and for issuing commands for actions (Cohen & Oviatt, 1995; Oviatt & Cohen, 1991).

A primary guideline is therefore to support modalities so that the representational systems required by users are available. The language that results when adequate complementary modalities are available tends to be simplified linguistically, briefer, syntactically simpler, and less disfluent (Oviatt, 1997), and it contains less linguistic indirection and fewer co-referring expressions (Oviatt & Kuhn, 1998). One implication of this is that the fundamental language models needed to design a multimodal system are not the same as those used in the past for processing textual language.

Structure the Interface to Elicit Simpler Language

A key insight in designing multimodal interfaces that lead to simpler, more processable language is that the language employed by users can be shaped very strongly by system presentation features. Adding structure, as opposed to having an unconstrained interface, has been demonstrated to be highly effective in simplifying the language produced by users, resulting in more processable language and fewer errors. A forms-based interface that guides users through the steps required to complete a task can reduce the length of spoken utterances and eliminate up to 80 percent of hard-to-process speech disfluencies (Oviatt, 1995). Similar benefits have been identified in map-based domains. A map with more detailed information displaying the full network of roads, buildings, and labels can reduce disfluencies compared to a minimalist map containing one-third of the roads (Oviatt, 1997).

Other techniques that may lead users toward expected language are guided dialogs and context-sensitive cues (Bourguet, 2006). These provide additional information that helps users determine what their input options are at each point of an interaction, leading to more targeted production of terms that are expected by the interface at a given state. This is usually implemented by having a prompt that explicitly lists the options the user can choose from.

Exploit Natural Adaptation

A powerful mechanism for transparently shaping user input relies on the tendency that users have of adapting to the linguistic style of their conversational partners (Oviatt, Darves et al., 2005). A study in which 24 children conversed with a computer-generated animated character confirmed that children's speech signal features, amplitude, durational features, and dialog response latencies spontaneously adapt to the basic acoustic-prosodic features of a system's text-to-speech output, with the largest adaptations involving utterance pause structure and amplitude. Adaptations occurred rapidly, bidirectionally, and consistently, with 70 to 95 percent of children's speech converging with that of their computer partners.

Similar convergence has been observed in users' responses to discrete system prompts. People will respond to system prompts using the same wording and syntactic style (Larson, 2003). This suggests that system prompts should be matched to the language that the system ideally would receive from users—usually presenting a simple structure, restricted vocabulary, and recognizable signal features.

Offer Alternative Modalities Users Can Switch to When Correcting Errors

There is evidence that users switch modalities they are using to correct misrecognitions after repeated failures (Oviatt & VanGent, 1996). Users correcting a spoken misrecognition will attempt to repeat the misrecognized word via speech a few times, but will then switch to another modality such as handwriting when they realize that the system is unable to accept the correction. This behavior appears to be more pronounced for experienced users compared to novices. The latter tend to continue to use the same modality despite the failures (Halverson et al., 1999). Therefore, a well-designed system should offer alternative modalities for correcting misrecognitions. The absence of such a feature may lead to "error spirals" (Halverson et al., 1999; Oviatt & VanGent, 1996)—situations in which the user repeatedly attempts to correct an error, but due to increased hyperarticulation, the likelihood of correct system recognition actually degrades.

Make Interpretations Transparent But Not Disruptive

One disconcerting effect of systems that rely on interpretation of users' ambiguous input is when users cannot clearly connect the actions performed by the system with the input they just provided. That is particularly disconcerting when the system display disagrees with the expectations that users had when providing the input, as would happen when a user commands a system to paint a table green and sees the floor turning blue as a result (Kaiser & Barthelmess, 2006). One technique to make system operation more transparent is to give users the opportunity to examine the interpretation and potentially correct it, such as via a graphical display of the interpretation.

One popular way of making users aware of a system's interpretation is to make available a list of alternate recognitions. These lists, sometimes called "*n*-best" lists, represent a limited number of the most likely ("best") interpretations identified by a system. This strategy needs to consider very carefully how easy it is to access and dismiss the list, and also how accurate the list is. In cases in which the accuracy is such that most of the time the correct interpretation is not present (for very ambiguous input), this strategy can become counterproductive (Suhm, Myers, & Waibel, 1999). When well designed, making alternative recognitions available can be effective, to the point that users may try the list first and just attempt to repeat misrecognized terms as a secondary option (Larson & Mowatt, 2003). One example of a simple interface that makes an alternative interpretation available is Google. A hyperlink lists a potential alternative spelling of query terms and may be activated by a single click.

Displaying the state of recognition can sometimes prove disruptive, such as during multiparty interactions. Displaying misrecognitions or requiring the users to choose among alternative recognitions in the course of a meeting has been shown to disrupt the interaction, as users turn their attention to correcting the system. A fruitful approach (explored, for example, in the Distributed Charter System) is to present the state of recognition in a less distracting or forceful way, such as via subtle color coding of display elements. Users are then free to choose the moment that is most appropriate to deal with potential issues (Kaiser & Barthelmess, 2006).

12.6.2   Choosing Modalities to Support

The choice of which modalities to support is naturally an important one when a multimodal interface is being designed. The appropriate modalities and characteristics of the language support within each modality are influenced by the tasks that users are to face, conditions of use, and user characteristics.

Context of Use

Conditions of use determine, for instance, whether a system is required to be mobile or whether it is to be used within an office environment or a meeting room. That in turn determines the nature and capability of the devices that can be employed. Mobile users will certainly not accept carrying around heavy loads and will therefore not be able to take advantage of modalities that require processing power or sensors that cannot be fit into a cell phone or PDA (portable digital assistant), as is the case for most vision-based techniques. Meeting rooms, on the other hand, may make use of a much larger set of modalities, even in cases in which several computers are required to run a system.

Most current mobile systems provide support for pen input and are powerful enough to execute at least a certain level of speech recognition. Speech interfaces are ideal for many mobile conditions because of the hands-and-eyes-free use that speech affords. Pen input is used in many devices as an alternative for keyboard input in small devices.

Other considerations associated with usage context that affect choice of modalities are privacy and noise. Speech is less appropriate when privacy is a concern, such as when the interface is to be used in a public setting. Speech is also not indicated when noisy conditions are to be expected (e.g., in interfaces for construction sites or noisy factories).

A well-designed interface takes advantage of multiple modalities to provide support for a variety of usage contexts, allowing users to choose the modality that is most appropriate given a particular situation.

User Characteristics

A well-designed interface leverages the availability of multiple modalities to accommodate users' individual differences by supporting input over alternative modalities.

Motor and cognitive impairments, age, native language (e.g., accent), and other individual characteristics influence individual choice of input modalities. For example, pen input and gesturing are hard to use when there is diminished motor acuity. Conversely, spoken input may prove problematic to users who have speech or hearing impairments or who speak with an accent.

A particular kind of temporary impairment occurs when users are mobile or are required to keep a high level of situational awareness, such as on a battlefield or in an emergency response situation, or even while operating a vehicle (Oviatt, 2007). Supporting spoken interaction then becomes an attractive option.

12.7 CASE STUDIES

Case studies for these kinds of multimodal interfaces can be found at *www. beyondthegui.com*.

12.8 FUTURE TRENDS

Multimodal interfaces depend ultimately on the quality of the underlying recognizers for multiple modalities that are used. Thus, these interfaces benefit from the advances of natural language processing techniques and advances in hardware capabilities that make it possible for more challenging recognitions to be successfully achieved.

12.8.1 Mobile Computing

Mobile computing is one of the areas in which multimodal interfaces are expected to play an important role in the future. Multimodal interfaces suit mobility particularly well because of the small factor requirements that are usually imposed by on-the-move operation. Multimodal interfaces provide expressive means of interaction to users without requiring bulky, hard-to-carry equipment such as keyboards or mice.

The flexibility provided by interfaces that support multiple modes also fits the demands introduced by mobility via the adaptation to shifting contexts of use. A multimodal interface can, for instance, provide spoken operation for users whose hands and eyes are busy, such as while driving, or maintaining situational awareness in dangerous environments, such as disaster areas. These interfaces can also adapt to noisy environments either by leveraging mutual disambiguation to enhance recognition or by providing means for commands to be given via nonspoken means such as a pen.

Speech and pen input are already supported by a variety of devices (smart phones, PDAs), and successful commercial implementations are available. Further computational power should lead to an era in which a majority of the mobile devices might offer multimodal capabilities, including increasing levels of video-based modalities.

12.8.2   Collaboration Support

Collaborative human interaction is intrinsically multimodal (Tang, 1991). Groups of people collaborating make ample use of each other's speech, gaze, pointing, and body motion to focus attention and establish required shared contexts. Collaborative multimodal interfaces are therefore a natural fit for groupware.

Interesting challenges are introduced by the shift from single-user to multi-user interfaces. The language used by humans to communicate among themselves can be considerably more complex than that employed when addressing computers (Oviatt, 1995). Systems that are based on observation of human–human interaction will need to employ novel techniques for extracting useful information.

The introduction of technology into collaborative settings has not been without problems. These interactions are also known to be brittle in the face of technology (Grudin, 1988) and the potential disruptions of subtle social processes that technology may introduce. Considerable care is therefore required to examine how systems that operate via natural language can best be integrated.

Pioneering systems have mostly employed a passive approach, in which observations performed by a system are collected with minimal direct interaction between the system and a group of users. System results are delivered after the interactions have concluded, in the form of browsable information (Ehlen, Purver, & Niekrasz, 2007) or MS-Project charts (Kaiser et al., 2004) or via a semiautomated "coach" that presents episodes of the meeting to participants during which dysfunctional behavior is detected (Zancanaro et al., 2006).

Work is ongoing to leverage para-linguistic cues such as amplitude to facilitate the detection of who is being addressed (humans or computer) in a multiparty setting (Lunsford et al., 2005, 2006) and to examine the impact of the introduction of computational assistants during and after meetings (Falcon et al., 2005; Pianesi et al., 2006; Rienks, Nijholt, & Barthelmess, 2006; Zancanaro et al., 2006).

Given the time and expense that are involved in collaborative interactions in general, advances in supporting technology have a potential for high payoffs. It is expected that this area will continue to attract attention in the future.

12.8.3   Conclusion

The continuous advances in hardware and of the basic recognition technologies upon which multimodal systems rely should make it increasingly possible for designers to take advantage of multimodality to provide users with more intuitive, easier-to-use interfaces that will impact user performance and satisfaction in a positive way.

Systems that take advantage of a broader range of natural human communication capabilities, such as speech, handwriting, and gesturing, are expected to become increasingly important as they transfer the burden of communication from users to the systems themselves. This shift toward systems that are better able to understand their users, rather than requiring them to operate in ways that fit the systems' narrow limitations, is expected to eventually result in unprecedented gains in productivity as performance becomes less hindered by technology and users become freer to concentrate on their tasks rather than on the intricacies of the technology.

REFERENCES

Adjoudani, A., & Benoit, C. (1995). Audio-visual speech recognition compared across two architectures. Paper presented at the 4th European Conference on Speech Communication and Technology.

Allegayer, J., Jansen-Winkeln, R., Reddig, C., & Reithinger, N. (1989). Bidirectional use of knowledge in the multi-modal NL access system XTRA. Paper presented at the 11th International Joint Conference on Artificial Intelligence.

Anderson, R. J., Anderson, R., Hoyer, C., & Wolfman, S. A. (2004). A study of digital ink in lecture presentation. Paper presented at the 2004 SIGCHI Conference on Human Factors in Computing Systems. Available at *http://doi.acm.org/10.1145/985692.985764*.

Anderson, R. J., Hoyer, C., Prince, C., Su, J., Videon, F., & Wolfman, S. A. (2004). Speech, ink and slides: The interaction of content channels. Paper presented at the 12th Annual ACM International Conference on Multimedia. Available at *http://doi.acm.org/10.1145/1027527.1027713*.

Andre, E. (2003). Natural language in multimedia/multimodal systems. In Mitkov, R., ed., *The Oxford Handbook of Computational Linguistics*. New York: Oxford University Press, 650–69.

Andre, E., Muller, J., & Rist, T. (1996). The PPP persona: A multipurpose animated presentation agent. Paper presented at the 1996 Workshop on Advanced Visual Interfaces. Available at *http://doi.acm.org/10.1145/948449.948486*.

Anoto. (2007). Anoto technology—*http://www.anoto.com/*.

Arthur, A. M., Lunsford, R., Wesson, R. M., & Oviatt, S. L. (2006). Prototyping novel collaborative multimodal systems: Simulation, data collection, and analysis tools for the next decade. Paper presented at the 8th International Conference on Multimodal Interfaces, Oral Session 4: Collaborative Systems and Environments. Available at *http://portal.acm.org/citation.cfm?id=1181039&coll=portal&dl=ACM&CFID=6362723&CFTOKEN=90890358*.

Baddeley, A. D. (1992). Working memory. *Science* 255(5044):556–59.

Baddeley, A. D. (2003). Working memory: Looking back and looking forward. *Nature Reviews Neuroscience* 44(10):829–39.

Banerjee, S., & Rudnicky, A. I. (2004). Using simple speech-based features to detect the state of a meeting and the roles of the meeting participants. Paper presented at the 8th International Conference on Spoken Language Processing. Available at *http://www.cs.cmu.edu/~banerjee/Publications/icslp2004.pdf*.

Bangham, J. A., Cox, S. J., Elliott, R., Glauert, J. R. W., Marshall, I., Rankov, S., et al. (2000). Virtual signing: Capture, animation, storage and transmission—an overview of the ViSiCAST project. Paper presented at the IEE Seminar on Speech and Language Processing for Disabled and Elderly people, London.

Barthelmess, P., & Ellis, C. A. (2005). The Neem Platform: An evolvable framework for perceptual collaborative applications. *Journal of Intelligent Information Systems* 25(2): 207–40.

Barthelmess, P., Kaiser, E. C., Huang, X., & Demirdjian, D. (2005). Distributed pointing for multimodal collaboration over sketched diagrams. Paper presented at the 7th International Conference on Multimodal Interfaces, Recognition and Multimodal Gesture Patterns. Available at *http://doi.acm.org/10.1145/1088463.1088469*.

Barthelmess, P., Kaiser, E. C., Huang, X., McGee, D. R., & Cohen, P. R. (2006). Collaborative multimodal photo annotation over digital paper. Paper presented at the 8th International Conference on Multimodal Interfaces, Poster Session 1. Available at *http://portal.acm.org/citation.cfm?id=1181000&coll=portal&dl=ACM&CFID=6362723&CFTOKEN= 90890358*.

Barthelmess, P., McGee, D. R., & Cohen, P. R. (2006). The emergence of representations in collaborative space planning over digital paper: Preliminary observations. Paper presented at the 2006 ACM Conference on Computer-Supported Cooperative Work, Workshop on Collaboration over Paper and Digital Documents (CoPADD).

Bellik, Y. (1997). Media integration in multimodal interfaces. Paper presented at the IEEE First Workshop on Multimedia Signal Processing, Princeton, NJ.

Bellik, Y., & Burger, D. (1994). Multimodal interfaces: New solutions to the problem of computer accessibilty for the blind. Paper presented at the 1994 SIGCHI Conference on Human Factors in Computing Systems, Conference Companion. Available at *http://portal.acm.org/citation.cfm?id=260482&coll=portal&dl=ACM&CFID=6362723&CFTOKEN =90890358*.

Bellik, Y., & Burger, D. (1995). The potential of multimodal interfaces for the blind: An exploratory study. Paper presented at the 18th Annual RESNA Conference. Available at *http://www.limsi.fr/Individu/bellik/publications/resna95.rtf*.

Bellik, Y., & Farcy, R. (2002). Comparison of various interface modalities for a locomotion assistance device. Paper presented at the 8th International Conference on Computers Helping People with Special Needs, ICCHP 2002, Linz, Austria.

Bellik, Y., & Teil, D. (1993). A multimodal dialogue controller for multimodal user interface management system application: A multimodal window manager. Paper presented at the 1993 SIGCHI Conference on Human Factors in Computing Systems, Amsterdam.

Benoit, C., & Le Goff, B. (1998). Audio-visual speech synthesis from French text: Eight years of models, designs and evaluation at the ICP. *Speech Communication* 26(1–2):117–29.

Benoit, C., Martin, J.-C., Pelachaud, C., Schomaker, L., & Suhm, B. (2000). Audio-visual and multimodal speech-based systems. In Gibbon, D., Mertins, I., & Moore, R., eds., *Handbook of Multimodal and Spoken Dialogue Systems: Resources, Terminology and Product Evaluation*. Boston: Kluwer Academic Publishers, 102–203.

Bernstein, L. E., & Benoit, C. (1996). For speech perception by humans or machines, three senses are better than one. Paper presented at the 4th International Conference on Spoken Language Processing, Philadelphia.

Bers, J., Miller, S., & Makhoul, J. (1998). Designing conversational interfaces with multimodal interaction. *Proceedings of DARPA Workshop on Broadcast News Understanding Systems*, Lansdowne, VA, 319–21.

Bolt, R. A. (1980). "Put-that-there": Voice and gesture at the graphics interface. Paper presented at the 7th Annual Conference on Computer Graphics and Interactive Techniques, Seattle.

Bourguet, M.-L. (2006). Towards a taxonomy of error-handling strategies in recognition-based multi-modal human–computer interfaces. *Signal Processing Journal* 86(12):3625–43.

Bourguet, M.-L., & Ando, A. (1998). Synchronization of speech and hand gestures during multimodal human–computer interaction. Paper presented at the 1998 SIGCHI Conference on Human Factors in Computing Systems. Available at *http://portal.acm.org/citation.cfm?id=286726&coll=portal&dl=ACM&CFID=6362723&CFTOKEN=90890358*.

Calvert, G. A., Spence, C. E. D. T., & Stein, B. E., eds. (2004). *The Handbook of Multisensory Processing*. Cambridge, MA: MIT Press.

Cassell, J., Sullivan, J. W., Prevost, S., & Churchill, E. F., eds. (2000). *Embodied Conversational Agents*. Cambridge, MA: MIT Press.

Cheyer, A. J., & Julia, L. (1995). Multimodal maps: An agent-based approach. Paper presented at the 1995 International Conference on Cooperative Multimodal Communication, Eindhoven, The Netherlands.

Cheyer, A. J., & Luc, J. (1998). MVIEWS: Multimodal tools for the video analyst. Paper presented at the 4th International Conference on Intelligent User Interfaces, San Francisco.

Codella, C., Jalili, R., Koved, L., Lewis, J. B., Ling, D. T., Lipscomb, J. S., et al. (1992). Interactive simulation in a multi-person virtual world. Paper presented at the 1992 SIGCHI Conference on Human Factors in Computing Systems, Monterey, CA.

Cohen, M. M., & Massaro, D. W. (1993). Modeling coarticulation in synthetic visual speech. In Magnenat-Thalmann, M., & Thalmann, D., eds., *Models and Techniques in Computer Animation*. Tokyo: Springer-Verlag, 139–59.

Cohen, P. R. (1992). The role of natural language in a multimodal interface. Paper presented at the 5th Annual ACM Symposium on User Interface and Software Technology, Monterey, CA.

Cohen, P. R., Dalrymple, M., Moran, D. B., Pereira, F. C. N., Sullivan, J. W., Gargan, R. A., Schlossberg, J. L., & Tyler, S. W. (1989). Synergistics use of direct manipulation and natural language. *Proceedings of Human Factors in Computing Systems (CHI '89)*. New York: ACM Press, 227–34.

Cohen, P. R., Johnston, M., McGee, D. R., Oviatt, S. L., Pittman, J., Smith, I. A., et al. (1997). QuickSet: Multimodal interaction for distributed applications. Paper presented at the 5th ACM International Conference on Multimedia, Seattle.

Cohen, P. R., & McGee, D. R. (2004). Tangible multimodal interfaces for safety-critical applications. *Communications of the Association for Computing Machinery* 47(1):41–46.

Cohen, P. R., McGee, D., & Clow, J. (2000). The efficiency of multimodal interaction for a map-based task. Paper presented at the 6th Applied Natural Language Processing Conference, Seattle.

Cohen, P. R., & Oviatt, S. L. (1995). The role of voice input for human–machine communication. *Proceedings of National Academy of Sciences of the United States of America* 92(22):9921–27.

Dalal, M., Feiner, S. K., McKeown, K. R., Pan, S., Zhou, M. X., Hollerer, T., et al. (1996). Negotiation for automated generation of temporal multimedia presentations. Paper presented at the Fourth Annual ACM International Conference on Multimedia, Boston.

Danninger, M., Flaherty, G., Bernardin, K., Ekenel, H., Kohler, T., Malkin, R., et al. (2005). The connector: Facilitating context-aware communication. Paper presented at the 7th International Conference on Multimodal Interfaces, Trento, Italy.

Demirdjian, D., Ko, T., & Darrell, T. (2003). Constraining human body tracking. Paper presented at the Ninth IEEE International Conference on Computer Vision, Nice.

Deng, L., Wang, K., Acero, A., Hon, H.-W., Droppo, J., Boulis, C., et al. (2002). Tap-to-talk in a specific field: Distributed speech processing in miPad's multimodal user interface. *IEEE Transactions on Computer Speech and Audio Processing* 10(8):605–19.

Duncan, L., Brown, W., Esposito, C., Holmback, H., & Xue, P. (1999). Enhancing virtual maintenance environments with speech understanding. Technical Report TECHNET-9903, Boeing Mathematics and Computing Technology.

Dupont, S., & Luettin, J. (2000). Audio-visual speech modeling for continuous speech recognition. *IEEE Transactions on Multimedia* 2(3):141–51.

Ehlen, P., Purver, M., & Niekrasz, J. (2007). A meeting browser that learns. Paper presented at the AAAI 2007 Spring Symposium: Interaction Challenges for Artificial Assistants, Stanford, CA.

Ellis, C. A., & Barthelmess, P. (2003). The Neem dream. Paper presented at the 2nd Tapia Conference on Diversity in Computing, Atlanta.

Epps, J., Oviatt, S. L., & Chen, F. (2004). Integration of speech and gesture inputs during multimodal interaction. Paper presented at the 2004 Australian International Conference on Computer–Human Interaction. Available at *http://www.cse.unsw.edu.au/~jepps/ozchi04.pdf*.

Falcon, V., Leonardi, C., Pianesi, F., Tomasini, D., & Zancanaro, M. (2005). Co-located support for small group meetings. Paper presented at the 2005 SIGCHI Conference on Human Factors in Computing Systems, Workshop: The Virtuality Continuum Revisited. Available at *http://portal.acm.org/citation.cfm?id=1057123&coll=portal&dl=ACM&CFID=6362723&CFTOKEN=90890358*.

Faure, C., & Julia, L. (1994). An agent-based architecture for a multimodal interface. Paper presented at the 1994 AAAI Spring Symposium, Palo Alto, CA.

Fell, H. J., Delta, H., Peterson, R., Ferrier, L. J., Mooraj, Z., & Valleau, M. (1994). Using the baby-babble-blanket for infants with motor problems: An empirical study. Paper presented at the First Annual ACM Conference on Assistive Technologies, Marina del Rey, CA.

Flanagan, J. L., & Huang, T. S. (2003). Scanning the issue: Special issue on human–computer multimodal interfaces. *Proceedings of the IEEE* 91(9):1267–71.

Fukumoto, M., Suenaga, Y., & Mase, K. (1994). Finger-pointer: Pointing interface by image processing. *Computers & Graphics* 18(5):633–42.

Grudin, J. (1988). Why CSCW applications fail: Problems in the design and evaluation of organizational interfaces. Paper presented at the 1988 ACM Conference on Computer-Supported Cooperative Work, Portland, OR.

Gruenstein, A., Niekrasz, J., & Purver, M. (2005). Meeting structure annotation: Data and tools. Paper presented at the 9th European Conference on Speech Communication and Technology, 6th SIGdial Workshop on Discourse and Dialogue. Available at *http://www.mit.edu/~alexgru/pubs/sigdial05.pdf*.

Gupta, A. K., & Anastasakos, T. (2004). Integration patterns during multimodal interaction. Paper presented at the 8th International Conference on Spoken Language Processing.

Halverson, C. A., Horn, D. B., Karat, C.-M., & Karat, J. (1999). The beauty of errors: Patterns of error correction in desktop speech systems. Paper presented at the 7th IFIP International Conference on Human–Computer Interaction, Edinburgh.

Hauptmann, A. G. (1989). Speech and gestures for graphic image manipulation. Paper presented at the 1989 SIGCHI Conference on Human Factors in Computing Systems, Austin, TX.

Hina, M. D., Ramdane-Cherif, A., & Tadj, C. (2005). A ubiquitous context-sensitive multimodal multimedia computing system and its machine learning-based reconfiguration at the architectural level. Paper presented at the Seventh IEEE International Symposium on Multimedia, Irvine, CA.

Horndasch, A., Rapp, H., & Rottger, H. (2006). SmartKom-Public. In Wahlster, W., ed., *SmartKom: Foundations of Multimodal Dialogue Systems.* Berlin: Springer, 471–92.

Huang, X., Acero, A., Chelba, C., Deng, L., Droppo, J., Duchene, D., et al. (2001). MiPad: A multimodal interaction prototype. Paper presented at the Acoustics, Speech, and Signal Processing Conference (ICASSP).

Huang, X., & Oviatt, S. L. (2005). Toward adaptive information fusion in multimodal systems. Paper presented at the Second Joint Workshop on Multimodal Interaction and Related Machine Learning Algorithms.

Huang, X., Oviatt, S. L., & Lunsford, R. (2006). Combining user modeling and machine learning to predict users' multimodal integration patterns. Paper presented at the 3rd Joint Workshop on Multimodal Interaction and Related Machine Learning Algorithms.

Johnston, M., & Bangalore, S. (2004). MATCHKiosk: A multimodal interactive city guide. Paper presented at the 42nd Annual Meeting of the Association of Computational Linguistics. Available at *http://acl.ldc.upenn.edu/acl2004/postersdemos/pdf/johnston.pdf*.

Johnston, M., & Bangalore, S. (2005). Finite-state multimodal integration and understanding. *Natural Language Engineering* 11(2):159–87.

Johnston, M., Bangalore, S., Vasireddy, G., Stent, A., Ehlen, P., Walker, M., et al. (2002). MATCH: An architecture for multimodal dialogue systems. Paper presented at the 40th Annual Meeting of the Association of Computational Linguistics, Discourse and Dialogue. Available at *http://portal.acm.org/citation.cfm?id=1073083.1073146*.

Johnston, M., Cohen, P. R., McGee, D. R., Oviatt, S. L., Pittman, J. A., & Smith, I. A. (1997). Unification-based multimodal integration. Paper presented at the 35th Annual Meeting of the Association for Computational Linguistics. Available at *http://portal.acm.org/ citation.cfm?id=976909.979653&dl=GUIDE&dl=%23url.dl.*

Jovanovic, N., op den Akker, R., & Nijholt, A. (2006). Addressee identification in face-to-face meetings. Paper presented at the 11th Conference of the European Chapter of the Association for Computational Linguistics, Trento, Italy.

Kaiser, E. C. (2005). Multimodal new vocabulary recognition through speech and handwriting in a whiteboard scheduling application. Paper presented at the 10th International Conference on Intelligent User Interfaces, San Diego.

Kaiser, E. C. (2006). Using redundant speech and handwriting for learning new vocabulary and understanding abbreviations. Paper presented at the 8th International Conference on Multimodal Interfaces, Oral Session 5: Speech and Dialogue Systems. Available at *http://portal.acm.org/citation.cfm?id=1181060&coll=portal&dl=ACM&CFID=6362723& CFTOKEN=90890358.*

Kaiser, E. C., & Barthelmess, P. (2006). Edge-splitting in a cumulative multimodal system, for a no-wait temporal threshold on information fusion, combined with an underspecified display. Paper presented at the 9th International Conference on Spoken Language Processing.

Kaiser, E. C., Barthelmess, P., Erdmann, C. G., & Cohen, P. R. (2007). Multimodal redundancy across handwriting and speech during computer mediated human–human interactions. Paper presented at the 2007 SIGCHI Conference on Human Factors in Computing Systems, San Jose, CA.

Kaiser, E. C., Barthelmess, P., Huang, X., & Demirdjian, D. (2005). A demonstration of distributed pointing and referencing for multimodal collaboration over sketched diagrams. Paper presented at the 7th International Conference on Multimodal Interfaces, Workshop on Multimodal, Multiparty Meeting Processing. Available at *http://www.csee. ogi.edu/~kaiser/distgest_demo.3.pdf.*

Kaiser, E. C., Demirdjian, D., Gruenstein, A., Li, X., Niekrasz, J., Wesson, R. M., et al. (2004). A multimodal learning interface for sketch, speak and point creation of a schedule chart. Paper presented at the 6th International Conference on Multimodal Interfaces, Demo Session 2. Available at *http://portal.acm.org/citation.cfm?id=1027992&coll= portal&dl=ACM&CFID=6362723&CFTOKEN=90890358.*

Katzenmaier, M., Stiefelhagen, R., & Schultz, T. (2004). Identifying the addressee in human–human–robot interactions based on head pose and speech. Paper presented at the 6th International Conference on Multimodal Interfaces, State College, PA.

Kendon, A. (1981). Gesticulation and speech: Two aspects of the process of utterance. In Key, M. R., ed., *The Relationship of Verbal and Nonverbal Communication.* The Hague: Mouton de Gruyter, 207–27.

Koons, D. B., Sparrell, C. J., & Thorisson, K. R. (1993). Integrating simultaneous input from speech, gaze, and hand gestures. In Maybury, M. T., ed., *Intelligent Multimedia Interfaces.* Cambridge, MA: AAAI Press/MIT Press, 257–76.

Larson, J. A. (2003). *VoiceXML: Introduction to Developing Speech Applications.* Upper Saddle River, NJ: Prentice Hall.

Larson, K., & Mowatt, D. (2003). Speech error correction: The story of the alternates list. *International Journal of Speech Technology* 8(2):183–94.

Lauer, C., Frey, J., Lang, B., Becker, T., Kleinbauer, T., & Alexandersson, J. (2005). AmiGram: A general-purpose tool for multimodal corpus annotation. Paper presented at the 2nd Joint Workshop on Multimodal Interaction and Related Machine Learning Algorithms.

Lucente, M., Zwart, G.-J., & George, A. (1998). Visualization Space: A testbed for deviceless multimodal user interface. Paper presented at the 1998 AAAI Spring Symposium on Applying Machine Learning and Discourse Processing, Palo Alto, CA.

Lunsford, R., Oviatt, S. L., & Arthur, A. M. (2006). Toward open-microphone engagement for multiparty interactions. Paper presented at the 8th International Conference on Multimodal Interfaces, Banff, Alb.

Lunsford, R., Oviatt, S. L., & Coulston, R. (2005). Audio-visual cues distinguishing self- from system-directed speech in younger and older adults. Paper presented at the 7th International Conference on Multimodal Interfaces, Trento, Italy.

Mankoff, J., Hudson, S. E., & Abowd, G. D. (2000). Interaction techniques for ambiguity resolution in recognition-based interfaces. Paper presented at the 13th Annual ACM Symposium on User Interface Software and Technology, San Diego.

Martin, J.-C., & Kipp, M. (2002). Annotating and measuring multimodal behaviour—TYCOON metrics in the Anvil tool. Paper presented at the 3rd International Conference on Language Resources and Evaluation. Available at *http://www.dfki.uni-sb.de/~kipp/public_archive/MartinKipp2002-lrec.ps.*

Massaro, D. W., & Stork, D. G. (1998). Speech recognition and sensory integration. *American Scientist* 86(3):236–44.

McCowan, I., Gatica-Perez, D., Bengio, S., Lathoud, G., Barnard, M., & Zhang, D. (2005). Automatic analysis of multimodal group actions in meetings. *IEEE Transactions on Pattern Analysis and Machine Intelligence* 27(3):305–17.

McGee, D. R. (2003). Augmenting Environments with Multimodal Interaction. PhD diss., Oregon Graduate Institute of Science and Technology.

McGee, D. R., & Cohen, P. R. (2001). Creating tangible interfaces by transforming physical objects with multimodal language. Paper presented at the 6th International Conference on Intelligent User Interfaces, Santa Fe, NM.

McGee, D. R., Pavel, M., Adami, A., Wang, G., & Cohen, P. R. (2001). A visual modality for the augmentation of paper. Paper presented at the 3rd Workshop on Perceptual/Perceptive User Interfaces, Orlando.

McGee, D. R., Pavel, M., & Cohen, P. R. (2001). Context shifts: Extending the meaning of physical objects with language. *Human–Computer Interaction* 16(2–4):351–62.

McGrath, M., & Summerfield, Q. (1985). Intermodal timing relations and audio-visual speech recognition by normal-hearing adults. *Journal of the Acoustical Society of America* 77(2):678–85.

McGurk, H., & MacDonald, J. (1976). Hearing lips and seeing voices. *Nature* 264:746–48.

McLeod, A., & Summerfield, Q. (1987). Quantifying the contribution of vision to speech perception in noise. *British Journal of Audiology* 21(2):131–41.

McNeil, D. (1992). *Hand and Mind: What Gestures Reveal about Thought.* Chicago: University of Chicago Press.

Meier, U., Hürst, W., & Duchnowski, P. (1996). Adaptive bimodal sensor fusion for automatic speechreading. Paper presented at the 1996 IEEE International Conference on Acoustics, Speech, and Signal Processing, Atlanta.

Miller, D., Culp, J., & Stotts, D. (2006). Facetop tablet: Note-taking assistance for deaf persons. Paper presented at the 8th International ACM SIGACCESS Conference on Computers and Accessibility, Portland, OR.

Morency, L.-P., Sidner, C. L., Lee, C., & Darrell, T. (2005). Contextual recognition of head gestures. Paper presented at the 7th International Conference on Multimodal Interfaces, Trento, Italy.

Morimoto, C. H., Koons, D. B., Amir, A., Flickner, M., & Zhai, S. (1999). Keeping an eye for HCI. Paper presented at the 12th Brazilian Symposium on Computer Graphics and Image Processing. Available at *http://doi.ieeecomputersociety.org/10.1109/SIBGRA.1999.805722*.

Mousavi, S. Y., Low, R., & Sweller, J. (1995). Reducing cognitive load by mixing auditory and visual presentation modes. *Journal of Educational Psychology* 87(2):319–34.

Naughton, K. (1996). Spontaneous gesture and sign: A study of ASL signs co-occurring with speech. Paper presented at the 1996 Workshop on the Integration of Gesture in Language & Speech. Available at *http://www.sign-lang.uni-hamburg.de/BibWeb/LiDat.acgi?ID=40674*.

Neal, J. G., & Shapiro, S. C. (1991). Intelligent multi-media interface technology. In Sullivan, J. W., & Tyler, S. W., eds., *Intelligent User Interfaces*. New York: Addison-Wesley, 11–43.

Negroponte, N. (1978). *The Media Room: Report for ONR and DARPA*. Cambridge, MA: MIT, Architecture Machine Group.

Nigay, L., & Coutaz, J. (1993). A design space for multimodal systems: Concurrent processing and data fusion. Paper presented at the 1993 SIGCHI Conference on Human Factors in Computing Systems, Amsterdam.

Nigay, L., & Coutaz, J. (1995). A generic platform for addressing the multimodal challenge. Paper presented at the 1995 SIGCHI Conference on Human Factors in Computing Systems, Denver.

Nijholt, A. (2006). Towards the automatic generation of virtual presenter agents. *Informing Science* 9:97–110.

Nijholt, A., Rienks, R. J., Zwiers, J., & Reidsma, D. (2006). Online and off-line visualization of meeting information and meeting support. *Visual Computer* 22(12):965–76.

Oliver, N., & Horvitz, E. (2004). S-SEER: Selective perception in a multimodal office activity recognition system. Paper presented at the First International Workshop on Machine Learning for Multimodal Interaction, Martigny, Switzerland.

Oulasvirta, A., Tamminen, S., Roto, V., & Kuorelahti, J. (2005). Interaction in 4-second bursts: The fragmented nature of attentional resources in mobile HCI. Paper presented at the 2005 SIGCHI Conference on Human Factors in Computing Systems, Portland, OR.

Oviatt, S. L. (1995). Predicting spoken disfluencies during human–computer interaction. *Computer Speech and Language* 9(1):19–35.

Oviatt, S. L. (1996a). Multimodal interfaces for dynamic interactive maps. Paper presented at the 1996 SIGCHI Conference on Human Factors in Computing Systems. Available at *http://portal.acm.org/citation.cfm?id=238438&coll=portal&dl=ACM&CFID=6362723&CFTOKEN=90890358*.

Oviatt, S. L. (1996b). User-centered modeling for spoken language and multimodal interfaces. *IEEE Transactions on Multimedia* 3(4):26–35.

Oviatt, S. L. (1997). Multimodal interactive maps: Designing for human performance. *Human–Computer Interaction* 12(1–2):93–129.

Oviatt, S. L. (1999a). Mutual disambiguation of recognition errors in a multimodal architecture. Paper presented at the 1999 SIGCHI Conference on Human Factors in Computing Systems, CHI Letters. Available at *http://portal.acm.org/citation.cfm?id=303163&coll= portal&dl=ACM&CFID=6362723&CFTOKEN=90890358.*

Oviatt, S. L. (1999b). Ten myths of multimodal interaction. *Communications of the Association for Computing Machinery* 42(11):74–81.

Oviatt, S. L. (2000). Taming recognition errors with a multimodal interface. *Communications of the Association for Computing Machinery* 43(9):45–51.

Oviatt, S. L. (2002). Breaking the robustness barrier: Recent progress on the design of robust multimodal systems. *Advances in Computers* 56:305–41.

Oviatt, S. L. (2006a). Human-centered design meets cognitive load theory: Designing interfaces that help people think. Paper presented at the 14th Annual ACM International Conference on Multimedia, Santa Barbara, CA.

Oviatt, S. L., ed. (2006b). *Encyclopedia of Multimedia*, 2nd ed. Boston: Kluwer Academic Publishers.

Oviatt, S. L., (2007). Multimodal interfaces. In Chen, F., & Jokinen, J., eds. *New Trends in Speech-Based Interactive Systems*, 2nd ed. New York: Springer.

Oviatt, S. L., & Adams, B. (2000). Designing and evaluating conversational interfaces with animated characters. In Cassell, J., Sullivan, J. W., Prevost, S., & Churchill, E. F., eds., *Embodied Conversational Agents*. Cambridge, MA: MIT Press, 319–43.

Oviatt, S. L., Arthur, A. M., & Cohen, J. (2006). Quiet interfaces that help students think. Paper presented at the 19th Annual ACM Symposium on User Interface Software and Technology, Montreaux, Switzerland.

Oviatt, S. L., & Cohen, P. R. (1991). Discourse structure and performance efficiency in interactive and noninteractive spoken modalities. *Computer Speech and Language* 5(4):297–326.

Oviatt, S. L., Cohen, P. R., Fong, M. W., & Frank, M. P. (1992). A rapid semi-automatic simulation technique for investigating interactive speech and handwriting. Paper presented at the 2nd International Conference on Spoken Language Processing, Banff, Alb.

Oviatt, S. L., Cohen, P. R., & Wang, M. Q. (1994). Toward interface design for human language technology: Modality and structure as determinants of linguistic complexity. *Speech Communication* 15:283–300.

Oviatt, S. L., Cohen, P. R., Wu, L., Duncan, L., Suhm, B., Bers, J., et al. (2000). Designing the user interface for multimodal speech and gesture applications: State-of-the-art systems and research directions. *Human–Computer Interaction* 15(4):263–322.

Oviatt, S. L., Coulston, R., & Lunsford, R. (2004). When do we interact multimodally? Cognitive load and multimodal communication patterns. Paper presented at the 6th International Conference on Multimodal Interfaces, Pittsburgh.

Oviatt, S. L., Coulston, R., & Lunsford, R. (2005). Just do what I tell you: The limited impact of instructions on multimodal integration patterns. Paper presented at the 10th International Conference on User Modeling, Edinburgh.

Oviatt, S. L., Coulston, R., Tomko, S., Xiao, B., Lunsford, R., Wesson, R. M., et al. (2003). Toward a theory of organized multimodal integration patterns during human–computer interaction. Paper presented at the 5th International Conference on Multimodal Interfaces, Vancouver, BC.

Oviatt, S. L., Darves, C., Coulston, R., & Wesson, R. M. (2005). Speech convergence with animated personas. *Spoken Multimodal Human–Computer Dialogue in Mobile Environments* 28:379–97.

Oviatt, S. L., DeAngeli, A., & Kuhn, K. (1997). Integration and synchronization of input modes during multimodal human–computer interaction. Paper presented at the 1997 SIGCHI Conference on Human Factors in Computing Systems. Available at *http://portal. acm.org.liboff.ohsu.edu/citation.cfm?id=258821&coll=portal&dl=ACM&CFID=6362723& CFTOKEN=90890358.*

Oviatt, S. L., & Kuhn, K. (1998). Referential features and linguistic indirection in multimodal language. Paper presented at the 5th International Conference on Spoken Language Processing, Sydney.

Oviatt, S. L., Lunsford, R., & Coulston, R. (2005). Individual differences in multimodal integration patterns: What are they and why do they exist? Paper presented at the 2005 SIGCHI Conference on Human Factors in Computing Systems, Portland, OR.

Oviatt, S. L., & Olsen, E. (1994). Integration themes in multimodal human–computer interaction. Paper presented at the 3rd International Conference on Spoken Language Processing.

Oviatt, S. L., & VanGent, R. (1996). Error resolution during multimodal human–computer interaction. Paper presented at the 4th International Conference on Spoken Language Processing, Philadelphia.

Pavlovic, V. I., Berry, G. A., & Huang, T. S. (1997). Integration of audio/visual information for use in human–computer intelligent interaction. Paper presented at the 1997 IEEE International Conference on Image Processing, Santa Barbara, CA.

Pianesi, F., Zancanaro, M., Falcon, V., & Not, E. (2006). Toward supporting group dynamics. Paper presented at the 3rd IFIP Conference on Artificial Intelligence Applications and Innovations, Athens.

Poppe, P. (2007). Special Issue on Vision for Human–Computer Interaction. *Computer Vision and Image Understanding* 108(1-2):4–18

Potamianos, G., Neti, C., Luettin, J., & Matthews, I. (2004). Audio-visual automatic speech recognition: An overview. In Bailly, G., Vatikiotis-Bateson, E., & Perrier, P., eds., *Issues in Visual and Audio-Visual Speech Processing*. Cambridge, MA: MIT Press.

Purver, M., Ehlen, P., & Niekrasz, J. (2006). Detecting action items in multi-party meetings: Annotation and initial experiments. In Renals, S., Bengio, S., & Fiscus, J., eds., *Machine Learning for Multimodal Interaction*. New York: Springer-Verlag, 200–11.

Purver, M., Körding, K. P., Griffiths, T. L., & Tenenbaum, J. B. (2006). Unsupervised topic modelling for multi-party spoken discourse. Paper presented at the 21st International Conference on Computational Linguistics, 44th Annual Meeting of the ACL, Sydney.

Reithinger, N., Alexandersson, J., Becker, T., Blocher, A., Engel, R., Löckelt, M., et al. (2003). SmartKom: Adaptive and flexible multimodal access to multiple applications. Paper presented at the 5th International Conference on Multimodal Interfaces, Vancouver, BC.

Reithinger, N., & Herzog, G. (2006). An exemplary interaction with SmartKom. In Wahlster, W., ed., *SmartKom: Foundations of Multimodal Dialogue Systems*. Berlin: Springer, 41–52.

Rienks, R. J., & Heylen, D. K. J. (2005). Dominance detection in meetings using easily obtainable features. Paper presented at the 2nd Joint Workshop on Multimodal Interaction and Related Machine Learning Algorithms, Edinburgh.

Rienks, R. J., Nijholt, A., & Barthelmess, P. (2006). Pro-active meeting assistants: Attention please! Paper presented at the 5th Workshop on Social Intelligence Design, Osaka.

Robert-Ribes, J., Schwartz, J.-L., Lallouache, T., & Escudier, P. (1998). Complementarity and synergy in bimodal speech: Auditory, visual, and auditory-visual identification of French oral vowels in noise. *Journal of the Acoustical Society of America* 103(6): 3677–89.

Rogozan, A., & Deléglise, P. (1998). Adaptive fusion of acoustic and visual sources for automatic speech recognition. *Speech Communication* 26(1–2):149–61.

Ruiz, N., Taib, R., & Chen, F. (2006). *Examining the redundancy of multimodal input*. Paper presented at the 2006 Australian International Conference on Computer–Human Interaction, Sydney.

Salber, D., & Coutaz, J. (1993a). Applying the Wizard of Oz technique to the study of multimodal systems. Paper presented at the Third International Conference on Human–Computer Interaction, Moscow.

Salber, D., & Coutaz, J. (1993b). A Wizard of Oz platform for the study of multimodal systems. Paper presented at the 1993 SIGCHI Conference on Human Factors in Computing Systems, Amsterdam.

Sellen, A. J., & Harper, R. H. (2003). *The Myth of the Paperless Office*. Cambridge, MA: MIT Press.

Siroux, J., Guyomard, M., Multon, F., & Remondeau, C. (1995). Modeling and processing of the oral and tactile activities in the Georal tactile system. Paper presented at the First International Conference on Cooperative Multimodal Communication, Eindhoven.

Stock, O., & Zancanaro, M., eds. (2005). *Multimodal Intelligent Information Presentations*. New York: Springer.

Stotts, D., Bishop, G., Culp, J., Miller, D., Gyllstrom, K., & Lee, K. (2005). *Facetop on the Tablet PC: Assistive Technology in Support of Classroom Note-Taking for Hearing-Impaired Students*. Technical Report. Chapel Hill: University of North Carolina.

Suhm, B. (1998). Multimodal Interactive Error Recovery for Non-Conversational Speech User Interfaces. PhD diss., University of Karlsruhe (Germany).

Suhm, B., Myers, B. A., & Waibel, A. (1999). Model-based and empirical evaluation of multimodal interactive error correction. Paper presented at the 1999 SIGCHI Conference on Human Factors in Computing Systems, Pittsburgh.

Sumby, W. H., & Pollack, I. (1954). Visual contribution to speech intelligibility in noise. *Journal of the Acoustical Society of America* 26(2):212–15.

Summerfield, Q. (1992). Lipreading and audio-visual speech perception. *Philosophical Transactions of the Royal Society of London: Biological Sciences* 335(1273):71–78.

Tang, A., McLachlan, P., Lowe, K., Saka, C. R., & MacLean, K. (2005). Perceiving ordinal data haptically under workload. Paper presented at the 7th International Conference on Multimodal Interfaces, Trento, Italy.

Tang, J. C. (1991). Finding from observational studies of collaborative work. *International Journal of Man–Machine Studies* 34(2):143–60.

Tomlinson, M. J., Russell, M. J., & Brooke, N. M. (1996). Integrating audio and visual information to provide highly robust speech recognition. Paper presented at the 1996 IEEE International Conference on Acoustics, Speech, and Signal Processing, Atlanta.

Turk, M., & Robertson, G. (2000). Perceptual user interfaces (introduction). *Communications of the Association for Computing Machinery* 43(3):32–34.

Van Leeuwen, D. A., & Huijbregts, M. A. H. (2006). The AIM speaker diarization system for NIST RT06's meeting data. Paper presented at the NIST Rich Transcription 2006 Spring Meeting Recognition Evaluation.

van Turnhout, K., Terken, J., Bakx, I., & Eggen, B. (2005). Identifying the intended addressee in mixed human–human and human–computer interaction from non-verbal features. Paper presented at the 7th International Conference on Multimodal Interfaces, Trento, Italy.

Vatikiotis-Bateson, E., Munhall, K. G., Hirayama, M., Lee, Y., & Terzopoulos, D. (1996). The dynamics of audiovisual behavior of speech. *Speechreading by Humans and Machines: Models, Systems and Applications* 150:221–32.

Verbree, D., Rienks, R. J., & Heylen, D. K. J. (2006a). Dialogue-act tagging using smart feature selection: Results on multiple corpora. Paper presented at the First International IEEE Workshop on Spoken Language Technology, Palm Beach, FL.

Verbree, D., Rienks, R. J., & Heylen, D. K. J. (2006b). First steps towards the automatic construction of argument-diagrams from real discussions. *Frontiers in Artificial Intelligence and Applications* 144:183–94.

Vergo, J. (1998). A statistical approach to multimodal natural language interaction. Paper presented at the AAAI Workshop on Representations for multi-modal Human–Computer Interaction, Madison, WI.

Verlinden, M., Zwitserlood, I., & Frowein, H. (2005). Multimedia with Animated Sign Language for Deaf Learners. Paper presented at the World Conference on Educational Multimedia, Hypermedia & Telecommunications, ED-MEDIA, Montréal.

Vo, M. T. (1998). A Framework and Toolkit for the Construction of Multimodal Learning Interfaces. PhD diss., Carnegie Mellon University.

Vo, M. T., & Wood, C. (1996). Building an application framework for speech and pen input integration in multimodal learning interfaces. Paper presented at the 1996 IEEE International Conference on Acoustics, Speech, and Signal Processing, Atlanta.

Wahlster, W. (2006). Dialogue systems go multimodal: The SmartKom experience. In Wahlster, W., ed., *SmartKom: Foundations of Multimodal Dialogue Systems*. Berlin: Springer, 3–27.

Wahlster, W., Andre, E., Finkler, W., Profitlich, H.-J., & Rist, T. (1993). Plan-based integration of natural language and graphics generation. *Artificial Intelligence* 63:387–427.

Wainer, J., & Braga, D. P. (2001). Symgroup: Applying social agents in a group interaction system. Paper presented at the 2001 International ACM SIGGROUP Conference on Supporting Group Work, Boulder, CO.

Wang, J. (1995). Integration of eye-gaze, voice and manual response in multimodal user interfaces. Paper presented at the 1995 IEEE International Conference on Systems, Man and Cybernetics, Vancouver, BC.

Wang, K. (2004). From multimodal to natural interactions. Paper presented at the W3 Workshop on Multimodal Interaction. Position Paper. Available at *http://www.w3.org/2004/02/mmi-workshop/wang-microsoft.html*.

Wang, S., & Demirdjian, D. (2005). Inferring body pose from speech content. Paper presented at the 7th International Conference on Multimodal Interfaces.

Wickens, C. D., Sandry, D., & Vidulich, M. (1983). Compatibility and resource competition between modalities of input, central processing, and output. *Human Factors* 25(2):227–48.

Xiao, B., Girand, C., & Oviatt, S. L. (2002). Multimodal integration patterns in children. Paper presented at the 7th International Conference on Spoken Language Processing, Denver.

Xiao, B., Lunsford, R., Coulston, R., Wesson, R. M., & Oviatt, S. L. (2003). Modeling multimodal integration patterns and performance in seniors: Toward adaptive processing of individual differences. Paper presented at the 5th International Conference on Multimodal Interfaces, Vancouver, BC.

Zancanaro, M., Lepri, B., & Pianesi, F. (2006). Automatic detection of group functional roles in face to face interactions. Paper presented at the 8th International Conference on Multimodal Interfaces, Banff, Alb.

Zhai, S., Morimoto, C. H., & Ihde, S. (1999). Manual and gaze input cascaded (MAGIC) pointing. Paper presented at the 1999 SIGCHI Conference on Human Factors in Computing Systems, Pittsburgh.

Zoltan-Ford, E. (1991). How to get people to say and type what computers can understand. *International Journal of Man–Machine Studies* 34(4):527–47.

Index